Institutions and Environmental Change

Institutions and Environmental Change
Principal Findings, Applications, and Research Frontiers

edited by Oran R. Young, Leslie A. King, and Heike Schroeder

Institutional Dimensions of Global Environmental Change
A Core Research Project of the International Human Dimensions
Programme on Global Environmental Change (IHDP)

The MIT Press
Cambridge, Massachusetts
London, England

HC
79
.E5
I5I59
2008

MIT Press books may be purchased at special quantity discounts for business or sales promotional use. For information, please email special_sales@mitpress.mit .edu or write to Special Sales Department, The MIT Press, 55 Hayward Street, Cambridge, MA 02142.

This book was set in Sabon on 3B2 by Asco Typesetters, Hong Kong. Printed and bound in the United States of America.

Printed on recycled paper.

Library of Congress Cataloging-in-Publication Data

Institutions and environmental change : principal findings, applications, and research frontiers / edited by Oran R. Young, Leslie A. King, and Heike Schroeder.
    p. cm.
Includes bibliographical references and index.
ISBN 978-0-262-24057-4 (hardcover : alk. paper) — ISBN 978-0-262-74033-3 (pbk. : alk. paper) 1. Environmental policy. 2. Environmental protection. 3. International relations—Environmental aspects. 4. Globalization—Environmental aspects. 5. International cooperation. I. Young, Oran R. II. King, Leslie A. III. Schroeder, Heike.
HC79.E5I5159 2008
333.7—dc22   2008007304

10 9 8 7 6 5 4 3 2 1

Contents

Preface

This book constitutes the culmination of a long and rewarding experience with scientific research involving participants from many disciplines and many countries. The project on the Institutional Dimensions of Global Environmental Change (IDGEC) was one of the four original core projects of the International Human Dimensions Programme on Global Environmental Change (IHDP). Planning for IDGEC actually got under way in 1995 with an invitation to Arild Underdal of the University of Oslo and Oran Young, then of Dartmouth College, from IHDP's predecessor, the Human Dimensions Programme, to prepare a feasibility study identifying opportunities for engaging in cutting-edge research on the roles that institutions play both in causing and in addressing large-scale environmental problems. This was followed by the appointment of a Scientific Planning Committee with a mandate to prepare a Science Plan for the project, a series of workshops organized to develop this plan, a process involving reviews from outside experts together with revisions to address the suggestions of the reviewers, and finally the formal approval of the plan by IHDP's Scientific Committee (SC) in November 1998. IDGEC's own Scientific Steering Committee (SSC) met for the first time in Japan during June 1999 and continued to meet regularly throughout the life cycle of the project, which came to a close formally in March 2007.

The purpose of this volume is to distill the principal findings of IDGEC and to highlight the scientific legacy of the project. Three other volumes dealing with what IDGEC treats as the problems of interplay and scale, as well as with international bodies created to administer environmental regimes, complete this record of IDGEC's accomplishments. Together the four volumes provide a substantial account of the results arising from research carried out under the auspices of IDGEC.

We have learned a lot about the conduct of large-scale scientific research on issues relating to global environmental change in the course
of our work with IDGEC. Like other projects dealing with global environmental change, IDGEC has not been blessed with a research fund
and the authority to allocate it among individual researchers. Rather,
we had a mandate from IHDP to announce the launching of a concerted
program of research on institutional issues, to encourage a large number
of researchers to join our common journey, to establish infrastructure in
the form of an International Project Office (IPO) to develop and service a
community of researchers working in this realm, and to create opportunities for interested researchers to interact with one another and to bring
their findings to the attention of policy makers and what we have come
to call "knowledge brokers." We discovered early on that this was a tall
order. Not being in a position to distribute significant material resources,
we learned at once about the need to rely on intellectual leadership or, in
other words, on the effort to establish and energize a vibrant community
or network of researchers who would be motivated to participate largely
because of the intellectual stimulation and feelings of efficacy to be
derived from interacting with colleagues who share both interests and
values. Needless to say, this process was not always straightforward,
but it did give rise to a situation in which we can say with assurance
that the whole of the project was substantially greater than the sum of
the parts.

Early on, we also realized the importance both of forging a strong link
to high-priority concerns in the social sciences and of applying our work
to salient issues in the realm of global environmental change (e.g., climate change, losses of biodiversity, the degradation of ocean ecosystems). To this end we deliberately grounded the work of IDGEC in
matters of central concern to students of social institutions, while at the
same time turning our attention to cutting-edge issues in the realm of
environmental governance. Thus, what IDGEC calls research foci—the
questions of causality, performance, and design—are generic concerns
of interest to scholars interested in institutions. What IDGEC calls analytic themes—the problems of fit, interplay, and scale—are issues that
have emerged as high-profile concerns among those interested in
human-environment interactions. To strengthen the links between the
research foci and the analytic themes, we launched a series of flagship
activities: one dealing with atmospheric issues and emphasizing the topic
of climate change; a second dealing with marine systems and especially

recent changes in the law of the sea; and a third dealing with terrestrial issues and particularly matters pertaining to the use and conservation of forests. Integrating these individual pieces of our puzzle into a coherent picture turned out to be a challenge requiring focused attention on the part of members of the SSC and the IDGEC executive officer. Meeting this challenge fully and effectively became an exciting and productive effort, giving rise to strong and lasting ties among all the key players in the project.

Along the way we have accumulated many debts of gratitude that we are happy to acknowledge publicly at this time. We are grateful to Eckert Ehlers, the first chair of the IHDP SC, and to Larry Kohler, the first executive director of IHDP, who guided IDGEC through its formative stages. Larry, in particular, turned out to be a demanding taskmaster but one whose high expectations regarding performance played an important role in making IDGEC a success. We are also pleased to acknowledge the support of Hans Opschoor, who acted as liaison during the early years between the IHDP SC and the IDGEC SSC. A number of individual staff members located in the IHDP Secretariat in Bonn played constructive roles over the years as science coordinators responsible for maintaining a connection between IDGEC as an autonomous core project and the central administration of IHDP.

Three outside experts—Abram Chayes, Joke Waller-Hunter, and Narpat Jodha—served as external reviewers of the draft IDGEC Science Plan. Bringing very different perspectives to this task, they provided excellent feedback and recommendations that made it possible to strike a proper balance between analytical and applied concerns and to strengthen the final text of the Science Plan considerably.

Perhaps our largest debt of gratitude goes to all those who have served over the years as members of the Scientific Planning Committee and of the IDGEC SSC. The architects of the IDGEC Science Plan—including Arun Agrawal, Peter Sand, Arild Underdal, and Merrilyn Wasson, as well as two of us (King and Young)—produced a road map for the project that has withstood the test of time remarkably well. Some twenty-two individuals, including practitioners as well as scientists, served at one time or another as members of the IDGEC SSC. They played a crucial role in providing guidance for the collective efforts of IDGEC. In this connection we would like especially to recognize Agus Sari, who took over from Oran Young as chair of the IDGEC SSC at the beginning of 2006 and guided the project successfully to its close.

IDGEC would not have been able to function without an effective IPO and especially without the dedicated service of those who occupied the role of executive officer during the period 1999–2007. One of us (Schroeder) held this position from the fall of 2003 through the end of the project in March 2007. Her predecessors included Nicholas Flanders, Virginia Walsh, and Syma Ebbin. Here we want to pay special tribute to Virginia, whose tragic death in 2004 cut short a promising career in the academic world.

The U.S. National Science Foundation (NSF) played an essential role in granting the bulk of the funding needed to operate the IPO throughout the life of the project. We extend our thanks publicly not only to the foundation itself but also to Tom Baerwald, our program officer at NSF, who has held us to high standards and supported our efforts loyally from the outset. IHDP awarded an annual subvention that made it possible to organize regular meetings of the IDGEC SSC in locations in Asia, Europe, and North America. We are grateful also to Dartmouth College, which hosted the IPO and provided contributions in kind from 1999 through 2002, as well as to the Donald Bren School of Environmental Science and Management at the University of California, Santa Barbara, which played the same role from 2003 through the formal close of the project in March 2007. This is also the place to acknowledge with thanks the crucial support provided by Nicki Maynard, our office manager at Dartmouth, and Maria Gordon, our office manager at the Bren School, whose efforts made all the difference on a day-to-day basis.

An IDGEC Synthesis Planning Group guided our efforts throughout the synthesis process. The three of us were joined as members of this committee by Frank Biermann, Song Li, and Agus Sari. We had the good fortune to be able to organize a major IDGEC synthesis conference during December 2006 in Bali, which not only played a central role in helping us to crystallize IDGEC's scientific legacy but also provided an initial opportunity to engage in a consultative process regarding future directions in this field of research. Under the leadership of Frank Biermann, the latter effort has given rise to a planning process that we expect will eventuate in the launching of a new project on Earth system governance. The Bali Conference was made possible by generous support from the following sources: the Asia-Pacific Network for Global Change Research; the Donald Bren School of Environmental Science and Management at the University of California, Santa Barbara; Charles Darwin University; the Clayton H. Riddell Faculty of Environment, Earth, and

Resources at the University of Manitoba; the Global Carbon Project; the Inter-American Institute for Global Change Research; the International Human Dimensions Programme on Global Environmental Change; the Land-Ocean Interactions in the Coastal Zone project; the Research Council of Norway; the System for Analysis, Research, and Training; the United Nations University Institute of Advanced Study; the U.S. National Science Foundation; and Yayasan Pelangi Indonesia.

Finally, we want to acknowledge the help of Maria Gordon, our office manager and in-house editor at the Bren School; Clay Morgan, our editor at MIT Press; and the anonymous reviewers the Press engaged to evaluate our manuscript. One of the reviewers, in particular, provided lengthy and sophisticated comments that have made us think hard and work to clarify the main messages of this book.

IDGEC has been a formative experience for us in human as well as scientific terms. We come away with strong bonds of friendship across many scientific, political, and geographical boundaries that have emerged in conjunction with the work of our project.

Oran R. Young
Chair, IDGEC SSC, 1999–2005
Member, IDGEC Synthesis Planning Group, 2004–2006
Chair, IHDP SC, 2006–

Leslie A. King
Member, IDGEC SSC, 1999–2005
Chair, IDGEC Synthesis Planning Group, 2004–2006

Heike Schroeder
Executive Officer, IDGEC IPO, 2003–2007
Member, IDGEC Synthesis Planning Group, 2004–2006

Summary for Policy Makers

This book summarizes a decade of research on the question of *whether institutions matter* in tackling environmental problems. Institutional analysis is on the cutting edge of the social sciences today. Institutions have been critical forces in shaping "real world" environmental governance systems, and therefore are significant not only for scientific but also for policy advances. Institutions are defined in the research as systems of rights, rules, and decision-making procedures. The studies indicate that institutions play a role in both causing and addressing problems that arise from human-environment interactions but that the nature of this role is complex. This book aims to promote further scientific inquiry. It aims also to inform policy makers in their efforts to address the challenges posed by problems such as loss of biological diversity, degradation of forests and oceans, and the overarching issue of climate change.

Overview

Institutions give rise to social practices, assign roles to the participants in these practices, and govern the interactions among the occupants of the various roles. Environmental and resource regimes are types of institution. They address situations in which actions can degrade ecosystems through overuse of natural resources (e.g., fish stocks) or because of unintended side effects (e.g., air pollution). Approached in this way, regimes constitute important components of governance systems at levels of social organization ranging from the local to the global. Institutions are distinct from organizations, which are material entities typically possessing personnel, offices, budgets, a legal personality, and so forth. Organizations play important roles in the administration and management of regimes dealing with a wide range of topics (e.g., the U.S. Environmental Protection Agency, the International Maritime Organization). The work reported in this volume constitutes the scientific legacy of the International Human Dimensions Programme (IHDP) core project on the Institutional Dimensions of Global Environmental Change (IDGEC). The project's Science Plan highlighted two sets of issues: (i) research foci, or questions about the causality, performance, and design of institutions; and (ii) the analytic themes of institutional fit, interplay, and scale. It also focused attention on the science-policy interface, reviewing ways in which research could influence policy and policy makers could help shape research agendas. The synthesis phase of the project harvested the scientific findings, investigated policy implications of those findings, and explored new directions in research. The main topics and policy-relevant findings of the research are outlined below.

Causality (see chapter 2)

Institutions are important to consider in policy making, but the roles they play are complex and hard to decipher.

• *Interactive causes and effects*   Effects of institutions are typically *nonlinear*, as they are characterized by thresholds and tipping points and are often *contingent* upon a set of other factors. In environmental governance, institutions often work as integral components of complex responses, owing much of their influence to other elements, such as nongovernmental organization (NGO) activities, but in turn themselves facilitating and focusing collective learning and action.
• *Assignment of institutional cause and effect*   Regimes designed specifically for the purpose of environmental governance are not necessarily the institutions most important in causing or addressing environmental change.
• *Motives for implementation and compliance*   Utilitarian motives (the "logic of consequence," where sanctions and rewards change the cost-benefit calculations of the players) and normative motives (the "logic of appropriateness," where rules are followed because they are seen as rightful and legitimate and obligations are encapsulated in the identity and social collectivity generated by the institution) are most often both at work and simultaneously so. The relative importance of normative motives tends to increase with the density and stability of inter- and transnational relationships. The more dense and stable the inter- or

transnational relationship, the more likely it is that normative motives will prevail.

Performance (see chapter 3)

The criteria best suited for evaluation of institutional performance by policy makers are efficiency, equity, and sustainability. Performance assessment involves the following categories and key actions:

• *Stated goals* Evaluate to what extent these are achieved.
• *Unstated goals* Assess whether any have been fulfilled, intentionally or unintentionally.
• *Comparison* Compare the performance of two or more different regimes dealing with related issues (e.g., which fisheries agreements or river pollution agreements are more and which are less effective in achieving their goals).
• *Trade-offs* Estimate trade-offs between different criteria (e.g., equity versus efficiency).
• *Cost-effectiveness* Evaluate the cost-effectiveness of alternative ways to achieve agreed-upon goals.
• *Baseline and operating circumstances* Take these into account in the assessment of similar regimes, since they may produce different results under varying conditions (one size or type does not fit all).

Design (see chapter 4)

Given that one size or type does not fit all, policy makers are more likely to succeed in creating an institution that proves effective in solving (a) specific problem(s) by using a diagnostic method.

• *Diagnostic approach* Because institutions interact with a range of other factors, this approach works better in designing institutions than a search for design principles or generalizations applicable to the full range of environmental and resource regimes.
• *Major factors* Diagnostic queries seek to probe the nature of the problem, the overarching political setting, the character of the actors or players, and the prevailing practices. A composite picture needs to be built of all major factors contributing to a specific issue in order to provide insight into (i) the scope of the biophysical system to be addressed by the institution, (ii) the appropriate goal(s) and its/their nature—

environmental and/or behavioral—to set for an institution, (iii) the rights to be conferred by the institution, (iv) the rules to be implemented, (v) the decision-making procedures to be followed, (vi) key agencies responsible for implementation of the institution, (vii) bodies with which the institution needs to be in communication, and (viii) the hierarchy of administration in which the institution will operate.

• *Feasibility*   Recommendations for institutional design must emphasize proposals that are realistic or feasible within the relevant sociopolitical setting. Yet changes that seem utopian under normal conditions may become feasible during "windows of opportunity" brought about by economic, political, and social changes.

Fit (see chapter 5)

Misfits between institutions and biophysical and socioecological systems are common; they are often extremely difficult to eliminate, even when they are well known, such as water rights in the western United States. Environmental regimes need to account for the fact that such systems are highly dynamic and multilevel, entailing periods of both incremental and abrupt change as well as considerable uncertainty.

• *Biophysical and socioeconomic diversity*   Case-by-case assessment and allowance for biophysical and socioeconomic diversity are the best ways to avoid institutional misfits.
• *Results of promotion of multilevel governance*   Such promotion does not always produce an enhanced fit between ecosystem dynamics and governance in environmental regimes.
• *Quality of interaction among institutional players*   This is important in the way learning is stimulated, how different interests are bridged and common goals worked out, and how polycentric institutions are used to ensure political, legal, and financial support for the sustained existence of the institutional framework.

Interplay (see chapter 6)

Institutional interplay occurs when the operation of one set of institutional arrangements affects the results of another or others. Given the rapid growth of institutional arrangements at all levels of social organization, interplay is an increasingly common occurrence, one that can produce positive as well as negative results for environmental governance.

• *Issue area overlap* To the extent that overlap occurs, actors can choose the most suitable existing institution(s) for a policy initiative.
• *Integrated strategies* Actors can pursue preferences that take into consideration the potential of the varying institutions affecting an issue area both for establishing new norms and for policy implementation.
• *"Strategic inconsistency"* This has been successfully created by several environmental institutions in terms of their scope, for example, by regulating particular areas of international trade or employing trade measures as an enforcement tool. As a result they have limited the implications of existing free-trade rules and have carved out certain areas of the regulatory authority of the World Trade Organization (WTO).
• *Potential strength of institutional fragmentation* Fragmentation may constitute a strength rather than a weakness in international environmental governance. Institutions with large regulatory overlaps appear to create substantial added benefit if they employ complementary governance instruments, represent different memberships, or provide for significantly different decision-making procedures.

Scale (see chapter 7)

There is no optimal level of sociopolitical organization from which to address a problem. Rather, levels are identified through a political framing process, a process that itself changes the nature of the problem, the menu of possible solutions, and the way in which the results are evaluated.

• *Scalar analysis* This type of analysis helps to ensure recognition of complexity in the way a problem is defined and of the appropriate levels for the application of solutions or specific components of solutions.
• *Administrative levels and/or times* Tasks relating only to one solution may need to be assigned at different levels and/or times. An emphasis on comanagement, integrated management, and adaptive management is a natural corollary.
• *Subsidiarity* Although this principle of governance has become widely accepted, research about how subsidiarity translates into practice shows it is difficult to guarantee its aim of local control over local issues. Subsidiarity has become unworkable and is often an illusory panacea offered to local and national governments in return for loss of sovereignty.
• *Scaling* Scaling can be used to address equity concerns and to help bring about coordinated, consistent, and effective efforts at all appropriate levels to solve problems, including those prioritized globally.

Science-Policy Interface (see chapter 8)

There is a serious communications gap between science and policy in the global environmental change arena. In the face of major environmental changes, the need is greater than ever for scientists and policy makers to engage in two-way communication.

• *Greater policy maker input*   Such input could help meet the need to refine present research findings in depth, detail, and range, and to define new research agendas.

• *Joint research projects or processes*   If both sides are well connected in expanded and deeper research, this will in turn help to produce more useful and more directed advice to policy makers on creating effective institutions.

• *New institutions and institutional redesign*   These are needed to confront emerging environmental problems and sets of interdependent problems.

• *Obstacles to improved science-policy interaction*   These include vastly different time horizons, lack of opportunities for scientists and policy makers to interact informally, and lack of knowledge about the policy-making process and opportunities for scientific input.

• *Ways to overcome obstacles*   These include the identification and support of knowledge brokers to link the two communities and increased opportunities in the research process and in funding decisions for policy makers to make their knowledge needs known to scientists and funding bodies.

New Directions (see chapter 9)

Part of the legacy of the IDGEC project is a process of identifying new research themes. A new program of research on Earth system governance has emerged that is looking at the role of institutions in a broader governance framework and that addresses issues of governance from the local to the global level. This new program will be policy relevant through its development of a new paradigm that reflects the current political context, that is, one that acknowledges the transformation presently occurring from dedicated, single-institution environmental policy to governance systems that encompass all aspects of the Earth system—geosphere (land), atmosphere (air), hydrosphere (water and ice), and biosphere (life, particularly human life as a primary agent of environmental change). Specific themes include:

• *Earth system governance architectures*   Investigating how specific governance systems fit together.
• *Agency of actors*   Understanding the way power is exerted in the roles played by both public and private actors in Earth system governance.
• *Adaptive Earth system governance*   Analysis of institutional change will be conducted with the objective of developing *adaptive* forms of Earth system governance.
• *Accountability and legitimacy*   Examining how to ensure these are created in governance systems.
• *Access to goods and their allocation*   Inquiry into the distribution of material and nonmaterial values.

An Invitation

Those engaged in past and new research welcome input from members of the policy community. What is or is not helpful about the research so far? Can future research become more relevant to the needs of policy makers? What improvements in communication are feasible and desirable? Those who participated in IDGEC as well as those responsible for developing the new Earth System Governance Project are interested in comments and ideas and would be pleased to engage in a dialogue about these issues in any convenient format.

IDGEC Glossary

The IDGEC Glossary includes terms that are of central importance to the research carried out under the auspices of IDGEC or inspired by the project. Members of the project coined some of these terms. Others are in common use, but the definitions provided here reflect the way the terms are used by members of the IDGEC community.

causality IDGEC research focus regarding the roles that institutions play in producing and addressing global environmental change processes

complex causality A form of causality in which two or more independent variables interact with one another

cross-level interaction An interaction across levels of a single scale (e.g., an interaction between national and international levels on the scale of jurisdiction)

cross-scale interaction An interaction across different scales (e.g., an interaction between spatial and temporal scales)

design IDGEC research focus regarding the (re)formation of institutional arrangements to address environmental problems

fit IDGEC analytic theme that addresses the (in)compatibility or (mis)match between the properties of biophysical systems and the attributes of institutions

governance The process of steering or guiding societies toward collective outcomes that are socially desirable and away from those that are socially undesirable

governance system An institutional arrangement created to perform the function of governance with regard to a specific society and sometimes a specific issue

IDGEC analytic themes A set of research topics focusing on relationships between institutions and the biophysical world (fit), between or among distinct institutions (interplay), and between or among different levels of social organization (scale)

IDGEC research foci A series of substantive topics dealing with issues of interest to all those studying institutions, including the effectiveness of institutions (causality), the extent to which the effects of institutions meet criteria of evaluation (performance), and the (re)formation of institutions to achieve desirable ends (design)

institution   A cluster of rights, rules, and decision-making procedures that gives rise to a social practice, assigns roles to participants in the practice, and guides interactions among occupants of these roles

interplay   IDGEC analytic theme that scrutinizes interactions among institutional arrangements, horizontal or vertical and political or functional, that significantly influence institutional outputs, outcomes, and impacts

knowledge broker   An individual well versed in both the policy and scientific worlds who facilitates communication between the two

misfit/mismatch   An incompatibility between institutions and biophysical systems

multilevel governance   Governance that operates at two or more levels of social organization (e.g., local, regional, national levels)

new institutionalism   A school of thought that explores the role of social institutions as sources of governance

organization   A group of people joined together to achieve a specific purpose. Typically, an organization has personnel, offices, equipment, a budget, and, often, legal personality.

performance   IDGEC research focus that explores the effectiveness of institutions evaluated in terms of criteria such as efficiency, equity, and sustainable development

portfolio approach   A methodology featuring the use of multiple techniques of analysis to explore the role of institutions in causing and addressing environmental issues

regime   A type of institution that focuses on a specific issue of concern to society

regime complex   Two or more regimes linked together to form a larger system of governance

resource or environmental regime   A regime dealing with an issue relating to human-environment interactions

scalar politics   Shifting levels on a particular scale to achieve political advantage

scale   IDGEC analytic theme that explores the generalizability of findings across levels of a specific scale (e.g., across local, national, and international levels on the scale of social organization)

I

Introduction

1

Institutions and Environmental Change: The Scientific Legacy of a Decade of IDGEC Research

Oran R. Young

Introduction

How does current thinking about the institutional dimensions of environmental change differ from the way researchers and practitioners thought about this subject a decade ago (Young et al. 1999/2005)? Can research produce scientifically valid claims about conditions determining the success of environmental and resource regimes? What insights can we derive from this effort that will prove helpful to policy makers responsible for creating institutional arrangements dealing with the most pressing environmental issues of our times (e.g., the impacts of climate change, the accelerating loss of biological diversity, the depletion of marine living resources)? Can research provide the basis for practical advice to those responsible for administering environmental governance systems?

This volume addresses these questions through an assessment of the scientific contributions of the long-term, international research project on the Institutional Dimensions of Global Environmental Change (IDGEC). In the process it seeks to distill and appraise the project's legacy in a manner accessible to a variety of audiences. Individual chapters evaluate the contributions of the project both to generic issues relating to all governance systems and to issues that are more specific to environmental governance. Separate chapters explore the policy relevance of research carried out under the auspices of the project and consider cutting-edge questions that will be of interest to researchers working in this field in the coming years. Uncertainty remains a prominent feature of knowledge regarding the institutional dimensions of large-scale environmental change; there is no shortage of priority topics for future research in this field. But we will endeavor to demonstrate in this volume that the work of members of our scientific community together with that of many others engaged in related research is advancing knowledge in this domain substantially.

IDGEC emerged as one of the original core projects of the International Human Dimensions Programme on Global Environmental Change (IHDP), itself a member of the Earth System Science Partnership (ESSP). The project's history is more or less coterminous with that of IHDP. The Scientific Committee of the then Human Dimensions Programme (HDP) authorized a feasibility study for a potential project on institutional issues at its final meeting in September 1995. The process of developing what became the IDGEC Science Plan featured a number of phases, including a rigorous review process mandated by IHDP, which replaced the HDP in 1996. The Scientific Committee of IHDP formally approved the Science Plan in November 1998. The project hit the ground running with the appointment of a Scientific Steering Committee (SSC), the inaugural meeting of that committee, and the establishment of an International Project Office, all taking place during the first half of 1999.

Like other global change research projects, IDGEC has passed through a well-defined life cycle lasting approximately ten years. Now we are engaged in a synthesis process designed to capture the scientific legacy of the project and to evaluate future directions in research in this field. Apart from the project on Land Use and Land Cover Change which was already under way when IHDP came into existence and which was sponsored from the outset jointly with the International Geosphere-Biosphere Programme (IGBP), IDGEC is the first IHDP core project to pass through a focused and comprehensive synthesis process. The results will therefore be of interest to all members of the IHDP community and to members of the broader global change research community, as well as to those whose primary interests are the institutional dimensions of environmental change.

Our research proceeds within the overarching milieu of the "new institutionalism" in the social sciences, treating institutions as sets of rights, rules, and decision-making procedures that give rise to social practices, assign roles to the participants in these practices, and guide interactions among the occupants of these roles (North 1990; Young 1999a). Viewed in this way, institutions are not only important in efforts to solve problems; they also can play a role in the onset and impact of environmental problems. The "tragedy of the commons," for instance, is basically a story about missing or inappropriate rights and rules governing the actions of users of renewable but depletable resources (G. Hardin 1968).

Most proposals for avoiding or overcoming this problem focus on intro-ducing changes in prevailing rights and rules, whether they prescribe a transition to private property, a shift to public property, or the develop-ment of some form of restricted common property (Baden and Noonan 1998; Ostrom et al. 2002). These are precisely the sorts of issues that lie at the heart of the project's research agenda. When and how do prevail-ing institutional arrangements influence subjects to give them reasons to behave in a manner that is unsustainable, whether this takes the form of depleting renewable resources (e.g., stocks of fish or mammals) or emit-ting pollutants (e.g., sulfur dioxide or greenhouse gases) into the Earth's atmosphere? Under what circumstances can institutional reform solve or alleviate these problems—or even prevent them from occurring in the first place—and what are the prospects for initiating such reforms and implementing them successfully (Young 1999a)?

Structure of the Book

We have organized this volume to facilitate efforts to identify and assess the scientific legacy of our project. This opening chapter provides an overview of a decade-long research effort and a taste of things to come. Parts II and III address the substantive concerns of the project, grouped under the headings of research foci and analytic themes. Part IV then turns to a discussion of the policy relevance of the project's findings and to an initial assessment of emerging opportunities for the next phase of research on environmental institutions.

Addressing the core of the IDGEC research agenda as laid out in the project's Science Plan, individual chapters in parts II and III identify and evaluate in depth the contributions of the project to the generic questions of causality, performance, and design and to the specific problems of fit, interplay, and scale. In each case we start from the most fundamental question articulated in the Science Plan: what do we know now that we did not know at the time the project was launched in the 1990s? There is no simple way to quantify or determine the exact role our project has played in stimulating the emergence of insights relating to these research foci and analytic themes. We have tried to exercise caution in evaluating the project's role in stimulating specific advances in knowledge relating to institutions. Contributions dealing with the research foci are particu-larly difficult to attribute, since these findings are relevant to understand-ing institutions in generic terms and emerge from a much broader stream of research encompassing work in a variety of disciplines and issue areas. Because the project has been a leader in prioritizing the analytic themes of fit, interplay, and scale, contributions in these areas are somewhat easier to attribute. Where the project has championed the importance of a particular theme, there is reason to believe that it has made a difference in the growth of research on that theme.

Part IV deals with policy relevance and future directions. There is no straightforward way to assess the interactions between scientific research and the policy process. Nevertheless, we have sought in chapter 8 to illuminate the implications of our major findings for policy. We also regard the issue of future directions as vital. There is widespread agreement within the global change research community that a continued emphasis on issues of governance is essential for research on the human dimensions of environmental change. Although we offer some initial observations about research opportunities in chapter 9, we expect the issue of future directions to be a topic of lively discussion in the broader community. There can be no doubt that this debate can and will be informed by the track record of our project.

Foundational Choices

We have sought from the beginning to set our work on the institutional dimensions of environmental change within a broader stream of research of interest to leading social scientists. This effort has led to conceptual, methodological, and substantive choices that define the overarching character of our research program.

The New Institutionalism

Although the project emphasizes the roles that institutions play with regard to environmental change, our research has sought from the outset to take advantage of the intellectual capital of the *new institutionalism* in formulating our research agenda and bringing our findings to the attention of those who are interested in the role of institutions more generally. To take a single prominent example, the project shares with the new institutionalism a strong interest in what are known as collective-action problems, or situations in which seemingly rational choices on the part of individual members of a group lead to societal results that are undesirable from the perspective of all the members of the group (Schelling

1978; R. Hardin 1982). We have known for some time, for instance, that the tragedy of the commons exhibits the defining features of what is known to those who analyze collective-action problems as the prisoner's dilemma (Ostrom 1990). It is apparent as well that efforts to address many environmental problems involve the supply of collective goods and, as a result, often give rise to what is known as the free-rider problem (Olson 1965). Under the circumstances it makes sense to think about the creation of institutional arrangements designed to solve or alleviate environmental problems as exercises in overcoming collective-action problems.

The new institutionalism has become influential throughout the social sciences and in law. An interest in institutions treated as clusters of rights, rules, and decision-making procedures constitutes the glue that holds those who work in this realm together and gives this movement a distinctive "personality" that is well known not only to practitioners of the new institutionalism but also to the movement's critics. As one would expect from such a wide-ranging movement, however, the new institutionalism encompasses a number of analytic strands that are quite distinct (March and Olsen 1989; Rutherford 1994; Scott 1995). Researchers on environmental institutions have taken a particular interest in two of these strands, which we call the *collective-action perspective* and the *social-practices perspective* on the nature and role of institutions (Young 2002b).

It will come as no surprise that the collective-action perspective is the better known of the two. This perspective assumes that individuals have preferences that are exogenous to their membership in groups, that they act on the basis of some sort of utilitarian calculation, and that they endeavor to maximize payoffs to themselves as individuals. Institutions form through a process—explicit or implicit—of developing social contracts. The prisoner's dilemma, the free-rider problem, and, more generally, problems of burden sharing and compliance loom as critical concerns among collective-action thinkers (Barrett 2003). The social-practices perspective, by contrast, assumes that the identities of individuals are shaped in part by group membership, that actors are influenced by what is known as the logic of appropriateness as opposed to the logic of consequences, and that compliance with institutional rights and rules often becomes a matter of second nature or habit (Hart 1961; March and Olsen 1998). As we would expect, economists and many political scientists are attracted to the collective-action perspective, whereas sociologists and many anthropologists find the social-practices perspective more appealing.

The collective-action and social-practices perspectives existed prior to the initiation of our research. During the course of our work, a third outlook on the links between institutions and environmental change has emerged. Less crisply articulated than the preexisting perspectives, this way of thinking, which we characterize as the *knowledge-action perspective*, stresses agency, individual leadership, and the role of governance systems in shaping the way environmental problems are understood (Breitmeier, Young, and Zürn 2006). On this account, prevailing discourses underpin institutions, and institutional change often reflects shifts in pertinent discourses. Knowledge brokers play particularly prominent roles in this perspective (Litfin 1994). So do "champions" who have the ability to move issues to the top of the policy agenda and to make sure that they do not get relegated to the backwater of the policy process. We expect future research in this realm to make a concerted effort to enhance understanding of this perspective.

A hallmark of our research program is an effort to marry—or at least to deploy in tandem—the three perspectives to illuminate the roles that institutions play both in causing and in addressing particular environmental problems. Difficulties in (re)forming institutions, for instance, can be attributed both to the transaction costs associated with institutional bargaining and to the "stickiness" of institutions once they are firmly entrenched and embedded in the thought processes or standard operating procedures of actors as a matter of second nature. Compliance with sets of rights and rules can be explained, then, both in terms of calculations regarding the expected costs of noncompliance and in terms of the influence of socialization or the habit of obedience. Sluggishness in responding to major environmental problems may reflect either opposition on the part of influential interest groups or the absence of clear characterizations of the problems or of champions needed to make sure they are not ignored. We are not in a position at this point to merge the three perspectives fully to create a single, overarching theory of environmental institutions. But researchers studying these institutions regularly make use of all three perspectives, often in efforts to explain the success or failure of specific institutional arrangements (e.g., the successful ozone regime as articulated in the Vienna Convention of 1985 and the Montreal Protocol of 1987 as amended, in contrast to the limp climate regime embedded in the 1992 UN Framework Convention on Climate Change and the Kyoto Protocol of 1997 as operationalized in the subsequent Marrakech Accords).

Complex Causality

Beyond the effort to situate our research within major social science perspectives to provide context for the analysis of environmental institutions, our research addresses fundamental questions regarding the roles that institutions play as determinants of societal outcomes. Mirroring broader perspectives in the social sciences, many observers of institutions approach this issue in terms of the idea of causal chains and draw a distinction between what are typically called underlying factors and proximate or intervening variables. From this perspective, the underlying forces in human affairs are factors like population growth, increases in affluence and shifts in consumption patterns associated with affluence, and the emergence of new technologies. Those who think in these terms typically treat institutions as intervening variables in the sense that they affect the impact of underlying forces but are not such forces themselves (Krasner 1983). Thus, institutions may play some role in channeling or guiding demographic forces or patterns of consumption and therefore in steering interactions among the members of societies. But they do not account for the nature and causal impact of the underlying forces.

From a methodological point of view, this perspective makes life easier for students of institutions. As researchers have discovered time and again, the most recent links in causal chains are easier to identify and analyze rigorously than links located farther back in these chains. Clues regarding causal connections grow cold quickly as we move backward from one link in the causal chain to another. By contrast, it is often comparatively easy to identify the links in such chains located closest to outcomes of interest to the analyst. As an example, it is easy to see the causal connection between the 1987 Montreal Protocol and the title of the U.S. Clean Air Act Amendments (CAAA) of 1990 dealing with the implementation of the Montreal Protocol. It is far more challenging to probe the economic and political sources leading to the adoption of the CAAA themselves (Bryner 1995).

At the same time, research on environmental institutions has raised profound questions about the usefulness of the simple view of causal chains outlined above (Young 2002a; Lambin and Geist 2006; Young, Lambin, et al. 2006). Systems of rights and rules (e.g., arrangements regarding taxes and subsidies) can and often do serve to guide the choices individuals make regarding consumption. The operation of rules dealing with patents and copyrights can influence substantially the incentives of those endeavoring to develop new technologies. Even demographic trends are influenced by prevailing rights and rules. Compare China with its one-child-per-family rule, for instance, with India, which has no such rule. Restrictive rules regarding family size not only affect overall trends in population—India will soon surpass China as the world's most populous country; they also affect things like the sex ratio of children added to the population.

What inferences can we draw from these observations? Institutions certainly can operate as proximate forces. Arguably this is an appropriate way to think about the arrangements set up to curb emissions of greenhouse gases or to preserve stocks of fish that move in and out of the jurisdictions of a number of coastal states. But institutions can also operate as underlying forces. Even more fundamental is the observation that institutions often form elements of interactive causal clusters in contrast to the mainstream conception of causal chains. Systems of land tenure, for instance, often interact both with patterns of social stratification and with biophysical forces like patterns of rainfall and soil types to produce changes in land use and land cover over time (Lambin and Geist 2006). Emissions trading schemes interact with broader investment opportunities, tax policies, and technological advances to determine the results of efforts to use incentive mechanisms to curb greenhouse gas emissions.

Causal clusters made up of a number of interacting variables pose methodological challenges, a point examined in some detail later in this chapter. For now it is sufficient to note that the shift from research into causal chains to the study of causal clusters has major implications for how we think about the roles that institutions play in steering societies toward desirable outcomes and away from harmful outcomes. A focus on causal chains leading from deep structure to intervening variables and on to outcomes is perfectly appropriate in some settings. But in analyzing the institutional dimensions of environmental change, we regularly find ourselves seeking to understand the impacts of these clusters as composite drivers rather than engaging in frustrating attempts to assign weights to individual elements in these clusters as determinants of collective outcomes. One important consequence is that it is often helpful to make use of the idea of complex systems in studying institutions and to approach outcomes generated by causal clusters in terms of the concept of emergent properties.

Crosscutting Applications

A third starting point centers on the observation that institutions constitute a crosscutting theme in research on issues of environmental change. Most projects launched under the auspices of the global environmental change research programs—the World Climate Research Programme and DIVERSITAS as well as IGBP and IHDP—focus on more or less bounded issues. These include matters like industrial transformation, urbanization, coastal zone processes, the carbon cycle, and food systems. They strive to bring an extensive collection of tools to bear in efforts to enhance our understanding of matters like transitions from industrial to postindustrial societies, the extraordinary growth of cities during the twentieth century, or changes in the concentration of greenhouse gases in the Earth's atmosphere. By contrast, researchers analyzing institutions seek to understand the roles that institutions play in all these realms. How do rules affecting the use of the atmosphere as a repository for wastes or residuals resulting from the burning of fossil fuels affect rates of emissions of greenhouse gases? How do systems of taxes and subsidies influence decisions about investments that have consequences for the introduction of new technologies or the development of new products involved in the transition from industrial to postindustrial society? Can the creation of quasi-markets help to control greenhouse gas emissions or to avoid severe depletions of living marine resources? In each of these cases, will the results be favorable from the perspective of various conceptions of fairness or equity?

The crosscutting nature of the role of institutions is both an opportunity and a potential pitfall for analysts interested in environmental institutions. It has provided no end of requests for collaboration with those engaged in other projects, whether they involve issues relating to the allocation of carbon allowances, the development of entry barriers designed to conserve fish stocks, the protection of coastal wetlands and mangrove forests, or the degradation of dryland ecosystems. At the same time, researchers studying environmental institutions are acutely aware that the investigation of institutional issues of interest to other global change projects could easily divert attention from research on environmental institutions per se. The need to establish priorities does not preclude mutually beneficial collaboration between those focusing on institutions and those concerned with climate change, the loss of biological diversity, the allocation of fresh water to different uses, and so forth. But it does set up a tension within the global change research community that is worth considering carefully and reflecting on regularly.

A decision made at the first meeting of our SSC in 1999 has had long-term consequences regarding collaboration with other projects. To provide empirical grounding for studies of a wide range of institutional issues, the SSC created three flagship activities known respectively as the Political Economy of Forests (Contreras, Lebel, and Pasong 2001), the Performance of Exclusive Economic Zones (A. Hoel 2000), and the Carbon Management Research Activity (Sewell, Wasson, and Yamagata 2000). These activities have turned out to be useful for a number of purposes (Young 2003b). Still, it is only fair to observe that the existence of the flagship activities has constrained interactions with other global environmental change research projects. Built-in sources of rich empirical materials have weakened incentives to go farther afield in search of interesting applications than would have been the case in the absence of the flagship activities.

Institutional Discourses

One of the most far-reaching and powerful contributions a coherent school of thought or paradigm can make arises from its role in structuring mental maps and framing the questions asked in contrast to providing answers to specific questions (Kuhn 1962). The rise of the Keynesian approach to fiscal policy growing out of the experiences of the Great Depression and its subsequent displacement by an approach placing greater emphasis on monetary policy is a well-known case in point. With respect to the environment, both the rise of interest in incentive systems, in contrast to command-and-control regulations, as a means of channeling behavior and the shift from a focus on reaping sustainable yields from discrete populations or stocks of living resources to the idea of ecosystem-based management have profoundly changed our ways of thinking about human-environment interactions.

The emergence of a stream of research rooted in the new institutionalism and centered on the idea of environmental governance constitutes another major shift in discourse among those working on human-environment relations. The rise of a new discourse is a complex process involving both perceptions and judgments regarding probable payoffs for individual actors. Assessments of the roles that specific actors play in this process are notoriously subjective. Nonetheless, paradigmatic change is an important part of the legacy of recent work on the institutional dimensions of environmental change. As a result, it is worth describing this development in some detail and paying careful attention to the conceptual building blocks that serve to fix our project's place in the resultant discourse regarding governance.

Institutions versus Organizations
The new institutionalism draws a clear distinction between *institutions*, treated as clusters of rights, rules, and decision-making procedures that give rise to social practices, and *organizations*, construed as material entities that typically have personnel, offices, equipment, financial resources, and often legal personality (Young 1989a, 1994a; North 1990). The political system set up under the terms of the U.S. Constitution, with its emphasis on federalism together with checks and balances, is an institution; the U.S. Congress is a large and highly complex organization whose purpose is to select policies through a legislative process spelled out in the Constitution and to ensure that these policies are properly implemented. The world market for oil is an institution; British Petroleum, Royal Dutch Shell, and ExxonMobil are all organizations formed to take advantage of opportunities for producing, refining, and marketing petroleum products through the operation of this market. For shorthand purposes, we often say that institutions are the rules of the game and organizations are the players in these institutions.

The introduction of a distinction between institutions and organizations is not meant to downgrade the importance of understanding organizations. Not only are organizations the key players in many institutions; they also, as the example of the U.S. government in the preceding paragraph suggests, can and often do become important as bodies responsible for administering the rights, rules, and decision-making procedures that constitute the defining features of institutions. In this light the new institutionalism highlights the relationships between institutions and organizations as a prominent topic for research. Studies of small-scale, traditional societies have made it clear that the establishment of organizations is not a necessary condition for the creation and operation of effective institutions. Many small-scale societies, for instance, have developed sophisticated arrangements governing the appropriation of living resources and competing uses of land without creating a government in the conventional sense to administer these arrangements (Ellickson 1991; Ostrom et al. 2002; Berkes 2007). Nor is the creation of organizations sufficient to ensure that institutions are implemented effectively and fairly in more complex social settings. The world is full of failed states along with organizations that greedy leaders have established and operated largely as vehicles for acquiring power and wealth for themselves. Nonetheless, the link between institutions and organizations is an important one.

Governance versus Government

Institutions arise in many settings and play a wide variety of roles. Institutions that emerge in response to a demand for steering mechanisms to guide societies toward outcomes that are socially beneficial and away from outcomes that are harmful can become elements of *governance systems*. Of course, institutions may fail to satisfy the demand for governance in specific situations, and they are subject to corruption when they fall under the influence of actors not motivated by a commitment to social welfare. Still, our focus here is on the contributions that institutions can and do make to meeting the demand for governance.

Our project belongs to a broad stream of research concerned with conditions determining the success or failure of governance systems in a variety of settings and with the unintended consequences or social costs of arrangements designed to solve specific problems. The development of thinking about governance—in contrast to government—has become a growth industry among those interested in a wide range of substantive issues. Several of those who crafted the IDGEC Science Plan were influenced by the emerging discourse on governance before assuming their roles in the project (Young 1994a); many scholars having no affiliation with the project have played influential roles in the development of this discourse in recent years (Rosenau and Czempiel 1992; Kooiman 2003). It is helpful to place the work of the project in an intellectual setting featuring the emergence of new thinking about governance. Environmental issues have triggered some of the most innovative experiments with new forms of governance over several decades (von Moltke 1997; Young 1997). The project has enjoyed the good fortune of operating within a remarkably vibrant intellectual setting (Keck and Sikkink 1998; Reinecke 1998; Reinecke and Deng 2000; Cashore 2002; Kaul and Le Goulven 2003; Bellamy 2005).

A recent development, stimulated in part by work carried out under the auspices of our project, involves a clarification of the relationship between institutions and governance. Institutions play critical roles in meeting the demand for governance, but they are not the only factors that contribute to the supply of governance in most settings. Belief systems, norms, culture, and a sense of community typically operate alongside institutions as mechanisms guiding the behavior of actors toward collectively desirable outcomes and away from social snares. This is not to downplay the roles that institutions play; they are necessary to the supply of governance. Yet the performance of institutions is conditioned by the character of the sociocultural environment in which they operate. The presence, for example, of a broader culture of compliance can alleviate or even eliminate the need to build elaborate compliance mechanisms into institutions created to address specific problems. It follows that efforts to design effective governance systems must pay attention to the compatibility of institutional arrangements and the principal features of the relevant sociocultural setting.

Resource and Environmental Regimes

Although governance systems come in many forms, researchers working in this field have found it useful to draw a distinction between broad, overarching arrangements or institutions designed to address a wide range of substantive issues and issue-specific institutions focusing on a particular issue area and often created to address a particular problem. The U.S. Constitution, the UN Charter, and the Law of the Sea Convention are all examples of broad constitutive arrangements; they provide mechanisms for arriving at collective choices about all sorts of issues. The international arrangement created to protect stratospheric ozone, the procedures established under the terms of the U.S. Fishery Conservation and Management Act as amended, and informal practices that arise to handle disputes between neighboring landowners, by contrast, are all specific governance systems (Young 1982b; Ellickson 1991; Parson 2003).

Our project has adopted the usage of those who employ the term *regime* to refer to the large universe of these issue-specific arrangements as they arise and operate at levels of social organization ranging from the local to the global. This has produced a strong interest in interactions—both horizontal and vertical—between and among regimes as institutions dealing with specific issue areas (e.g., the interaction between the international trade regime and the regimes embedded in multilateral environmental agreements or MEAs [Young et al. 2008]). There are obvious links as well between broad constitutive arrangements and regimes. Regional fisheries regimes, for instance, operate within the overarching framework of the law of the sea (Ebbin, Hoel, and Sydnes 2005); the U.S. regime created to curtail emissions of sulfur dioxide operates within the broader framework of the American political system (Tietenberg 2002).

Given our focus on environmental concerns like global warming and the loss of biological diversity, researchers associated with the project have taken a strong interest in the creation and performance of regimes. This strategic choice has played a role in bounding the scope of our efforts, producing a body of research that casts an intense light on issues like the formation and effectiveness of regimes, while directing less attention to issues like the links between regimes as issue-specific arrangements and overarching governance systems. In our view this choice has proved fruitful, a judgment that later sections of this chapter and the substantive chapters to follow endeavor to justify.

Our community generally refers to regimes created to address issues relating to natural resources and the environment as *resource and environmental regimes* (Young 1982b). It is common to speak of resource regimes managing human uses of renewable and nonrenewable resources (e.g., fish, hydrocarbons) and environmental regimes managing anthropogenic pollutants and the disposal of wastes or residuals (e.g., air pollution, greenhouse gases). Our working hypothesis is that the two categories are sufficiently similar to justify treating them as a single universe of cases. Taken together, such regimes are common across the spectrum from local to global arrangements, and they deal with a wide range of substantive issues. The potential universe of cases is large, though researchers are confronted immediately by questions relating to generalizability both across levels of social organization and across issue areas (Young 2005b). The project joins others who ask whether it is possible to scale conclusions up or down across levels of social organization or to transfer conclusions derived from the study of regimes operating in one issue area to understand what is happening in other issue areas.

Our project shares the tendency of much regime analysis to look first at institutions that are governmental or intergovernmental in nature (Young 2005a; Breitmeier, Young, and Zürn 2006). Yet we now know that such arrangements are special cases of a broader category of environmental regimes that include private governance systems (e.g., the Chicago Climate Exchange in the United States), systems in which actors located in civil society play prominent roles (e.g., codes of conduct), and hybrid arrangements in which several distinct types of actors emerge as prominent players (e.g., the Forest Stewardship Council and the Marine Stewardship Council). This is good news not only in the sense that it expands the scope of efforts to address problems of environmental governance in today's world but also in the sense that it increases the size of the universe of cases available to researchers seeking to answer fundamental questions about the formation and effectiveness of environmental governance systems. Work on systems featuring important roles for customary practices, markets, and various types of networks is now going on in many quarters (Keck and Sikkink 1998; Reinecke 1998). Researchers working on our project have joined others in pursuing this line of thinking.

Research Foci: Causality, Performance, and Design

From the outset, one of the project's goals has been to address a set of generic questions of interest to all analysts of institutions in order both to take advantage of the broader stream of thinking about institutions and to elicit attention from scholars conducting research on institutions who have no special interest in resource and environmental regimes. This goal is reflected most clearly in the Science Plan's emphasis on research foci and, more specifically, on the questions of causality, performance, and design. The question of causality concerns the extent to which institutions influence the course of human affairs in a variety of social settings. The question of performance examines that subset of institutions that are effective in the sense that they make a difference; it seeks to evaluate institutional consequences in terms of well-defined criteria, including sustainability as well as efficiency and equity. When institutions do play a role of some importance in solving problems, questions of design also arise. Can we hope to (re)form regimes in ways that will enhance the prospects for achieving collective outcomes that are socially desirable or avoiding outcomes that are harmful? All those who think about institutions in the context of governance systems are concerned with these questions in one form or another. The first substantive section of this volume includes separate chapters devoted to each of these research foci. In each case we address the question: what do we know now that we did not know at the inception of the project? Here the aim is only to introduce the issues at stake and to capture the flavor of the contributions of our work to current understanding of these matters.

The Question of Causality
Despite the rise of the new institutionalism as a powerful force throughout the social sciences, there are lingering doubts regarding the roles that institutions play in influencing the course of human affairs (Young 1989c). Partly this is a matter of criticism launched by those who argue that other drivers account for most of the variance in collective or societal outcomes. Analysts who emphasize the central role of power, for instance, regularly assert that institutions are epiphenomena reflecting the political bargains underlying them and changing or adapting readily when the distribution of power in society shifts (Strange 1983; Mearsheimer 1994/1995). In part, doubt arises as a reflection of methodological problems confronting those who seek to develop and test propositions regarding the roles institutions play (Underdal and Young 2004). Like ecosystems, governance systems have fuzzy boundaries, a fact that can make it hard to separate individual regimes cleanly and that can lead to disagreements about what to include in the universe of cases. And because opportunities to engage in natural experiments—much less controlled experiments—are limited in research on social institutions, testing hypotheses about the formation and effectiveness of regimes requires a lot of ingenuity.

Two distinct messages regarding causality are worth noting in this introductory account. The first centers on the distinction among outputs, outcomes, and impacts that many analysts working in this field have adopted (Underdal and Young 2004).[1] It is a relatively easy task to demonstrate causality at the level of outputs or immediate products (e.g., treaties, statutes, regulations) of the policy process. No one doubts the persuasiveness of counterfactuals of the following sort: The U.S. government would not have adopted implementing legislation in the form of a major title of the CAAA of 1990 if there had been no 1987 Montreal Protocol. The Environmental Protection Agency would not have promulgated regulations covering phaseouts of certain chlorofluorocarbons

(CFCs) and related ozone-depleting substances in the absence of the CAAA. Efforts to apply them to specific cases would not have occurred in the absence of these regulations appearing in the U.S. Code of Federal Regulations. In this sense it is a simple matter to demonstrate the occurrence of some cause-and-effect relationships regarding environmental institutions.

This is not a trivial observation; it would be a mistake to dismiss the significance of such relationships out of hand. Still, it is obvious that focusing only on outputs in examining the causal significance of institutions will not do. We want to know something about outcomes or the effects of institutions on the behavior of key actors in the relevant systems. Even more to the point, information is needed about impacts or the extent to which resource and environmental regimes play influential roles in solving or at least alleviating the concerns leading to their creation (Miles et al. 2002; Breitmeier, Young, and Zürn 2006). This is where the problems begin to mount. There is, in short, a direct relationship between the importance of the issue addressed by a regime and the methodological challenges involved in analyzing the role of the regime in dealing with it. Although the causal role of institutions in producing outputs is easy to demonstrate, outputs are relatively unimportant as measures of the significance of institutions. Demonstrating the influence of institutions in terms of impacts, by contrast, is extremely hard. Yet this is precisely what our research aims to illuminate regarding the roles that institutions play in causing and mitigating environmental problems.

An understandable, though less than fully satisfactory, response to this tension is to focus attention on the effects of institutions on the behavior of key actors (Young and Levy 1999; Young 1999a). Guiding the behavior of human actors is essential to the success of any effort to solve environmental problems. Increasingly we have come to realize that human actions are critical drivers in the onset of these problems. Human behavior is easier to analyze rigorously than the impacts of institutions measured in terms of the degree to which they cause or solve problems. Researchers can examine large universes of cases to develop propositions about human behavior; there is even considerable scope for the conduct of controlled experiments dealing with such matters as the prevalence of behavior conforming to the logic of appropriateness in contrast to the logic of consequences (March and Olsen 1998). It is well worth noting, for instance, that individuals do not always choose the option known as *defect* in situations exhibiting the structure of the prisoner's dilemma, that cooperation increases when subjects are allowed to communicate even in the absence of enforcement mechanisms, and that many subjects are risk averse in the sense that they will choose the certainty of a fixed payoff over a probabilistic payoff yielding a higher expected return (Kahneman 2003).

Still, this does not allay concerns regarding the causal significance of resource and environmental regimes. This is where our second message relating to causality comes into focus. As discussed in the previous section, complex causality is common rather than exceptional in relation to institutions. This means that institutions typically form elements of interactive clusters of driving forces that determine collective outcomes in social settings (Young 2002a; Lambin and Geist 2006). For the most part these clusters include biophysical forces (e.g., changes in the length of the growing season, the extent of seasonal sea ice, the temperature of ocean water) as well as socioeconomic forces (e.g., the emergence of new harvesting technologies, shifts in human consumption patterns, changes in dominant political coalitions, movements of human populations). The fact that these forces interact—movements of human populations may follow environmental changes and affect political coalitions; new technologies may play a role in altering consumption patterns—means that it is always hard and sometimes impossible to separate out the signals of individual elements in causal clusters and assign specific weights to them in terms of explaining or predicting the character of collective outcomes.

This observation is sobering. It complicates—though it does not preclude—the use of many familiar statistical procedures (e.g., various forms of regression) in evaluating the causal significance of institutions (Young, Lambin, et al. 2006). It also explains why institutional arrangements that yield satisfactory results in some settings (e.g., systems of land tenure based on private ownership) may work poorly or even fail miserably in other settings (Komesar 2001; Cole 2002; Rajesh and Lebel 2006). It is a simple exercise to compile a long list of instances in which recommendations regarding institutional arrangements that worked well elsewhere have failed or, in some cases, even generated negative results for those who have adopted them. This observation has profound consequences for the question of design to be addressed later in this section. But for now it helps to emphasize that the question of causality with regard to the effects of resource and environmental regimes is one that often calls for a high order of sophistication in the analysis of causal clusters rather than the application of reductionist procedures designed to tease out the significance of individual factors through the use of some form of statistical inference.

The Question of Performance

As the preceding account suggests, there is a clear link between causality and performance. It is pointless to worry about the performance of a resource or environmental regime unless there are good reasons to believe the arrangement makes a difference in causal terms. But the analysis of performance differs fundamentally from the study of causality. An institution that alters a problem without solving it or that engenders new problems as unintended side effects of efforts to address a preexisting problem makes a difference in causal terms. The question of performance comes into focus, by contrast, when researchers ask not only whether a regime makes a difference but also whether it produces results that meet the requirements of criteria of evaluation involving standards like efficiency, equity, sustainable development, robustness, or any other standard deemed appropriate (Young 1982b). There is an essential normative component in the assignment of standards to the question of performance that is not present in the question of causality.

Some researchers in this field—especially those concerned with the performance of international environmental regimes—have sought to address this question in one bold stroke (Helm and Sprinz 1999; Sprinz and Helm 2000; Young 2001b, 2003a; Hovi, Sprinz, and Underdal 2003a,b; Bernauer and Siegfried n.d.). They propose a scale ranging from the outcome that would have occurred in the absence of a regime (i.e., no regime as the counterfactual) to some outcome that is deemed by relevant players to be the social or "collective" optimum. Performance, on this account, is a measure of the location of the actual outcome on a well-defined continuum ranging from the counterfactual to the optimum. A particularly attractive feature of this approach is that it offers a means of comparing and contrasting the performance of different regimes.

The appeal of this way of thinking, sometimes described as the Oslo-Potsdam solution, is obvious. But there are also grounds for questioning the usefulness of this appealing stream of analysis. It is hard to deal with the counterfactual embedded in most efforts to predict what would have happened in the absence of a regime. Observers can and often do differ in their conception of the social optimum, and the idea is difficult to operationalize, even for those who agree on the optimum in conceptual terms. Judging the state of affairs at any given time with regard to a regime's location on this scale is by no means straightforward. Further work on the development of the Oslo-Potsdam solution and other approaches to the evaluation of performance constitutes a high priority for those interested in the institutional dimensions of environmental change.

As explored in chapter 3, other approaches exist for evaluating the performance of regimes (R. Mitchell 2004; Zaelke, Kaniaru, and Kruzikova 2005; Zürn and Joerges 2005). One alternative is to apply familiar concepts of efficiency, equity, and even sustainability to the consequences of specific regimes. It is hard to judge the efficiency of regimes except in terms of very weak standards like Pareto optimality. This is largely a consequence of problems in calculating the benefits associated with steering clear of harmful outcomes or increasing the probability that desirable outcomes will occur over long time periods. As a result, it is easier to think in terms of cost-effectiveness. Could a concrete goal or objective, such as curbing intentional oil pollution at sea, have been met through the use of some means other than the equipment standards that constitute a core element of the regime established under the terms of the International Convention for the Prevention of Pollution from Ships 1973–78 (R. Mitchell 1994a)? Are the claims regarding cost-effectiveness of those who favor cap-and-trade approaches to curbing emissions of airborne pollutants persuasive?

Beyond these issues of efficiency, we have come to believe that many studies of resource and environmental regimes have not placed enough emphasis on evaluating their consequences in terms of standards of fairness (A. Hoel and Kvalvik 2006). The fact that this criterion encompasses a range of issues framed as matters of equity and justice complicates this approach. But there is a need to devote more systematic consideration to the roles that regimes play in determining who gets what and the extent to which subjects regard regimes as fair or just in procedural terms, regardless of the existence of winners and losers. This suggests that there is a compelling case for paying more attention in future research to the politics of the (re)formation and operation of regimes. Such an emphasis would naturally lead to more sustained work relating to a range of topics, including leadership, coalition formation, rent-seeking behavior, and the roles of various nonstate actors.

Evaluation of resource and environmental regimes in terms of problem solving offers a cruder, less easily analyzed, but in some respects more tractable approach (Miles et al. 2002; Breitmeier, Young, and Zürn 2006). Researchers may be able to arrive at relatively clear-cut conclusions regarding the roles that regimes play in solving problems, even when there is no consensus regarding specific criteria such as efficiency, cost-effectiveness, fairness, justice, and so forth. It is relatively easy to gain consensus around the propositions that the Antarctic regime has worked well in alleviating jurisdictional conflicts; the ozone regime has played a significant role in reducing the production and consumption of ozone-depleting substances; and the cap-and-trade system established under the CAAA of 1990 has been successful in cutting emissions of sulfur dioxide, regarded as a precursor to acid rain. Nor would anyone disagree with the judgment that the regime articulated in the United Nations Framework Convention on Climate Change (UNFCCC) and the Kyoto Protocol has failed so far to mitigate the problem of climate change. These are crude judgments; they tell us little about whether some alternative approach might have produced equally good results at a lower cost or generated an outcome that would seem preferable in terms of some conception of fairness or justice. Still, it would be a mistake to underrate the significance of this mode of thinking about performance. Even a four- or five-point nominal scale ranging from no effect on the problem to decisive resolution can provide the basis for exploring and even "testing" a range of hypotheses regarding factors identified in theoretical work as significant determinants of success or failure in efforts to address environmental problems.

Using the International Regimes Database (IRD), participants in our project have taken some initial steps in this direction (Breitmeier, Young, and Zürn 2006). It turns out, for example, that consensus decision making does not always give rise to the law of the least ambitious program (Hovi and Sprinz 2006). A number of regimes that rely on consensus (e.g., the regime for the protection of stratospheric ozone) have performed well in terms of the criterion of problem solving; even the use of a unanimity rule (e.g., the Antarctic Treaty System) turns out to be compatible with problem solving under some conditions. Although enforcement is clearly important in efforts to maximize compliance, such utilitarian mechanisms cannot account for all the variance in compliance. Our studies point to factors like juridification and legitimacy as important determinants of compliance (Zürn and Joerges 2005). What is more, regimes often achieve results in terms of problem solving by refining and deepening actors' understanding of the problem rather than by the adoption and implementation of conventional regulatory arrangements (Young 1999b). And these findings are merely illustrative of a wide range of relationships that come into focus and can be investigated empirically in the context of this problem-solving approach to performance.

The Question of Design

One reason why policy makers and scientists alike exhibit an intense interest in institutions arises from the presumption that these determinants of the course of human-environment relations are more malleable than other determinants (e.g., population, consumption patterns). This leads to hope that we can exercise some control over human destiny by crafting the provisions of resource and environmental regimes and adjusting key provisions of existing regimes in order to improve their performance in the light of experience (Young 2002b). For many analysts this is the fundamental justification for devoting time, energy, and resources to the study of institutional arrangements. Practitioners seek to take charge of fate by creating governance systems that guide behavior in such a way as to generate both socially desirable outcomes and greater knowledge about how to promote the public interest or the common good. Ideally this should lead over time to the development and refinement of a set of tools for designing institutions as part of governance systems applicable to a wide range of specific cases.

Research on environmental and resource regimes makes it clear that successful crafting of the provisions of these arrangements is easier said than done. Some cases appear to exemplify what Hayek (1973) and others have called spontaneous or self-generating regimes. Here, institutions take the form of emergent properties of complex interactions among a number of self-interested actors. On this account, actors do not design resource and environmental regimes in order to solve specific problems (e.g., climate change, the loss of biological diversity); shifts in the character of the institutional arrangements that arise are largely by-products of the efforts of numerous actors to pursue their own interests. The views of the so-called neorealists who regard institutions as surface manifestations of underlying power relations and shifts in institutional arrangements as reflections of deeper changes in power relations illustrate this line of thinking. But there are other, more benign views that also belong to the Hayekian perspective on institutions and institutional change (Fiori 2006). One example is the environmental Kuznets curve, with its emphasis on the proposition that rights and rules shift to favor environmental protection as a society passes through the process of industrialization and on to modernization (Deacon and Norman 2004).

Still, research on issues of environmental governance indicates that this is by no means the whole story (Dietz, Ostrom, and Stern 2003). Actors from the public sector, the private sector, and civil society all devote large amounts of time and resources to negotiations that give rise to specific regimes, and they pay a great deal of attention to the processes through which these constitutive agreements are implemented on the part of individual subjects (Chayes and Chayes 1995). The resultant institutional bargaining has been a topic of great interest to members of the research community. Unlike ordinary legislative bargaining, which exhibits a tendency to revolve around efforts to forge minimum winning coalitions (Riker 1962), institutional bargaining often aims for the creation of larger coalitions and, in many cases, something approaching the coalition of all the actors involved (Young 1994a). The reason for this is simple. The success of regimes requires behavioral change (though not necessarily compliance in the normal sense) on the part of subjects; change of this sort is easier to obtain from subjects who have participated in the process of regime formation and believe that the key provisions of a regime are fair or at least legitimate.

This feature of regime (re)formation can and often does lead to the crafting of important provisions that are opaque and difficult to interpret; it also accounts for the fact that many constitutive agreements include distinct sections that appear to be at odds with one another (e.g., fisheries regimes that promote the conservation of stocks and at the same time subsidize fishers to increase their fishing power) (Young 1982b).

Such results are hard to square with the idea of institutional design as a means of solving social problems. Even so, it is understandable that interest in institutional design remains high and that policy makers and researchers alike expend a lot of effort on issues of (re)design. A significant contribution that has emerged from work carried out under the auspices of our project centers on what we characterize as the diagnostic method (Young 2002b). This method reflects the central role of complex causality as discussed above. Institutions are only one of a number of driving forces that interact with one another to determine the outcomes of human-environment relations. This means that researchers are unlikely to be able to develop design principles in the sense of propositions that identify necessary—much less sufficient—conditions for the creation and operation of regimes that are effective and durable and that apply across the full universe of cases of environmental and resource regimes (Ostrom 1990). A factor that is important in one setting may be of marginal significance or even irrelevant in others. Researchers can hope to identify and evaluate the key characteristics of individual situations and craft specific constitutive agreements to fit the circumstances at hand. To take a couple of simple but illustrative examples: compliance is not a concern when dealing with coordination problems in contrast to collaboration problems (Stein 1982); the problem of burden sharing to pay for the production of collective goods is much less severe in privileged groups than in groups that lack a dominant member (Olson 1965).

Our point of departure in developing the diagnostic method is the view that we need to frame the provisions of regimes to fit the characteristics of specific problems. This is one reason why those negotiating the terms of individual statutes or treaties expend so much time and energy crafting the content of these agreements. Even so, there are many traps in this realm awaiting the unwary and especially those who tend to reason by analogy, borrowing provisions from previous efforts to form regimes and applying them to new cases without careful scrutiny.

Analytic Themes: Fit, Interplay, and Scale

In addition to its research foci addressing topics relating to institutions in general, our Science Plan highlights a set of analytic themes that are more specific to resource and environmental regimes. The drafters of the plan judged these themes to be cutting-edge concerns during the late 1990s. The problem of fit is a matter of the match or congruence between biophysical systems and governance systems. Here, researchers investigate the determinants of fit and how to improve fit as part of the process of regime building and adaptation. The problem of interplay grows out of a perception that discrete regimes can interact with one another and that such interactions become both more common and more significant as the number of discrete governance systems grows in any given social setting. For its part, the problem of scale highlights the extent to which institutional arrangements are similar and exhibit comparable processes across levels of social organization ranging from the local to the global. Because our work pays particular attention to global changes, we have devoted a good deal of energy to analyzing governance systems operating at a large scale (e.g., the arrangements established under the terms of multilateral environmental agreements). Still, there is a need to consider arrangements operating at other levels as well, both because of the importance of what researchers call vertical interplay and because of the prospects for gaining significant insights by comparing and contrasting governance systems operating at different levels of social organization.

Researchers participating in our community make no claim that these three analytic themes encompass all the interesting questions ripe for consideration in a project dealing with resource and environmental regimes. Other questions deserve more concentrated attention, including those concerning the growth of knowledge, the role of political leaders, and the nature of institutional change (Ebbin 2004; Walsh 2004). Even so, the analytic themes embedded in the problems of fit, interplay, and scale have proved fruitful and given rise to significant insights.

The Problem of Fit

There is nothing new about the problem of fit. Creators and operators of resource and environmental regimes have struggled for a long time to create governance systems that are well matched to relevant biophysical systems. But it is worth noting at the outset that the importance of fit has increased along with the growing role of anthropogenic forces in biophysical systems. When human actions play no more than a minor role in the dynamics of biophysical systems, institutional arrangements are more relevant to the achievement of efficiency and equity than to the pursuit of sustainability as a standard for evaluating a regime. As anthropogenic forces rise and begin to take center stage, the problem of fit comes to the fore. Human actions have had far-reaching consequences for biophysical systems for hundreds—perhaps thousands—of years (Turner et al. 1990). Human drivers, however, and with them the importance of fit have become central concerns only in recent times. As some observers have put it, a no-analogue situation has arisen with regard to human-environment interactions (Steffen et al. 2004), a fact that has led prominent scientists in many quarters to argue that the Earth has made a transition from the era known as the Holocene to a new era best described as the Anthropocene (Crutzen and Stoermer 2000).

Misfits or mismatches between biophysical systems and institutional arrangements are common; they are often hard to eliminate or alleviate, even when their existence and negative consequences are widely known within the relevant community. Mismatches may be either spatial or temporal in character. The boundaries of legal and political jurisdictions often bear no relationship to the areal extent of ecosystems, and jurisdictional boundaries are usually hard to change. The rhythms of decision-making procedures frequently differ from the cycles of biophysical systems. A particularly important problem arises in cases where normally stable biophysical systems are affected by rapid-change events that produce nonlinear changes known as state changes or system flips (Gunderson and Holling 2002; Westley 2002). It is difficult for governance systems to match a pattern of this sort, operating in one mode most of the time but being able to switch quickly into a distinct crisis mode when the need arises. This is why governance systems often seem to be caught unprepared by the occurrence of large-scale and rapid biophysical changes (e.g., hurricanes, tsunamis) and to experience great difficulty in reacting in a timely manner when such nonlinear changes occur.

Our research has identified other types of misfits as well. An important case involves the connectivity or level of interdependence of biophysical systems and institutional arrangements. When internal links within a biophysical system tighten but the relevant governance system is highly decentralized or even fragmented, for example, efforts to deal with spreading or even cascading biophysical changes are apt to be tardy and uncoordinated (Crowder et al. 2006; Worm et al. 2006; Young, Berkhout et al. 2006).

Why are mismatches between biophysical systems and governance systems so hard to avoid in the first place, to recognize, and to eliminate even after their existence is widely known (Olson 1982)? A number of factors that typically operate together have been found to account for this phenomenon. Sometimes limited knowledge makes it hard to construct regimes that match biophysical systems. It is difficult to forecast the occurrence of nonlinear changes in complex systems, even when the basic character of the system is well understood (Ebbin 2004). A particularly serious obstacle arises from the rapid growth of anthropogenic drivers in large socioecological systems and the need to understand the dynamics of coupled systems. Not only do researchers lack experience in connecting the component parts of coupled systems; it is also hard to garner support for studies of coupled systems and to assemble diverse teams of scientists willing to devote time and energy to the analysis of these systems. Natural scientists have little experience with efforts to endogenize the human dimensions of these systems in their models. Social scientists, on the other hand, are inexperienced at incorporating biophysical aspects in their models. As the experiences of the Intergovernmental Panel on Climate Change (IPCC) and the Millennium Ecosystem Assessment make clear, there is a long way to go in the effort to bring together the scientific capacity, resources, and commitment needed to understand complex systems like the carbon-climate-human system or the global food system well enough to improve the fit between biophysical systems and governance arrangements with regard to most large-scale environmental problems.

Yet this is only part of the explanation for the persistence of misfits between biophysical systems and governance systems. Political factors also play an important role in this realm. Some actors or interest groups may well benefit, at least in the short run, from maintaining or even nurturing the growth of misfits. The infamous assertion that "rain follows the plow" relates to the desire of politicians located in arid western states to gain admission to the American union, so that they could occupy positions as senators, representatives, and governors in these states once they became members of the United States (Stegner 1954). As a result, ecological factors figured only accidentally in one of the major processes of regime formation in American history, resulting in severe mismatches in the governance of drylands and watersheds. Nor is there any cause for surprise in the observation that those who play central roles in individual MEAs (e.g., the climate regime, the biodiversity regime) generally resist efforts to integrate their activities into larger clusters, despite contentions that such a move could enhance the overall effectiveness of environmental governance systems (Biermann and Bauer 2005). This is a familiar problem of protecting turf and avoiding bureaucratic change that arises in all governance systems and that is just as pervasive in regimes designed to handle human-environment interactions as it is in other areas.

Another complication arises from the fact that efforts to eliminate or alleviate mismatches normally require acts of institutional reform. This leads back to the issue of institutional bargaining. There are good reasons to make the requirements for changing constitutions or constitutive arrangements relatively stringent. Creating institutions that are too flexible or easy to change increases the likelihood that the resultant governance systems will turn out to be epiphenomena. Yet making the requirements for reform too stringent will ensure that mismatches will be difficult or impossible to eliminate. It is easy to propose a general strategy of striking a balance between the marginal costs of misfits and the marginal costs of excessive flexibility, but it is hard to operationalize a grand principle like this for use in real-world settings. The best that can be done is to make a conscious effort to steer a middle course between these threats to the performance of specific regimes. In the meantime it is clear that the complexities of institutional bargaining can and often will thwart well-intentioned efforts to come to terms with mismatches between regimes and the relevant biophysical systems (Young 1994a). This remains true even when there is no secret about the existence of the mismatches and key players have been struggling to address them for long periods of time.

Our flagship activity on the Performance of Exclusive Economic Zones (PEEZ) has provided a particularly rich laboratory in which to study the problem of fit (Ebbin, Hoel, and Sydnes 2005). The creation of Exclusive Economic Zones (EEZs)—formalized in the 1982 UN Convention on the Law of the Sea—was one of the most dramatic and far-reaching institutional changes of the twentieth century. A critical argument emphasized by those in favor of this bold extension of coastal state authority focused on the need to manage marine resources—especially harvestable fish populations—sustainably and the inadequacy of preexisting arrangements to solve this problem. The addition of the 1995 Straddling Stocks Agreement has helped to alleviate some problems (e.g., jurisdictional boundaries that ignore the behavior of fish populations) arising from the creation of EEZs. Still, the state of many of the world's fisheries has continued to deteriorate (Worm et al. 2006). The shift from maximum sustainable yield (MSY) management practices to ecosystem-based management (EBM) has proved difficult at the applied level, despite widespread agreement among scientists and even many policy makers concerning the need for such a transition. All the factors identified in the preceding paragraphs are at work here. High levels of uncertainty plague efforts to understand the dynamics of marine systems. Entrenched interest groups are fearful of harm to their interests likely to flow from a transition from MSY to EBM. Although progress is occurring in some areas, agreement concerning how to structure management systems to achieve EBM is far off, even in cases where the need for a major change is acknowledged. As a result, mismatches prevail, and the crisis in ocean governance continues to grow (Crowder et al. 2006). Still, the work of PEEZ indicates that this situation is not hopeless. The secret to success appears to lie in coordinating distinct regimes at the local, state, regional, and global levels to produce positive or even synergistic effects (Ebbin

2005; Henriksen, Honneland, and Sydnes 2006) rather than attempting to create one overarching arrangement dealing with marine resources.

The Problem of Interplay

Studies of institutional interplay, also described as interactions between or among distinct governance systems, have developed into a cottage industry during the lifetime of our project. The explanation for this development is straightforward: interplay can be expected to increase—often at an exponential rate—as the density of discrete institutional arrangements increases in society. This has obviously occurred in recent times at the international level. MEAs, for instance, now number in the hundreds. But similar developments occur regularly at lower levels of social organization as well. As new activities come on stream and the interdependencies between new and preexisting activities increase, the demand for governance grows. Research on institutional interplay would have been on the rise even in the absence of our efforts. Nevertheless, the project has played a prominent role in shaping the development of this area of research. Many of those who study interplay are active members of the IDGEC community, and much of the resulting literature engages— sometimes critically—with the approach to interplay set forth in the project's Science Plan (Stokke 2001a, 2001b; Oberthür and Gehring 2006c; Cash et al. 2006; Young 2006; Young et al. 2008).

Our Science Plan differentiates between vertical and horizontal interplay, or interactions across or within levels of social organization, and between functional and political interplay, or, in other words, de facto and intentional interplay. The vertical/horizontal distinction is now widely accepted among those analyzing institutional interplay. Interactions between environmental regimes and the trade regime at the international level and interactions between rules relating to clean air or water and rules governing taxation at the domestic level are obviously important (von Moltke 1997). But so are interactions between national and even international governance systems and (often traditional) institutions operating at the local level (Berkes 2002; Young 2002c). An important finding in this regard is that small-scale systems based on traditional practices that work perfectly well on their own often fail when they are impacted heavily by the operation of large-scale resource and environmental regimes. So is the fact that trade regimes tend to affect ecosystems more through their general effectiveness in stimulating the expansion of trade than through specific clashes with the provisions of environmental regimes dealing with matters like ozone depletion, trade in hazardous wastes, or climate change.

The distinction between functional and political interplay, on the other hand, has come in for cogent criticism (Stokke 2001b; Oberthür and Gehring 2006c). As suggested above, it is certainly true that functional or unintentional interactions are important. But those working in this field have pointed to other factors, such as the behavioral mechanisms that give rise to interplay or the substantive features of the issue areas within which interplay occurs, as key distinctions needed to explain or predict the occurrence and consequences of institutional interplay. This is a healthy dialogue (Young et al. 2008). Research is at a relatively early stage in efforts to understand the sources, consequences, and dynamics of interplay in a variety of settings. Our project has figured prominently in the growth of interest in this area of analysis; movement beyond the crude road map for the study of interplay articulated in the planning process almost a decade ago is a sign of intellectual vigor.

Another point worthy of consideration concerns the sign—positive or negative—of institutional interplay. Much of the early interest in this phenomenon arose from a concern about negative effects and specifically about the prospect that regimes dealing with matters like international trade and finance would interfere with efforts to deal with problems like climate change and the loss of biological diversity through the creation of MEAs. But research in this area has raised searching questions about the prevalence of this perceived problem. Using a relatively large set of cases pertaining to the European Union as well as to international society, Oberthür and Gehring (2006c) have concluded that positive—sometimes even synergistic—interactions are at least as common as cases of interference. Much work remains to be done to flesh out these matters, ensuring that research on institutional interplay will continue to flourish during the near future.

Our flagship activity on the Political Economy of Forests, focused primarily on developments in Southeast Asia, has provided a helpful vehicle for interplay research (Lebel 2005; Garden et al. 2006; Rajesh and Lebel 2006). The most significant finding here relates to the effects of the devolution of authority from central governments to regional or even local governments that has been occurring in many countries in recent years (Pasong and Lebel 2000; Contreras 2003). The case for devolution is based on the logic of subsidiarity and the expectation that regional or local authorities not only understand better the dynamics of specific ecosystems and the needs of their constituents but also are less susceptible to corruption than those located in national capitals. For the most part these expectations or hopes have not fared well, at least in the forested areas of Southeast Asia. Our explanation for these disappointing results focuses on complex causality and especially on matters of interplay between local activities and the overarching processes of globalization. The growth of regional and increasingly global markets has shifted power to actors (e.g., multinational corporations in Japan) that have little knowledge of or interest in the fate of local communities in places like Indonesia, Malaysia, and the Philippines that are heavily dependent on renewable forest products (Dauvergne 1997). Pressures on central governments desperate to increase exports to service external debts or to encourage export-led growth have reinforced the influence of the multinational corporations in this realm rather than providing a counterweight for those seeking to stem the tide of globalization at the regional and local levels.

The Problem of Scale
As formulated in our Science Plan, the problem of scale centers on the transferability of knowledge regarding institutions from one level of social organization to another. If something is known about the determinants of effectiveness in resource and environmental regimes operating at the local or micro level, can these findings be scaled up and applied to the national and even the global levels? Conversely, if conclusions are reached about the relative significance of various compliance mechanisms at the global or macro scale, can these findings be scaled down to shed light on sources of the effectiveness of institutions operating at the national and even the local levels? The question of scale in this sense has long been a focus of attention in most of the natural sciences, but it is a comparatively unfamiliar topic for research among social scientists (Gibson, Ostrom, and Ahn 2000). In highlighting this question, the project has made a deliberate effort to alter this situation in regard to the institutional dimensions of environmental change.

A notable finding in this connection is that there are substantial similarities between resource and environmental regimes operating in small-scale, traditional societies and in international society (Ostrom et al. 1999; Young 2002b, 2005b). Both settings lack states in the conventional sense of the term. The evolution of governance systems on the basis of practice in contrast to a reliance on formal, constitutive agreements is important in both settings. Regimes in both domains are likely to depend heavily on stakeholder involvement and on the power of legitimacy in contrast to enforcement of a more conventional sort as a source of compliance. Needless to say, it is important not to ignore critical differences between these settings. Often small-scale systems can rely on a relatively high level of cultural homogeneity, so that the phenomenon of community can play a significant role in the success of issue-specific regimes. In international society, by contrast, the role of community seems less significant. Although some analysts do explore sources of community at this level (Bozeman 1960; Claude 1988), arguments regarding the role of community in this setting, in the sense of a group of actors who share beliefs, attitudes, and norms, seem thin. Similarly, the implementation of rights, rules, and decision-making procedures is far less complex in small-scale systems than in international society. As a number of analysts have pointed out, many resource users in small-scale settings can represent themselves in key decision-making processes, and the users themselves often play key roles in monitoring compliance with the provisions of regimes on the part of their peers (Ostrom 1990). Few opportunities of this sort arise in international society, where implementation is at least a two-step process in the sense that it requires efforts to incorporate the provisions of international agreements into domestic practices. And the monitoring of compliance is apt to be carried out by specialized public agencies that have little or no connection with those who are appropriators of living resources, users of nonrenewable resources, or beneficiaries of ecosystem services. In terms introduced in the discussion of regime consequences, outputs (e.g., the passage of implementing legislation) do not necessarily generate outcomes (e.g., behavioral change) where large-scale institutions are concerned.

A different—more political—take on scale has arisen in recent research carried out by a group of European and Asian researchers (Lebel 2004; Lebel, Garden, and Imamura 2005; J. Gupta and Huitema, forthcoming). In essence, the idea here is that problems are socially constructed, that they can be framed in such a way as to make them suitable for consideration at different levels of social organization, and that it often makes a difference in terms of the interests of key actors whether they are addressed at one level or another. Those concerned with the rights of indigenous peoples, for instance, are likely to prefer to address problems at a local scale, whereas those acting on behalf of multinational corporations will often have a preference for global arrangements that produce systems of rights, rules, and decision-making procedures that are uniform across the international system. Under these circumstances, actors should be expected to engage in "scale shopping" along with the more familiar activities known as "forum shopping." This research is at an early stage. So far it has proceeded inductively and identified a variety of reasons why individual actors may want to engage in scaling up or scaling down in addressing problems arising in human-environment relations. This is a promising initiative; further research on this theme seems likely to prove fruitful in adding to knowledge of environmental governance.

Our Carbon Management Research Activity, which directs attention to the design of the global climate regime both under the terms of the Kyoto Protocol and beyond Kyoto, has taken a particularly strong interest in the problem of scale. As we move to create cap-and-trade arrangements to reduce greenhouse gas emissions, what lessons can be drawn from experience with such institutions at different levels of social organization? Is the generally positive experience in the United States with the creation and operation of markets in allowances for sulfur dioxide and nitrogen oxide emissions under the provisions of the CAAA of 1990 transferable to the level of the European Union and even the global level (Tietenberg 2002; Morgenstern and Pizer 2007)? Can inferences be drawn from the operation of the EU (European Union) Emissions Trading Scheme that may prove helpful in the operation of a global regime and in redesigning the climate regime for implementation beyond the first commitment period ending in 2012 (Capoor and Ambrosi 2006)? Would a small number of linked regional arrangements prove more beneficial than a single, global market for allowances? It is already clear that key related issues will deal with matters like procedures governing the initial allocation of allowances, compliance and enforcement mechanisms, the volatility of the resultant markets, and the ability of these institutional arrangements to evolve through a process of adaptive management without losing any desirable impact on behavior (Sugiyama 2005). These are classic concerns that center on issues of scale. It may be some time before researchers are in a position to provide confident answers to questions of this sort, but the importance of tackling them now with regard to concrete issues like climate change is apparent.

Methodological Matters

Specific observations about methodology have arisen in previous sections. A more synoptic view of our project's contribution to methodology in studies of social institutions is served by a description of the nature and the magnitude of the challenges as well as the strategies and tactics devised to address them. The result is a cautionary tale but not one that should give rise to pessimism on the part of analysts seeking to answer questions about the (re)formation and effectiveness of institutions and especially those that govern human-environment interactions.

As social constructions, resource and environmental regimes have no existence outside the behavior of human actors—individual or collective—and are subject to change as a consequence of human actions (Onuf 1989; Wendt 1999). Some analysts, particularly those who describe themselves as constructivists, claim that the subjugation of institutions to human actions in contrast to the immutability of biophysical laws makes it difficult or even impossible to engage in normal scientific research on the (re)formation and effectiveness of institutions (Kratochwil and Ruggie 1986). But this claim is surely exaggerated. Few would deny the feasibility of conducting empirical research on markets, which are institutions themselves, and more specifically on the occurrence of market failures or the consequences for the operation of markets of variations in rules relating to contracts, liability, taxation, and so forth. Similar observations follow regarding research on the effects of laws. There is an extensive body of research, for instance, on matters like the economic consequences of alternative zoning systems and alternative interpretations of the commerce clause and the provisions regarding "takings" in the U.S. Constitution.

Still, institutions do have characteristics generating methodological pitfalls. Complex causality makes it hard to separate out the signal of institutions from the noise of a variety of other driving forces. In cases where it seems important to focus on the impacts of interacting clusters of drivers, it may even be ill advised to attempt to pull apart the individual factors included in these clusters for separate treatment. The fact that universes of cases are often (though not always) small in studies of institutions adds to the resultant difficulties. It means, for instance, that there may be little scope for subdividing the universe of cases in order to control for factors other than institutions themselves. As an example, those exploring the hypothesis that democracies are more likely to comply with the requirements of environmental regimes than countries whose political systems are nondemocratic would find it helpful to subdivide the overall universe of cases to test the validity of this hypothesis. But even under the best of circumstances, the number of cases in each of the resultant categories would be small.[2]

There are also problems with specifying and measuring the dependent variable(s) in efforts to evaluate the effectiveness or the success of resource and environmental regimes. Unlike familiar concerns relating to such matters as voter turnout or the frequency of wars, a causal judgment is embedded in the idea of effectiveness (M. Levy, Young, and Zürn 1995). So long as effectiveness per se is treated as our dependent variable, therefore, a way must be found to address causality. One way around this problem is to replace effectiveness with some other dependent variable for purposes of analysis. Some measure of problem solving, for instance, can be treated as the dependent variable and the role of a regime in accounting for success can be assessed in these terms. The Oslo-Potsdam solution, discussed earlier, provides an example of this strategy that is particularly elegant in analytical terms. Of course, this approach presents the risk of ending up with spurious correlations. If a problem goes away following the creation of a regime, can analysts safely assume that the regime has played a significant role in bringing about this result? This is exactly the sort of problem that statistical procedures, like various forms of regression, are designed to address (Young, Lambin, et al. 2006). Although the nature of these procedures precludes clear-cut results regarding matters of causality, they can and often do help researchers to identify and understand what is going on in complex systems. But here, too, significant limitations exist when it comes to the analysis of institutions. The small size of universes of cases, the pervasiveness of complex causality, and difficulties in finding a common measure of the dependent variable stand out in this connection.

Conceptual problems relating to the dependent variable prove a constant concern in the analysis of resource and environmental regimes. The beauty of the Oslo-Potsdam solution in this regard is that it offers an approach to measurement that can be applied to all regimes and that yields a normalized score for each regime on a scale ranging from 0 to 1. This procedure, or some similar approach to measuring the success of regimes, may well figure prominently in future research on environmental and resource regimes, but we are not there yet. For the moment the practice of approaching effectiveness in terms of problem solving and assessing it on a nominal scale containing four to five discrete points seems likely to attract the attention of those desiring to assess the consequences of environmental governance systems (Miles et al. 2002; Breitmeier, Young, and Zürn 2006). This methodological concern, however, remains fundamental; researchers will and should devote sustained attention to the issue in terms of analyzing institutional effectiveness.

How are we to respond to these challenges in the near future? Recent work on regime consequences (Underdal and Young 2004), as well as collaboration with our sister project on Land Use and Land Cover Change (Young, Lambin, et al. 2006), has highlighted the value of a *portfolio approach* to the analysis of systems that exhibit complex causality. The resultant tool kit contains a range of methods, including enhanced statistical procedures, comparative and meta-analyses, narratives and case studies, systems analyses, and simulations. There is much to be said for making use of a number of these individual tools in studies of particular institutions or categories of institutions such as resource and environmental regimes. When the results converge, analysts can have added confidence in the robustness of their findings. Divergence can be useful, too, in identifying important areas where more analysis is needed.

This approach to the methodological challenges arising in the study of institutions as components of governance systems does not offer neat solutions. It leaves researchers no worse off, though, than analysts concerned with other complex and dynamic systems like the Earth's climate system. Climate researchers have resorted to a range of methodological procedures, including the extensive use of proxies for key variables, natural experiments, and large-scale simulations (Linden 2006). Those working in this field have not been able to answer all the questions raised about climate change, but they have produced a stream of sophisticated insights that are improving understanding of the Earth's climate system at a rapid pace. There is no reason to conclude that the challenges facing those studying resource and environmental regimes are any more daunting.

Informing Policy

Researchers engaged in studies of global environmental change are receiving more and more requests to tease out and highlight the policy implications of their findings. Nowhere is the value of such an effort more apparent than in studies of the institutional dimensions of environmental change. Policy makers and administrators allocate large amounts of time to (re)forming and administering resource and environmental regimes. Any knowledge that can help them to design better regimes and to implement the resultant governance systems more successfully has obvious policy relevance. Moreover, a sizable proportion of those engaged in research on environmental regimes have experience in the policy world themselves or have observed policy processes closely enough to have a good grasp of the world of applications.

All this bodes well for efforts to highlight the policy relevance of studies of environmental and resource regimes. Yet the application of findings about the role of institutions to issues currently on the policy agenda is anything but straightforward. As both this chapter and the substantive chapters that follow make clear, researchers will not be able to develop simple and powerful prescriptive generalizations about institutions that provide surefire recipes for solving the day-to-day problems of those responsible for creating and managing regimes. It is apparent that one size does not fit all in this realm, and skill must be cultivated in crafting specific regimes in such a way that they are well matched to the major features of the relevant problems (Komesar 2001; Cole 2002). What is more, scientists working on institutional issues cannot expect to be close enough to the ups and downs of specific negotiating processes in either legislative or treaty-making settings to be able to jump in on the spur of the moment to suggest explicit provisions for incorporation into the texts of statutes or treaties. As a result, gaps are common between the needs of members of the policy community and the realities of what members of the research community are able to deliver.

Does this suggest gloomy conclusions about the policy relevance of IDGEC research? Not at all. The way forward is to distinguish among different phases or stages of the policy process and to identify those phases in which contributions from the research community are most likely to prove effective. Research findings relating to resource and environmental regimes seem most likely to prove useful during the initial framing of problems and the identification of solution concepts for consideration in the policy process, the provision of advice that is helpful to those charged with administering specific regimes on an ongoing basis, and evaluations of the performance of regimes as a basis for engaging in adaptive management. Each of these types of contributions deserves a few clarifying observations.

As many observers have noted, framing issues for consideration in policy processes and identifying options for policy makers to consider can have far-reaching consequences for efforts to solve problems, not only with regard to environmental matters but also with regard to the demand for governance more generally (Schattschneider 1960/1975; Kingdon 1995; Stone 2002). In this connection scientific research has already had a profound influence on policy processes relating to the institutional dimensions of environmental change by clarifying the distinction between governance and government and, as a result, directing attention to ways to supply governance without government (Rosenau and Czempiel 1992) as well as to the roles of nonstate actors—including corporations and groups operating in civil society—in meeting the demand for governance in a variety of settings. Less prominent but still important is the role of research in expanding the range of policy instruments available for consideration in addressing environmental problems. The Carbon Management Research Activity, for instance, has participated actively in assessments of the relative merits of "targets and timetables" versus "policies and measures" under the terms of the UNFCCC and the Kyoto Protocol and, in the process, in efforts to evaluate the transferability of insights about cap-and-trade arrangements derived from studies of domestic systems to the operation of governance systems at the international level (Sugiyama 2005).

When it comes to inputs into the day-to-day activities of those responsible for administering or operating regimes, the contributions of the research community take a different form. Here, the sources of compliance and, more generally, the roots of the behavior of the subjects of regimes come into focus as key concerns (Breitmeier, Young, and Zürn 2006). Statutes, treaties, and especially informal agreements typically provide administrators with considerable leeway when it comes to moving systems of rights, rules, and decision-making procedures from paper to practice (R. Mitchell 1994a; Underdal and Hanf 2000). Sometimes this reflects the fact that policy makers cannot agree on such matters in the course of institutional bargaining. But even when they do, there is little point in tying the hands of administrators who must cope with the complexity and dynamism of real-world situations. It follows that those engaged in systematic research on matters like burden-sharing arrangements, compliance mechanisms, and systems of implementation review can offer advice to managers about matters of administration that can make a difference in determining the success of regimes created to address specific problems as well as about implementation strategies that can produce the desired results while limiting the costs to society of the solutions.

Processes of evaluation and adaptive management offer another attractive point of intervention for research on environmental governance systems. Complex and dynamic systems require constant assessment to keep management systems in tune with changing circumstances. But policy makers and administrators seldom have the time to step back from day-to-day responsibilities to gain perspective on the performance of institutional arrangements and to adopt a comprehensive view in assessing what works or does not work in efforts to solve or alleviate specific problems. Such analyses are the stock-in-trade of the research community (Underdal and Young 2004). Policy makers periodically embrace innovations (e.g., cap-and-trade systems for controlling emissions of pollutants), but they and the agency personnel charged with implementing such arrangements are likely to be too busy to assess results systematically, much less to compare and contrast results arising in specific cases with those arising in other issue areas, in other countries, or even at other levels of social organization. Given an attitude of mutual respect and trust, the scientific community could do much along these lines to strengthen engagement in a step-by-step process designed to improve the fit between governance systems and environmental problems. In the process, researchers would benefit as well from enhanced opportunities to test theoretical ideas against evidence derived from real-world situations.

The opportunities for initiating mutually beneficial interactions between policy makers and administrators and members of the scientific community are substantial in respect to the creation and operation of resource and environmental regimes. Yet we have often failed to take advantage of these opportunities. Our experience points to two factors that are critical to success in fostering a productive dialogue between policy and science. Mutual respect can make a big difference. We have found repeatedly that once individuals on both sides of this relationship get to know each other as individuals and develop a sense of trust, communication improves dramatically. Once the ice is broken, members of the two communities find that they share many interests and have a lot to say to one another that is illuminating for all concerned. The second factor concerns the role of individuals sometimes called knowledge brokers, who are seldom researchers themselves but who have the capacity to understand complex scientific arguments and to communicate them in an accessible manner to policy makers and to members of the attentive public (Litfin 1994). The demand for knowledge brokers interested in the human dimensions of environmental change is rising rapidly as society moves deeper into an era of human-dominated ecosystems (Vitousek et al. 1997; Steffen et al. 2004) and as the science of coupled socioecological systems becomes more complex. Some steps have been taken already to meet this demand; many more lie ahead.

Future Directions: An Integrative Approach to Governance

The IDGEC project has completed its life cycle, a development that raises questions about future directions in this field of study. Efforts to address this topic as part of the synthesis process have produced substantial results. As described in some detail in chapter 9, there is clear consensus not only within IHDP but also throughout the ESSP that our project's research agenda is important and must be carried forward in some appropriate manner. This consensus suggests that future work in this field should be framed in terms of the overarching idea of governance and should explore interactions between and among institutions and a variety of other factors that play a role in the supply of governance to address specific problems. Already some hints of interesting research questions regarding such links are emerging. The role of discourses in shaping institutions and the reciprocal influence of institutions in shaping the content of discourses, for instance, has emerged as a rich domain for systematic analysis (Ebbin 2004; Agrawal 2005).

Our research foci addressing the generic issues of causality, performance, and design remain central concerns in any scenario for the future. But the planning team we have assembled to work on future directions has identified a number of more focused analytic themes to guide the ongoing efforts of the research community in much the same way that the themes of fit, interplay, and scale served to direct attention during the past decade. Among those that have evoked particular interest in the global change research community are matters of architecture, agency, accountability, allocation, and adaptation (Biermann 2007).

The idea of architecture refers to complex linkages among institutions or, more broadly, the elements that make up complex governance systems. Arising in part from our problem of interplay, the emerging interest in institutional architecture encompasses a broader set of developments. Individual regimes dealing with related issues are evolving into what some investigators have called "institutional complexes" (Raustiala and Victor 2004). Linkages across levels of social organization that have significant implications for efforts to address environmental issues have given rise to a rapid growth of interest in what analysts now call "multi-level governance" (Karlsson 2000; J. Gupta and Huitema, forthcoming). More generally, many have observed that individual environmental and resource regimes are embedded in broader or more general governance systems, whether these are states at the national level or the web of laws, norms, and practices in place at the international level. While there is much to be learned from additional research on individual regimes, an evident need exists to build out to encompass a range of issues associated with the theme of architecture.

The theme of agency directs attention to the constraints arising from IDGEC's focus on (inter)governmental institutions and calls for increased attention to what many have described as agency beyond the state. Partly this is a matter of directing attention to the growing roles that a variety of nonstate actors are now playing in the creation and operation of environmental and resource regimes (Betsill and Corell 2007). But the theme of agency also raises issues of a more fundamental nature. Increasingly, efforts are being made to bypass the state and, in the process, to develop a variety of new forms of governance, ranging from wholly private governance systems to a wide variety of hybrid arrangements (Delmas and Young, forthcoming). Beyond this, those who speak of agency beyond the state have noted the importance of leadership on the part of individuals both in the creation of individual regimes and in moving them from paper to practice under real-world conditions. Although this type of leadership is hard to analyze systematically, the evidence that individuals can and often do make a difference in these settings is strong (Young and Osherenko 1993).

Accountability is a matter of the extent to which governance systems must answer for their performance to one or more well-defined constituencies (Keohane and Nye 2003). The identity of the relevant constituencies may vary. We have heard a lot recently about alleged democracy deficits in arrangements like the European Union or various forms of private governance like the Forest Stewardship Council. But this does not mean that accountability to "the people" is the only form that accountability regarding environmental governance systems can take. There are cases in which accountability is a matter of satisfying experts or those with the qualifications to judge performance or agents who act on behalf of some higher authority. But in every case there is a pronounced tendency to link accountability with legitimacy. The assumption here is that a governance system is likely to be accepted as legitimate by those subject to its rules and decision-making procedures to the extent that it is accountable.

The theme of allocation picks up on an observation made above about our project's emphasis on the roles that institutions play in causing and addressing environmental problems in contrast to their influence in distributive terms or, in other words, in determining who gets what in relevant issue areas. But it is not necessary to ignore or even to downplay the importance of institutions in terms of problem solving in order to consider the allocative or distributive consequences of institutional arrangements. Sometimes this influence is quite specific, as in provisions governing the initial allocation of individual transferable quotas in various fisheries or the initial allocation of emissions allowances under the terms of emissions trading systems like the cap-and-trade arrangement set up under the EU's Emissions Trading Scheme (Raymond 2003). In other cases distributive effects are implicit or more diffuse but no less important for that. Systems that embrace the establishment of private property rights in contrast to various forms of public property or common property, for instance, affect the interests of various players in the system differentially, whatever that system's effectiveness in solving specific problems arising in human-environment relations.

Finally, there is the set of issues often addressed under the rubric of adaptation. Several components of this theme are worth noting here. Institutional adaptation comes into focus whenever the relevant problems are dynamic. Growing awareness of the nonlinear character of many environmental problems and the resultant prospect of abrupt changes has led to recognition of the importance of creating institutional arrangements that can monitor biophysical systems closely, provide early warning of the onset of dramatic changes, and respond to these changes in a timely and effective manner. Beyond this, researchers have come to realize that institutions—like the biophysical systems they address—are themselves dynamic and subject to many forms of change. Putting these observations together poses a particularly difficult challenge. There is a need to learn how to manage nonlinear systems through the development and operation of governance systems that are subject themselves to changes that are not well understood and that may have far-reaching consequences for the capacity of institutional arrangements to solve—or at least manage—the problems they are created to address.

Concluding Thoughts

The subject that we call the institutional dimensions of global environmental change encompasses a wide range of researchable topics. In our project we have focused on a few specific topics, including those like causality, performance, and design that are relevant to the analysis of institutional arrangements in all settings and those like fit, interplay, and scale that deal with cutting-edge issues pertaining to resource and environmental regimes. Although such claims are inevitably hard to demonstrate conclusively, we believe that our project has played a role of considerable importance in advancing our understanding of the complex interactions between human systems and biophysical systems or, as we now say, the dynamics of socioecological systems. Our work, as is commonly the case in scientific endeavors, has also highlighted a range of new issues that seem likely to become cutting-edge themes in the years to come. Even as we celebrate the accomplishments of IDGEC, therefore, we also welcome with enthusiasm the launching of a new project within the IHDP community on Earth system governance.

Notes

1. Alternative but roughly equivalent terminology differentiates among policy, behavioral, and environmental consequences.

2. One strategy for increasing the universe of cases is to treat distinct actions (e.g., compliance) by country per year as separate observations (R. Mitchell 2004b).

II

Research Foci

2

Determining the Causal Significance of Institutions: Accomplishments and Challenges

Arild Underdal

Introduction: The State of the Art

The first of the three focus questions specified in the Institutional Dimensions of Global Environmental Change (IDGEC) Science Plan asks, "What roles do institutions play in causing and confronting global environmental changes?" The order is appropriate, since interest in questions about performance, design, fit, or other institutional dimensions hinges very much on the assumption that institutions do play important roles in governing human behavior. At a basic level this assumption finds strong support in the research literature. Deductive and experimental analysis leaves no doubt that certain types of rules—including those specifying property rights, regulating access to a particular resource, and determining decision-making procedures—can make a substantial difference and will do so under many real-world circumstances (e.g., Conybeare 1980; Barrett 2003). A wide range of empirical studies have produced compelling evidence that specific institutions, or particular institutional forms, do in fact have at least some degree of success in serving the purpose for which they were established (e.g., Ostrom 1990; Miles et al. 2002; Gibson, Williams, and Ostrom 2005; Breitmeier, Young, and Zürn 2006). In addition, studies focusing on local management systems and studies examining international resource regimes converge on one very important conclusion: "governance without government" can indeed be effective, provided that certain conditions are met (compare, e.g., Ostrom 1990 and Breitmeier, Young, and Zürn 2006).

An equally clear message, however, is that the causal significance of specific regimes and organizations varies substantially, depending on the extent to which they influence human activities driving or mitigating environmental change. Since institutions cause effects by guiding or modifying human behavior, they can affect only those elements of environmental change that are open to human influence. And although institutions as a distinct category of social arrangements do play important roles in shaping behavior and outcomes, no extensive search is needed to find regimes and organizations that make, at best, only a marginal difference.

Important progress has been made over the past decade or two in understanding roles played by different types of institutions in causing and mitigating environmental change. Four achievements stand out as particularly significant. They are accomplishments made by the research community at large, but activities initiated by or in other ways related to the IDGEC research program have contributed to the advancement along all four frontiers.

1. Improved understanding of the causal *mechanisms* and *pathways* through which institutions shape behavior and outcomes. A number of studies published over the past ten to fifteen years have advanced our understanding of how institutions produce effects (examples include Ostrom 1990 and 2005; P. Haas, Keohane, and Levy 1993; Victor, Raustiala, and Skolnikoff 1998; Young 1999a). For the field at large this may well be the most important achievement made during this period—in part because progress has been substantial, in part because at least some causal mechanisms can be manipulated and used as tools. Thus the study of mechanisms can generate knowledge that can serve as premises for the design of regimes and organizations.

2. Improved understanding of *patterns of variance*, particularly with regard to regime effectiveness. Several major studies identifying and examining factors influencing institutional performance have been published (e.g., Ostrom 1990; Victor, Raustiala, and Skolnikoff 1998; Weiss and Jacobson 1998; Young 1999a; Miles et al. 2002; Breitmeier, Young, and Zürn 2006). More ambitious comparative studies have been undertaken. New databases have been developed, enabling researchers to search for patterns across a larger number of cases. These efforts have interacted productively with the study of causal mechanisms. As a result, we can now speak with greater confidence and precision about conditions for effectiveness and causes of failure.

3. Progress in the study of *institutional interplay* and institutional complexes (e.g., Young 1996, 2002b; Stokke 2001a; Raustiala and Victor 2004; Oberthür and Gehring 2006c). Ten years ago the relationship be-

tween or among institutions was of marginal concern to most students of environmental governance. IDGEC has played a pioneering role in setting a new agenda for research, developing conceptual frameworks, and initiating empirical studies. A new subfield is emerging, with interesting findings already reported and more to come.

4. More ambitious and sophisticated use of the *methodological repertoire* of social science (Underdal and Young 2004). Such a trend can be seen in more frequent use of demanding techniques for explicit, transparent, and rigorous measurement; in more systematic efforts to combine different modes of inquiry (such as intensive case studies and extensive statistical analysis); and in studies applying tools that have rarely been used in this field before (such as Boolean logic and agent-based simulation). Although by no means pervasive, these are important developments. Determining causality can be a major intellectual challenge, and the better use we can make of the methodological toolbox available to us, the more accurate and reliable will be the conclusions we reach.

Despite these and other achievements, outsiders turning to the research literature for guidance may well find a somewhat confusing diversity of messages. They will find differing taxonomies in use for distinguishing causal mechanisms and differing definitions of key concepts such as "regime effectiveness." They will also find areas of more or less clear disagreement at different levels of generality. In some instances, assessments of the same institution diverge. For example, whereas several case studies have concluded that the Long-Range Transboundary Air Pollution (LRTAP) regime has contributed to reducing air pollution in Europe—albeit to varying degrees for different pollutants—other studies, applying a different methodological strategy for separating effects, report that little or nothing of the change observed can be attributed to that particular institution (compare, e.g., Munton et al. 1999 with Ringquist and Kostadinova 2005). More disturbingly, the curious outsider will find that at the level of general theory the efficacy of institutions as determinants of behavior and outcomes is being questioned by at least two different strands of research. Realists argue that institutions are political constructs that reflect rather than change underlying and more important factors, notably configurations of interests and power, and basic ordering principles (e.g., Mearsheimer 1994/1995).[1] Where cooperation is a matter of voluntary participation it will therefore be confined to a "coalition of the willing," united in a mutual commitment to do essentially what they would have done anyway. Another and equally profound challenge stems from the sociological conception of (formal) regimes and organizations as embedded in and molded by some deep normative structure of the system in which they are situated (Ruggie 1982; Conca 2006). This proposition differs substantially from the realist notion of interests and power as the basic driving forces, but the two "schools" converge in arguing that "conditions shape institutions and institutions only transmit the causal effect of these conditions" (Przeworski 2004, 527).

What are we to make of all this? For someone deeply involved in the study of institutions, the most obvious implication may simply be that more research is needed! Although undoubtedly true, this statement immediately generates a new question: how can we best focus and design future research so that it can help fill important knowledge gaps, resolve controversies, and in other ways enhance our understanding of the roles that institutions play in causing and mitigating environmental change? What follows is premised on the assumption that in order to answer that question researchers must be able to distinguish between difference and incompatibility. Generally speaking, conceptual frameworks and methodological tools that differ from one another simply represent a range of options. They may well compete with each other, but concepts and techniques are tools and as such to be evaluated as more or less useful or appropriate for specific projects and purposes. Divergent substantive propositions may be mutually exclusive. If so, at least one of them must be wrong. A closer scrutiny of apparently conflicting hypotheses or conclusions, however, will indicate that a fair number in fact refer to incongruent (and perhaps poorly specified) scopes of validity or to different definitions of key variables. These propositions can—at least in principle—be integrated to provide a more comprehensive understanding or a higher-resolution picture. This is by no means to deny that there are instances of very real and substantively important disagreement. The point here is simply that a careful compare-and-contrast exercise is required to identify such instances correctly. A good way to start is to examine research questions and to relate hypotheses and conclusions to the questions that generated them.

Substance: Questions and Answers

Research on the role of institutions spans a wide range of specific questions. In order to determine how various hypotheses or conclusions relate to each other, we need to determine whether or not they answer the same question.

Mode of Inquiry: "Logic" or "Fact"?

A first step can be to distinguish questions framed and answered in terms of deductive logic from those framed and examined in terms of empirical evidence. The former explore whether a change in the type of institution will produce important effects, given a set of assumptions about the broader setting.[2] Empirical research explores whether a particular change made in one or more institutional variables did actually lead to a significant change. While the answer to the former question will often be a firm yes, the answer to the latter may well be no. Working with ideal type constructs, we can easily demonstrate that an open-access regime for common-pool resources can—and on plausible assumptions will—lead to more pressure on the resource than a regime imposing effective restrictions. Working with empirical evidence, what we observe in fact will often be change at the margins, for example, in the form of new regulations imposing modest cuts in catch quotas or emissions permits. Even a small change may produce a large effect, but only if it happens to cross some important threshold or tipping point or generates synergetic interaction with other driving forces. In other circumstances marginal change can be expected to produce, at most, marginal effects. Empirical studies confirming such a pattern do not necessarily challenge the basic proposition of institutional theory that institutions can be important causal factors. A more appropriate interpretation would be that the empirical evidence suggests that institutional engineering has in fact been confined to a narrow range of options and that change within that narrow range will rarely be enough to bring about significant change in the behavior or outcomes targeted. The take-home message for policy makers and stakeholders could simply be that substantial change in outcomes will require more radical moves—either in institutional design or by some other means.

Within each mode of inquiry, the specific questions addressed differ widely in terms of the type of institution or institutional change in focus, the kind of effect(s) mapped or measured (the dependent variable([s]), and the full explanatory model guiding the research effort.

The Main Independent Variable: Institution

For any research program a common definition of key concepts is essential to enable effective communication and facilitate integration. For the IDGEC project the most central concept is that of "institution." The Science Plan provides an elaborate definition (Young et al. 1999/ 2005, 27; see also chapter 1 in this volume). Others have offered other formulations (e.g., North 1990; Mearsheimer 1994/1995, 8; March and Olsen 1995, 6; Koremenos et al. 2001, 762; Ostrom 2005, 3; Duffield 2007, 2). These definitions differ from one other in terms of, inter alia, the degree to which constitutive elements are specified and their treatment of informal norms and practices as distinct from formal rules and procedures. Some of these differences are nontrivial in the sense that social arrangements can easily be found that would meet the defining characteristics of one but not qualify by those of another (Duffield 2007). Most of the definitions used in this field of research do nonetheless share a common core. Therefore the fact that assessments of the causal significance of environmental institutions sometimes come up with divergent conclusions cannot be interpreted as merely a consequence of conceptual incongruities.

Yet, probing deeper, one can find important clues in distinctions that serve to focus the analysis, often without being spelled out explicitly. One of these is the distinction between a (formal) institution seen as an accomplished edifice, or *structure*, and institution building seen as part of a more comprehensive *process* of responding to certain problems or opportunities. Another distinguishes institutions from other types of social orders or arrangements.

Structure or Process?   Questions about the causal significance of a particular institution often take the rules, norms, and procedures of that institution as established facts. Thus, the *strength* of a regime is usually defined in terms of the extent to which it constrains the freedom of legitimate or legal choice open to the individual member. A regime is considered *effective* to the extent it successfully solves a particular problem or performs a particular function. A regime is considered *robust* or *resilient* to the extent it is capable of surviving stress with its functioning capacity intact. And *compliance* is rated high if the behavior regulated is consistent with the institutional rules and norms. Researchers framing their questions in such terms do, of course, recognize that institutions are social constructions that often go through one or more periods of significant change during their lifetime. But the point of reference for the analysis is the institution or regime as it exists at a particular point in time.

Some studies take a broader perspective, focusing on the process of *institution building* rather than on institutions as established orders. Some also analyze institution building as an integral component of a more comprehensive process of social learning and collective response. This shift of focus has at least two important implications. First, the process of institution building may have significant consequences beyond those that can be attributed to the (formal) regime or organization that it may produce. It may, in fact, have important consequences even if it fails to produce any institutional arrangement (Underdal 1994). Such consequences can materialize in the form, for example, of new knowledge or ideas that can serve as a basis also for unilateral measures (e.g., P. Haas 1992b), the evolution or diffusion of new normative standards (R. Axelrod 1997; Risse, Ropp, and Sikkink 1999), or process-generated stakes leading into a game of "tote-board diplomacy" (M. Levy 1993). Similarly, the patterns of development described by the Social Learning Group (2001); the influence attributed to programmatic activities by Breitmeier, Young, and Zürn (2006); and several other findings reported in recent studies all indicate that processes of institutional development deserve attention in their own right.

Second, the process of institution building is itself an integral part of a more comprehensive drive to come up with an effective collective response to a joint problem. Other elements of that drive may involve a wide range of activities by nongovernmental organizations (NGOs), companies, citizens/consumers, governments, and other actors. New institutions are typically brought into existence through such activities and owe much of their subsequent influence to them. This is by no means to say that the institution itself is therefore unimportant. Rather, what this line of reasoning suggests is that its role, however important, can best be understood in this wider context; and that separating the influence of the institution itself from that of the more comprehensive response drive of which it is part will often be not only a methodologically difficult exercise but sometimes also substantively futile. What can most often be observed is the combined effect of this complex. Therefore, in assessments of regime effectiveness the change in outcomes attributed to the institution will often be such a combined effect, one to which the institution contributes along with other causal factors.

The main implication of these observations seems to be that supplementing the study of institutions as established structures with projects focusing on institution building as embedded in more comprehensive processes of collective learning and response can produce valuable insight. The two perspectives generate different questions that may well have different answers, but they both highlight important institutional dimensions of environmental change.

Institutions as Distinguished from Other Social Orders   The realist tradition makes a distinction between *institutions* and what Waltz (1979) refers to as *ordering* (or organizing) *principles*. The former are seen largely as epiphenomena, reflecting rather than shaping more important determinants, notably configurations of interests and power. By contrast, ordering principles—such as anarchy or hierarchy—are seen as profoundly important. The (neo)realist model of international politics is very explicitly proffered as a model of politics in an anarchical system and makes no claim to validity for systems ordered by a different principle. In this conceptual framework an ordering principle is not by itself an institution; it provides no system of rules prescribing or proscribing behavior.[3] Rather, it serves as the constitutional platform on which any institution will have to be built. As such it determines the range of feasible options and may also affect selection within that range. Thus, the only kinds of institutions that can be built on the platform of anarchy will be those that can be established and sustained through voluntary cooperation (consensus) and those that can be imposed through coercion by the most powerful actor or coalition of actors (hegemony).

Interestingly, the sociological tradition makes a somewhat similar distinction between what is often referred to as the deep, normative structure of a system or society on the one hand and more specific regimes and organizations on the other (e.g., Conca 2006). The latter are seen as embedded in the former and therefore as reproducing more than shaping the set of "powerful, overarching metanorms" on which they are built (Conca 2006, 26). This is certainly a different kind of order than the one described by (neo)realists. Yet the two approaches converge on the proposition that social systems are in a profound sense ordered, and that the truly important ordering principles or norms are to be found beneath the surface of specialized regimes or formal organizations.

This line of reasoning has at least one clear and simple implication for research: only programs that conceive of specialized institutions such as resource regimes as being embedded in and constrained by this deep layer of principles and norms can determine their "true" causal significance. An equally simple message can be derived for policy makers: for institutional engineering to provide truly effective instruments for environmental governance, it will have to penetrate this deep layer or somehow decouple the design of specific regimes and organizations from more basic ordering principles and norms of society.

The embeddedness of specific institutions in more basic or overarching structures has long been recognized by students of environmental governance (see, e.g., Krasner 1982b; Ruggie 1982; Young 1996; Agrawal 2002). Many studies designed to explain the establishment of international regimes very explicitly consider the basic ordering principle of the international system as a significant constraint. At the same time many have been eager to demonstrate that (formal) regimes and organizations deserve interest in their own right as partly "autonomous variables," to quote Krasner (1982a).[4] Although still sometimes challenged on theoretical or empirical grounds (e.g., Mearsheimer 1994/1995; von Stein 2005), this drive to "liberate" regimes and organizations from the grip of more basic orders has over the years produced substantial support for the general argument. It may even—at least by its own assessment— have been *too* successful. Autonomy is by no means absolute. To understand the roles that institutions play in shaping outcomes, the need arises also to understand the links that do exist between specific regimes on one hand and the kinds of deep structures highlighted in realist and sociological theory on the other.

IDGEC has taken an important and timely step in promoting the study of institutional interplay and institutional complexes. As currently developed, the theme of interplay focuses primarily on interaction at the level of specific regimes (Underdal and Young 2004; Young et al. 1999/2005, 60–65; Oberthür and Gehring 2006c). This may well be an appropriate order of priorities for a particular research program at a particular stage, but for the field at large, arguments can be made for extending the agenda to include research on links between specific regimes or other types of institutions on the one hand and more basic social orders on the other. Such an extension could be accommodated within the framework of the IDGEC Science Plan by adding another dimension of vertical interplay. One advantage of framing the study of links among layers in terms of interplay would be to conceptualize the relationship as one of interaction, with institutions not merely reflecting but—at least collectively and over time—also influencing more basic norms and principles. To come to grips with macro-level effects of institutional complexes, further steps will be required (Raustiala and Victor 2004). The network approach adopted by Ward (2006) seems to be one promising path.

Which Institutions Are Important?   Students of environmental governance have taken an interest in a wide range of institutional arrangements

—from local community schemes for resource management to global environmental regimes, from legally binding regulations administered by formal organizations to informal rules and norms, and from regulation by means of command-and-control directives to incentive- or market-based systems. Across this wide range, however, an agenda has emerged with certain common elements.

Most of the research effort has centered on institutions established specifically for the purpose of environmental governance or resource management. This is a sensible priority since these institutions are the dedicated tools of environmental policy. That does not, however, necessarily make them the most important institutions when it comes to influencing human interaction with the environment. A well-known formula conceives of the aggregate impact of human activities on the environment (I) as a function of population (P), affluence (A), and technology (T) $[I = P \cdot A \cdot T]$ (Ehrlich and Ehrlich 1990). To turn this crude formula into a useful tool for quantitative assessment, valid and precise answers are needed to a host of intriguing questions about, for example, the environmental impact of technological development and the relationship between population and wealth. Even in its basic form, however, the model can serve as a rough guide to key factors that institutions must influence in order to make a substantial difference. If population, affluence, and technology are the critical drivers, the institutions or institutional complexes most important to the environment are likely to be those that influence main economic activities (production and consumption, investment and trade); technological change; and collective systems of beliefs, values, and practices. By this standard any specialized environmental regime will face stiff competition from international regimes defining the basic rules for trade and investment, including the World Trade Organization (WTO) and the European Union; international law and norms regulating state sovereignty; and perhaps also transnational religious and cultural communities influencing beliefs, values, and practices of substantial segments of the world's population.

To see more clearly how the study of environmental institutions relates to this broader picture, it might be helpful to identify in more general terms what makes an institution "important." In theory, the causal significance of a particular institution for a particular outcome or state of affairs can be seen as a function of the importance of the domain of human activities influenced (D), the overall weight within this domain of the actors whose behavior is affected (W), and the extent to which the behavior of these actors is influenced (E). If we define each of these factors in relative terms, a standardized measure of the causal significance (S) of a particular institution (i) can be computed as $S_i = (D_i \cdot W_i \cdot E_i)$.[5]

Most environmental institutions are highly specialized in that they address a particular problem (such as the depletion of stratospheric ozone) and target particular activities (e.g., the emission of certain chemical substances). Some are (also) confined to a small geographical area (e.g., a small rural community). Within their narrow domains some of these institutions contribute significantly to alleviating the problem with which they were established to cope while others do not. Recent studies assessing the effectiveness of larger sets of environmental institutions report fairly encouraging conclusions. Thus, in their analysis of effectiveness, based on the twenty-three regimes included in the International Regimes Database, Breitmeier, Young, and Zürn (2006) find that where the conditions targeted improved "slightly" or "considerably," regimes had a "significant" or "very strong" influence in 52 percent of the cases and little or no influence in 9 percent. By contrast, where conditions deteriorated, regimes played a significant or very strong role in only 7 percent of the cases and little or no role in 40 percent. Similarly, in their analysis of fourteen international regimes Miles et al. (2002) report aggregate scores of .51 for positive behavioral change and .35 for effectiveness defined in terms of problem solving, on a scale ranging from 0 to 1. Interestingly, the former study reports significant effects also for regimes constrained by the most demanding decision rules (unanimity and consensus), whereas the latter found positive scores also for regimes dealing with politically malignant problems.[6] Both find that, up to a point, effectiveness tends to increase as regimes mature. From the rich case study literature on common property resource governance we have learned that many local communities or user groups have managed to establish and maintain self-governing institutions with success rates that defy Hardin's (1968) somber conclusion that only systems of centralized government or private property can avoid "the tragedy of the commons" (Dietz, Ostrom, and Stern 2003). Experimental research suggests that, at least under reasonably favorable circumstances, individuals are indeed able to come up with effective joint strategies for small-scale management of common-property resources (Ostrom, Walker, and Gardner 1992). Extensive empirical studies confirm that the strength of community-based institutions for resource management accounts for an important part of the variance observed in the state of the resource or ecosystem itself (see, e.g., Gibson, Williams, and Ostrom 2005; Agrawal and Chhatre 2006). And research on national environmental policies points to political systems and legal-administrative structures as variables to be included in explanatory models (see, e.g., Lundqvist 1980; Vogel 1986; Boehmer-Christiansen and Skea 1991; Lafferty and Meadowcroft 1996; Jänicke and Weidner 1997; Torras and Boyce 1998).[7]

That said, several studies offer cautionary remarks. For example, in summarizing the main findings of the International Institute for Applied Systems Analysis project on implementation and effectiveness of international commitments, Raustiala and Victor (1998, 698) report that "... the most important turning points and fundamental pressures that have caused regulatory action have not been institutions." Keohane, Haas, and Levy (1993, 14) say that "if there is one variable accounting for policy change, it is the degree of domestic environmental pressure in major industrial democracies, not the decision-making rules of relevant international institutions." Regression analysis of developments under the LRTAP regime points to noninstitutional variables as the most important determinants (Murdoch, Sandler, and Sargent 1997; R. Mitchell 2003b; Ringquist and Kostadinova 2005). And Agrawal and Chhatre (2006) find that biophysical variables stand out as the major determinants of forest conditions in the Indian Himalayas.

It would, furthermore, be a mistake to infer from positive findings in specific cases that the aggregate effect of specialized environmental institutions on human interaction with the environment is equally significant. Most of the studies cited suffer from a particular and perfectly understandable kind of selection bias; they focus on cases where such institutions have been established and can be observed "at work." In order to determine the aggregate causal significance of environmental institutions it would be necessary to study their *absence* as well as their presence. This would require a strategy of selecting cases of human activities that leave a significant imprint on the environment, or cases of environmental degradation caused to a significant degree by human activities (Underdal 2002a, 447).[8] Some of these cases will fall within the domain of institutions designed specifically for environmental governance, but others will not. For someone trying to determine the overall impact of such institutions on the state of the environment, the latter category is as important as the former. Since the two may differ systematically in important respects, there is no shortcut. For example, it may be suspected that most of the international regimes existing today respond to problems of moderate political malignancy and do so mainly by performing moderately demanding functions. If so, even a high score for each and every one of these regimes would not be sufficient to conclude that the current universe of regimes makes a truly substantial contribution to protecting nature's life-support systems. To determine the importance of international regimes for environmental governance at large, we would need not only an assessment of the effectiveness of each existing regime, but also an idea of how well the current universe and configuration of regimes fit the field seen as the aggregate configuration of governance challenges. To the extent that students of a particular type of institution focus only or primarily on cases where such institutions have been established, they may inadvertently foster an image of a world in which that particular type of institution appears as a more common and important phenomenon than it actually is.

Moreover, anyone trying to determine the generic role of institutions in causing and mitigating environmental change should keep in mind that institutions established specifically for the latter purpose make up only a subset of the universe of relevant cases, and that some of the most important institutions are likely to be found outside that subset. This is, of course, no news to students of environmental governance. The IDGEC Science Plan explicitly calls attention to the role of institutions such as trade and investment regimes and political systems and to regime interplay as an important new frontier of research (Young et al. 1999/2005). Nonetheless, it seems fair to say that the IDGEC project, as well as environmental change research more generally, has given higher priority to examining the efficacy of institutions as tools for environmental governance than to studying their role in influencing human activities that drive environmental change. Although the Science Plan points to functional interplay as an equally important topic, much of what has so far been published on regime interplay focuses primarily on linkages at the level of regime politics (see, e.g., Rosendal 2001a; Stokke 2001a; Young 2002b; Raustiala and Victor 2004; Oberthür and Gehring 2006c). A strong case can be made for this order of priorities, particularly for the early stages of a research program. But as the "human dimensions" community moves on to develop ambitious joint ventures with other global change programs, it faces new and demanding challenges to come up with a more comprehensive, precise, and dynamic understanding of the causal significance of basic types of institutions (such as markets or the sovereign state system) as well as of specific regimes based on these constitutive principles (such as the world trade regime or decision rules for international cooperation) for human interaction with the environment, broadly defined. What makes this challenge so demanding is that what is ultimately required is not merely a unidirectional flowchart showing how various institutions affect human activities, which in turn leave some imprint on the environment. What is needed is a dynamic model of a causal complex in which feedback loops and interaction effects are likely to be important elements (see Steffen et al. 2004).

We are not there yet, and we cannot even promise to deliver such a product in the near future. But progress is being made, and international programs like IDGEC may well play a useful role in guiding research as well as in integrating partial knowledge into a more comprehensive picture. Consider the current state of knowledge about the role of the global trade regime in governing human activities that cause environmental change. Since trade rules can affect critical activities such as production and consumption in multiple ways and through complex causal pathways, researchers cannot claim to have the full picture. They are, however, getting important bits and pieces of knowledge from various sources. Some contributions come through interesting case studies, such as Jahiel's analysis (2006) of the immediate effects of Chinese membership in the WTO. She identifies several positive effects in the form of efficiency gains from technology shifts and other developments but concludes that the sheer scale of growth—assumed in part to be a result of WTO membership—has overshadowed these gains, and that some of the improvements achieved at home have come at the expense of environmental harm abroad. China is an atypical country in many respects—including its size and its very high growth rate during this period—so more extensive empirical research is required to determine whether and to what extent Jahiel's conclusions can be generalized to other countries or other periods. But even in the absence of such studies, the means exist for assessing the general plausibility of the particular pattern reported for the Chinese case.

Important clues may be found in other types of research examining some of the key mechanisms assumed to be at work. For example, economists have produced a large number of studies examining the relationship between income on the one hand and pollution or resource depletion on the other. The environmental Kuznets curve hypothesis suggests that the environmental impact of human activities is an inverted U-shaped function of income per capita. Some empirical studies find such a pattern in specific cases, but the evidence is mixed. One critical assessment concludes that "... emissions of most pollutants and flows of waste are monotonically rising with income, though the 'income elasticity' is less than one and is not a simple function of income alone" (D. Stern 2004, 1420; see also Copeland and Taylor 2004). The pattern reported by Jahiel is basically consistent with this general proposition. Her finding that some of the environmental improvement China achieved in adjusting to WTO membership has been obtained by pushing more harmful activities over to poorer countries seems consistent with conclusions reported in several other studies, including Lofdahl's global systems analysis (2002) of the impact of trade on deforestation. In general, free trade enables high-consumption societies to leave some of their ecological footprints in other parts of the world. Other studies show, though, that this effect is sometimes mitigated by other trade- and investment-related mechanisms. Thus, trade can serve also as a vehicle for transmitting high environmental standards established by and for rich importing countries—particularly standards pertaining to product quality and production processes—to poorer countries exporting goods to these markets (Vogel 1995; Prakash and Potoski 2006). Similarly, foreign investment often leads to transfer of new technology, enabling firms in host countries to make more efficient use of scarce resources or to operate with lower pollution rates.

The net balance of all this will depend on the size of the scale effect relative to that of the efficiency effect. This ratio seems to vary with a wide range of intervening factors, such as initial level of income, growth rate, type of pollution, and type of political system. Such complexity permits, at best, differentiated and contingent statements about the impact of the global trade regime on the state of the environment. This should not deter further research. Contingent and differentiated statements can, in fact, be at least as useful to decision makers and stakeholders as more sweeping generalizations. There is no reason to assume that the methodological tools used by (social) science to examine the effectiveness of regimes and organizations designed to promote sustainable use of environmental resources cannot be used also for the purpose of studying environmental side effects of institutions established for other purposes.

The Dependent Variable(s): Effects

The choice and definition of the effect(s) to be measured—that is, the dependent variable—may have important implications for the conclusions generated. Two examples may suffice to prove the point.

First, the effects of a particular institution typically materialize in the form of a causal chain and can be measured at different points. For example, in examining the effectiveness of international environmental regimes, research can focus on compliance in the form of ratification or other formal steps of implementation taken by national governments *(output)*, change in the human behavior targeted by the regime *(outcome)*, or consequences defined in terms of change in the biophysical environment itself *(impact)*. The effectiveness score tends to decline the farther along this causal chain it is assessed. Methodological difficulties accumulate as research follows the chain, but in this case it can be assumed that the declining pattern reflects substantive realities, not (merely) measurement error. One reason to expect such a pattern is that government control over the behavior of corporate actors, groups, and individuals within its jurisdiction is imperfect. Moreover, domestic policy adjustment (output level) tends to be easier the less stringent the demands. The same is true for full compliance. This implies, as Downs, Rocke, and Barsoom (1996) remind us, that good news about compliance may well be bad news for the environment. With regard to studying consequences for the biophysical environment itself, it must be recognized that institutions can be a significant factor only insofar as human activities count as an important cause of environmental change. Where causal mechanisms inherent in nature account for a much larger proportion of the variance, studies measuring regime effectiveness in terms of environmental impact will normally produce lower, often substantially lower, scores than studies measuring effectiveness in terms of change in human behavior.

Second, in designing studies to assess institutional effects, it is necessary to determine the point of reference from which change is to be measured. Two principal options exist. One is the hypothetical state of affairs that would have arisen without the regime, sometimes referred to as the business-as-usual scenario. This is the appropriate approach to learn whether or to what extent a particular regime makes a difference. The other option is to assess the actual outcome observed with the regime in place against some notion of what qualifies as a "good" or "optimal" solution. This helps to determine the degree to which a particular problem is solved under present arrangements (Underdal 1992). These two approaches may be combined to produce a standardized measure of the contribution made by a particular institution to solving a particular problem (Sprinz and Helm 1999). The standards they provide are, however, substantively different, and there is no reason to assume that they will generally yield the same score. Sometimes a significant improvement from the business-as-usual scenario will not be enough to "solve" the problem. In other instances a small change will be all that is needed.

The Explanatory Model

Any study designed to map and measure effects brought about by institutions must somehow take into account other factors that can produce the same effects or in some other way leave an imprint on the outcome in focus. The number and range of potentially relevant variables increases along the causal chain, most sharply as focus shifts from effects on human activities to consequences for the biophysical environment itself. Models developed to guide research are constructed to capture the essence of this complex system in a simplified format. Since all models select and simplify, the decisions made about which independent and intervening variables to include, and the way relationships are specified among these variables, may have important implications for the kind of conclusions that can be legitimately inferred as well as for the specific findings of the analysis.

The most obvious of such implications is that the more independent and intervening variables included in the explanatory model, the lower will be the proportion of variance accounted for by each of them, other things being equal. There are exceptions to this rule, notably where institutional mechanisms interact synergistically with other types of mechanisms to "coproduce" outcomes. This discussion will return to one implication of this observation a little further on. The point here is that the findings reported in a particular study are to some degree products of the model that guided the analysis and should be interpreted in that context. Such contextual interpretation is particularly pertinent in comparing and contrasting conclusions from research relying on nonexperimental designs and a small number of cases, as is true of most studies of environmental governance.

Questions about the causal significance of institutions are most often framed in terms requiring separation of effects, that is, efforts to distinguish what is brought about by institutional mechanisms from what must be attributed to other factors. No research program aimed at advancing understanding of the roles institutions play in guiding or governing human activities can escape questions of separation. Students of institutions, however, have to wrestle also with another set of questions, about *joint* effects. Institutional mechanisms often interact with other mechanisms outside the purview of the institution to coproduce outcomes.[9] Assume that a local community introduces a new set of rules for the purpose of achieving more sustainable rates of extraction from a common-pool resource. The pattern of use found with this regime in place will be the product of a larger causal complex, in which rules and norms interact with attributes of the user group (e.g., mobility, social capital), market conditions (relationship of demand to supply), and characteristics of the ecosystem from which the resource is extracted (e.g., vulnerability). Jacobson and Brown Weiss (1998) bring together characteristics of the activity, the accord, the international environment, and the country in a complex model designed to explain compliance with international environmental agreements. In such instances questions about the relative contribution of each of these components may be less interesting and important than questions about their interaction in coproducing the outcome. The general lessons to be inferred about the effectiveness of a regime will most often be contingent in format, seeing effectiveness as a function of the *conjunction* of institutional characteristics and characteristics of the setting in which the institution is embedded.

Causal Mechanisms: *How* Do Institutions Produce Effects?

In order to establish causality, as distinct from mere correlation, it is necessary to identify at least one causal mechanism that *can* produce the effect, and—for specific instances—demonstrate that it in fact *did*. And in order to explain a particular outcome, it must be determined how it came about.

Students of institutions have tried to answer the *how* question by pointing to one or more causal mechanisms, or by describing a chain of events—a causal pathway—leading to a particular outcome. A survey of the literature will reveal that the term *mechanism* is used for different concepts (see Hovi 2004). Yet the common denominator is a search for the nuts and bolts of causation, for what links cause and effect. Researchers trying to determine the role of institutions in environmental governance have engaged in a search for such links at different levels of specificity. At the most fundamental level this is a quest for what might be called the generative sources of human behavior (Young and Levy 1999, 21). This leads into questions about, inter alia, whether or in what circumstances the behavior of different (types of) actors is driven by rational calculation aimed at maximizing utility or by internalized norms and routines.[10] Answers to such questions are most often framed in terms of correspondence with "reality" (validity), but the rational choice approach is often promoted (also) as giving "... a grip on the subject that is peculiarly conducive to the development of theory" (Schelling 1960, 4). It is easy to see that assumptions made about the generative sources of human behavior can have very important implications for policy. A practitioner assuming that most actors are motivated primarily by self-interest and capable of basically rational calculation will easily agree with Barrett (2003, 355) that "the principal task of a treaty is to restructure incentives." Someone believing that behavior is governed mainly by the logic of appropriateness will instead focus on the social construction of identities, beliefs, and norms and target processes such as social learning, socialization, and arguing.

The debate between "rationalists" and "constructivists" has sometimes been cast as a contest between two incompatible systems of ontological and epistemological premises. Such a contest may spur further refinement of each paradigm, but since it is abundantly clear that both identify mechanisms through which institutions can influence outcomes, it is highly unlikely to enhance understanding of how institutions work. Much to its credit, the study of environmental governance has largely avoided sterile posturing. Instead, some studies have relied essentially on one approach (e.g., Ostrom 1990; Barrett 2003; Hoffmann 2005), while others have pursued an eclectic strategy that combines causal mechanisms or pathways from both (e.g., Hasenclever, Mayer, and Rittberger 1997; Young 1999a). In recent years international relations scholars from both camps have begun exploring possibilities of combining the two on a more general basis, for example, by defining a set of "scope conditions" specifying their respective scopes of validity (Checkel 2001), by coupling them in a particular sequence (Risse, Ropp, and Sikkink 1999), or by using insights provided by one to fill in gaps left by the other (Katzenstein, Keohane, and Krasner 1998; Fearon and Wendt 2002). Skeptics certainly argue that little is gained by these efforts. Downs (2000), for instance, concludes, on the basis of an analysis of three major studies, that the mechanisms highlighted by the constructivist approach appear to have had only a marginal impact compared to those at the core of rational choice theory.[11] Even those who interpret available evidence as providing a stronger rationale for combining the two will have to admit that the integrative frameworks developed so far hardly compete with rational choice models when it comes to analytic virtues such as parsimony, stringency, or conclusiveness. This is by no means a fatal criticism; some conceptual frameworks and propositions aspire to higher scores on other important dimensions, such as empirical validity and policy relevance (Wendt 2001). Yet even the latter criteria require moving beyond unspecified propositions and nominal-level inventories of variables or mechanisms.

The complexity entailed in integrating a constructivist and a rationalist approach can be illustrated with the question of how to translate into specific form the vague proposition that actor behavior is driven by some combination of the logic of consequence and the logic of appropriateness. A first step is to determine which interests are pursued and which norms are recognized as valid and relevant. The next challenge is to determine how actors combine these criteria in evaluating alternative options. Decision theory offers two principal procedures. One—which Braybrooke and Lindblom (1963) referred to as the "synoptic" approach—is premised on the assumption that actors can and in fact do transform "scores" on different criteria into one integrated measure of utility or value. In order to obtain such integration, actors must standardize scores for alternative options across the set of relevant criteria and assign a certain vector to each criterion (v_i) describing its relative weight $(0 < v_i < 1)$. The overall utility or value of an option can then be calculated as a weighted aggregate of scores on the set of criteria included in the evaluation. Applied to policy problems such as global climate change, this recipe requires that economic costs and benefits be measured in the same format as the value of correspondence with salient norms, and that scores be weighted and aggregated into one integrated measure of attractiveness. No straightforward method for making such transformations has been described in the literature. More importantly, we have no evidence to suggest that this is the kind of intellectual exercise actors normally perform in coping with multicriteria decision problems.

The other procedure is premised on the constitutive ideas of cybernetics (see, e.g., Steinbruner 1976) and Simon's concept of "bounded rationality" (1957). The basic idea here is that evaluation is undertaken through a process of *sequential satisficing*. In practice this means that an actor first identifies a set of dimensions or criteria considered to be important characteristics of a "good" solution and then ranks these criteria in order of importance. Next, the actor determines what would qualify as a "satisfactory" score on each of these criteria. Considering the most important criteria first, available options are then evaluated in dichotomous terms, as meeting or not meeting the requirement specified. The first solution that meets all requirements is selected. If none of the options considered satisfies all requirements, the search continues in one or both of two directions: new options are added, and/or one or more of the criteria are relaxed or abandoned. In this procedure each criterion functions as a filter, a necessary condition that a solution must meet in order to be selected.

One attractive feature of the latter approach in this particular context is that it does not "discriminate" against rationalists or constructivists at the outset. A rationalist is free to start from the assumption that actors consider norms only where significant self-interest is not at stake (i.e., within a zone of indifference) and where outcomes are indeterminate (i.e., behind a veil of ignorance). Similarly, constructivists can start their analysis from the assumptions that norms come first and that self-interest enters the equation only when salient norms provide no (firm) guidance. Empirical evidence can be used to determine which model performs better in a particular context. But note also that the sequential satisficing procedure requires the ability to specify with a fair amount of precision the interests pursued and the norms recognized as valid and salient, the threshold of satisfaction for each criterion, and the ranking of different criteria on an ordinal scale.[12] This is a tall order, and much useful work can be done at more modest levels of ambition to develop research designs, conceptual frameworks, and empirical propositions combining rationalist and constructivist mechanisms. Yet it is hard to see how research can move from good inventories, interesting stories, and single propositions to potent theory without eventually coming to grips with these kinds of questions.

At a more operational level, the study of causal mechanisms focuses largely on particular functions that institutions can and often do perform, such as raising concern, enhancing the contractual environment and increasing capacity (P. Haas, Keohane, and Levy 1993), extending domestic policies of front-runners to a larger group of actors (Héritier, Knill, and Mingers 1996; DeSombre 2000), bestowing authority and legitimacy (Young and Levy 1999), or influencing the internal configuration of support and opposition (Boehmer-Christiansen and Skea 1991; Dai 2005). No common taxonomy has yet emerged. What we have at this stage is a rather diverse mix of typologies, some developed inductively (e.g., that of P. Haas, Keohane, and Levy 1993), others deductively (e.g., Young and Levy 1999); some building on rationalist premises, others on constructivist foundations, and still others cutting across that divide to highlight other distinctions (such as that between unitary and complex actors). This diversity makes it hard to compare, contrast, and integrate findings from different studies. In the early exploratory stages, however, a diverse multitude of categories may prove productive in detecting possibilities and generating hypotheses. Moreover, since much of this literature explores mechanisms defined in more or less operational terms, decision makers and stakeholders may well find that it provides a rich source of interesting ideas about levers that might be used for purposes of "political engineering." Looking behind the different labels and descriptions offered, they can also find some common notions of what are important institutional mechanisms and pathways. With an increasing number of studies published, and more systematic comparing and contrasting of findings, these notions could serve as a basis for a common taxonomy. This could be a significant step forward, perhaps providing the minimum of conceptual congruity needed for effective accumulation of knowledge. For the immediate future, a more urgent challenge is to link the various institutional mechanisms and functions to the nature of the problem and the wider operational setting. It makes no more sense to investigate the causal efficacy of a particular institutional mechanism without reference to these critical codeterminants than to engage in clinical testing of a medical cure with no reference to the particular expression of the disease in the patient and his or her general state of health. The observation made by Miles et al. (2002) that there can be more than one path to effective problem solving does not imply that these paths are functionally equivalent or randomly distributed. On the contrary, they describe a particular pattern. Fortunately, social science theory offers a number of propositions that can serve as useful guides for describing how mechanisms and problems are linked. A quick look at rational choice theory shows that the function of "restructuring incentives" will be basically irrelevant in games of pure coordination but critical in games of cooperation. Similarly, constructivist literature demonstrates that at least some of the mechanisms highlighted are assumed to be more active and effective in some circumstances than in others (e.g., Wendt 1999; Checkel 2001). A more precise understanding of such contingencies would be a significant achievement of considerable interest to practitioners as well as students of institutional impact on environmental change.

Preliminary Conclusions

This discussion assumes that some of the confusion that may arise from the rich diversity of propositions and findings on the efficacy of institutions can be cleared by distinguishing those propositions and findings that are (merely) different from those that are also incompatible. Three important observations have been made so far.

First, in a generic sense institutions do play important roles in steering human behavior and determining outcomes of social processes. This is not a real issue. Where human behavior leaves a significant imprint on the environment, a comprehensive and general causal model of environmental change would therefore have to include institutional variables. This does not, however, imply that all changes made in institutional arrangements are important. Many such changes are marginal adjustments that make, at best, an equally marginal difference. The answer to the theory-level question of the generic causal efficacy of institutions will therefore be more "positive" than the answers to many empirical questions concerning the effect of a particular change in a specific regime or organization.

Second, some institutions cause more important effects than others, and we can therefore expect comparative studies to produce findings reflecting a wide range of "true" variance. Some of the divergence in efficacy may, however, be traced back to different notions of what kinds of social orders qualify as institutions. Particularly outside research communities and networks specializing in institutional analysis, conceptual differences can be found with regard to, inter alia, the degree of formalization required and the boundary line between institutions on the one hand and more basic ordering principles and overarching norms on the other. To help bridge these and other gaps, a general formula for determining the causal significance of a particular institution for a particular outcome might be a useful tool. Such a formula is suggested above, deriving causal significance as a function of the importance of the actual domain of the institution, the weight of the actors whose behavior is affected, and the extent to which the behavior of these actors is influenced. Applied to the study of environmental change, this formula suggests that at least some of the most important institutions will be regimes defining the basic rules for key economic activities (such as trade and investment) and regimes specifying degrees of autonomy and decision rules at different levels of political organization. Good reasons exist for using other criteria to set research priorities. The effectiveness of different institutions as tools specifically for environmental governance could be one such criterion. Yet no program designed to help understand the role of institutions in driving or mitigating environmental change can confine itself to regimes and organizations designated to govern resource use or control environmental side effects of human activities. For this reason IDGEC adopted a broad agenda. So far, however, the research project itself and the wider research community have in fact mainly studied environmental institutions. As we are making progress in understanding the role of such institutions, the time has come to consider a drive to include more investigation into other institutions governing human activities that are significant driving forces of environmental change.

Third, institutions typically coproduce outcomes, as integral components of larger causal complexes. Efforts to determine the net causal influence of a particular environmental regime should therefore be supplemented with research to determine to what extent and how it interacts with other institutions, how it is embedded in more basic social orders, and how it interacts with the social and biophysical setting in which it operates. The complexity of the challenge increases with each topic in this sequence. Following the sequence can also be expected to reveal a decline in the proportion of the variance in a particular outcome attributed specifically to institutional arrangements. At the same time, interaction effects involving institutions will loom larger.

Method: How to Detect and Measure Effect?

Since some institutions are more important than others, studies of regimes will produce different conclusions, reflecting the full range of variance that actually exists. In some instances, however, different studies arrive at different assessments of the same institution. In some cases what appears to be disagreement can easily be traced back to the application of different standards. Thus, someone evaluating the performance of the International Whaling Convention in terms of preservationist values will see it as a failure through at least most of its lifetime, whereas an assessment in terms of its official purpose will lead to effectiveness scores ranging from very low to overshoot for different time periods (Andresen 2002). Breitmeier, Young, and Zürn (2006, 186) found that more progress had generally been made toward attaining officially stated goals than toward solving the substantive problems targeted. On the same track, Miles et al. (2002, 435) found higher scores for effectiveness defined in terms of behavioral change than for effectiveness defined in terms of functional problem solving. Sometimes, however, conclusions diverge even where researchers seem to apply basically the same standard(s) to the same institution. Studies assessing the effectiveness of the LRTAP convention provide an interesting case in point.

On the basis of in-depth case studies, M. Levy (1993), Munton et al. (1999), and—in more muted fashion—Wettestad (2002) conclude that the regime has contributed to an overall reduction in pollutant emissions, although varying substantially across protocols (substances) as well as countries. The results obtained by Ringquist and Kostadinova (2005) and by Murdoch, Sandler, and Sargent (1997) on the basis of multivariate regression analysis indicate, however, that little or none of the reductions observed in sulfur dioxide (SO_2) emissions in Europe can be attributed to LRTAP. In the words of Ringquist and Kostadinova (2005, 99), "... differences in emission reduction rates between ratifying and non-ratifying nations were virtually identical before and after ratification." R. Mitchell (2003b), also using regression analysis to separate effects, reports positive effects for LRTAP protocols, except for the first SO_2 agreement, but effects that are generally weak.[13] The fact that divergent conclusions are obtained through different methodological approaches raises one substantive and one methodological question. The former is simply which, if any, of the assessments reported can be considered correct. The latter asks whether at least part of the explanation for the divergence can be found in characteristics of the methods applied. Is there something about the intensive and largely qualitative case-study approach that makes it prone to "detecting" effects that would not appear through a more formal, statistical mode of analysis (or vice versa)? Does it matter whether we change the research question from "What has this institution accomplished?" to "What causes variance in this particular outcome?"

It very well might. Since both approaches essentially rely on the same time-series data to describe what has happened with the regime in place, the critical juncture at which they can take different routes is to be found in their estimates of what would have happened in its absence (i.e., in the construction of what Sprinz and Helm [1999] refer to as "no-regime counterfactuals"). A common approach of researchers—barred from experimental designs and without sufficiently large databases for extensive multivariate analysis—is to begin by comparing developments after the establishment of the institution with the situation existing prior to that event, and then to try to determine how much of the change observed can be attributed to the particular institution in focus (Young 2003a). This can be a perfectly sensible solution, provided that the analysis is guided by at least some crude theory-based model including a wider range of factors that can play a significant role and whose impacts can be systematically examined or controlled for.[14]

Case studies try to meet this challenge in different ways, but most of them focus primarily on detecting and describing institutional mechanisms and the pathways through which these mechanisms can influence events. The potential importance of other factors is clearly recognized, but their effects are not often estimated explicitly through some kind of systematic multivariate analysis. The acid rain study by Munton et al. (1999) is a good example. The authors confidently conclude that "the reductions [in SO_2 and NO_x emissions] would not have happened to the extent they did in the absence of the LRTAP and the MOI" (233).[15] They draw this conclusion largely on the basis of an in-depth analysis of a set of institutional mechanisms. Their account of events—particularly the section on developments in different countries—does include elements of counterfactual reasoning. But no explicit and systematic analysis is provided of emissions trajectories likely in a no-regime scenario.

By contrast, Ringquist and Kostadinova (2005) and Murdoch, Sandler, and Sargent (1997) develop explicit, multivariate models and use statistical tools to estimate the effects of each of the independent variables included in the model. In metaphorical terms we might say that while Munton et al. (1999) and many other case studies concentrate on identifying the institutional "engine" and understanding how it works, the kind of statistical studies referred to above go directly at estimating effect, more precisely the contribution of the engine relative to that of other forces (such as wind and currents) to the overall supply of energy moving the vessel. As modes of inquiry they are clearly complementary, each delivering something the other cannot, at least not equally well. The former approach cannot by itself answer the question addressed by the latter; no reliable method exists for inferring (relative) causal significance from existence proofs. Conversely, investigating relative causation cannot by itself help us understand how institutions produce effects. Explicit, multivariate models are therefore less well equipped to provide knowledge that can guide institutional design. There is, accordingly, much to be said for creatively combining these (and other) approaches. Relying on only one makes findings vulnerable where the choice of methodological approach predetermines substantive conclusions to a significant extent.

Further study is needed to determine whether such predispositions can explain the divergent estimates reported for the LRTAP regime. But the pattern found in the LRTAP case raises concern, at least in a field that has so far relied heavily on one mode of inquiry.[16] Although exceptions can be found (e.g., Stokke, forthcoming), it seems fair to say that most case studies aimed at identifying institutional mechanisms and determining how they work do not examine other kinds of mechanisms and causal pathways in equal depth and with equal fervor. This is by no means to say that they neglect other factors. On the contrary, intensive process tracing will often discover factors and causal pathways overlooked in more formal or extensive modes of inquiry. The critical question is how such studies go about estimating, or controlling for, the impact of these factors.

Many reports fail to provide such information. This may simply reflect the fact that the kind of qualitative counterfactual reasoning undertaken is much harder to specify and explain than numerical measurement and statistical analysis. At worst it may reflect a practice of meeting a major intellectual challenge through low-priority, ad hoc exercises in "counterfactuals light." In any case, specification is required for transparency, and without transparency replication and pointed critique become very difficult. In this perspective the fact that at least some multivariate statistical analyses have failed to confirm positive conclusions drawn in intensive case studies should not be dismissed lightly as reflecting only incidental anomalies. Rather, such discrepancies underline the importance of bringing the full methodological repertoire of social science to bear on this field of research and of comparing and contrasting findings obtained through different modes of inquiry.

In earlier sections I have suggested that specific institutions can best be understood as embedded in more basic social orders as well as in more comprehensive processes of social learning and collective response. This perspective leads to the expectation that the effects attributed to regimes and organizations will often be coproduced, contingent, and characterized by thresholds and tipping points (Jervis 1997; Steffen et al. 2004). The tasks of detecting, disentangling, and measuring effects would obviously be easier if this were otherwise. The most promising methodological strategy for coping with such complexity seems to be one of utilizing a range of different approaches according to what each does best. For example, in-depth case analysis can be particularly useful for identifying causal mechanisms and tracing causal pathways. Regression-type analysis has important advantages when it comes to separating and measuring effects in settings where multiple variables can influence the outcome. Boolean logic can help do the opposite, namely, detect particularly potent conjunctions of driving forces or roadblocks (Ragin 1987). And (agent-based) simulation offers unique opportunities for exploring the dynamics of complex systems, including consequences of hypothetical events and interventions that have never been made in "the real world" (R. Axelrod 1997). Taking advantage of these and other approaches within their proper domains seems a far more productive strategy than engaging in hot disputes over which is generally "best." Research on the role of institutions in environmental change has made progress toward such a differentiated and combinatorial strategy over the past ten to fifteen years, but much work remains ahead to take full advantage of the methodological repertoire of social science.

Concluding Remarks

Determining the causal significance of institutions means encountering and coping with the complexity, contingency, and nonlinearity that often characterize institutional effects and recognizing the important role of "nonenvironmental" institutions in shaping human behavior that drives environmental change. Significant gains may be harvested by combining insights from competing theoretical approaches and making good use of the full methodological repertoire of social science. These key points raise very demanding challenges but ones that emerge out of substantial progress made over the past decade or two in the study of institutional dimensions of environmental change. Demand is growing steadily for more precise and confident answers to increasingly complex questions about institutions as part of the problem and as part of the solution. An effective response cannot skirt the hard questions or confine itself to familiar designs and procedures. IDGEC set ambitious goals for itself and seems—through its own network and through its involvement with other global change programs—to have positioned participating researchers well to contribute continuing leadership and coordinating services.

Acknowledgments

I gratefully acknowledge useful comments to an earlier draft from Oran R. Young, Regine Andersen, Antonio Contreras, Thomas Gehring, and other participants in the IDGEC Synthesis Conference, Bali, Indonesia, December 6–9, 2006. I am grateful to Lynn Nygaard and Maria Gordon for language and style editing.

Notes

1. A different version of this argument can be found in Marxist literature and in so-called structural models of dominance.

2. Computer-based simulation is included in this category.

3. Following Thomas Hobbes, some see the organizing principle of anarchy as a social "state of nature" characterized by the *absence* of institutional arrangements. A similar interpretation is sometimes made of Hardin's (1968) notion of "commons" as a resource subject to no restrictions on access or use. Viewed in these terms, institutions would indeed be important arrangements; if the absence of something is profoundly important, then its presence must also be. For a more inclusive concept of "institution," see Wendt (1999).

4. See also Keohane 1984; Hasenclever, Mayer, and Rittberger 1997; B. Simmons 2000; Keohane and Martin 2003; and Haftel and Thompson 2006.

5. Similarly, the formal capacity of a particular institution can be seen as a function of the importance of the domain of activities regulated, the overall role or weight of its members, and the amount of change required in their behavior (or, for regulation by means of incentives, the change in costs and/or benefits of alternatives). Note that as specified each factor is a *necessary* condition for causal significance. Note also that this is a formula for causal efficacy in general; it does not distinguish positive from negative contributions.

6. The term *political malignancy* was introduced by Miles et al. (2002, 15–16) and defined as a function of incongruity between the interests of the individual actor and those of the group at large, and asymmetry (conflict of interests) among the actors in that group.

7. For a comprehensive review of regime performance studies, see Mitchell, chapter 3 in this volume.

8. This is a principle of selection that is actually used in some extensive studies, at least in studies of local, community-based systems (see, e.g., Agrawal and Chhatre 2006).

9. At the very least we have to recognize that an institution coproduces outcomes with the setting in which it functions. The effectiveness of a resource regime is a function of the regime-problem "complex" or "fit" (Jervis 1997, 39). See also chapter 5.

10. These basic orientations are often referred to by the labels coined by March and Olsen (1989): "the logic of consequentiality" (or "consequence") and "the logic of appropriateness" respectively.

11. The studies reviewed are those by Victor, Raustiala, and Skolnikoff (1998); Weiss and Jacobson (1998); and Young (1999a).

12. One possibility being, of course, that all criteria are considered *equally* important.

13. A similar difference is reported in some other cases as well, such as Siegfried and Bernauer's study of water management in the Naryn/Syr Darya basin: "A comparison of these results with results from a conventional compliance assessment revealed that the more sophisticated method produced much more negative performance estimates" (Siegfried and Bernauer 2006, 35).

14. In some instances this would ideally involve even counterfactual analysis of other kinds of processes, the most obvious candidate being that of policy diffusion (see, e.g., Elkins and Simmons 2005).

15. The acronym MOI stands for the Canada-U.S. Memorandum of Intent (on transboundary air pollution), signed in 1980.

16. Also some of the databases developed and used for more extensive analysis—such as the International Regimes Database and the regimes database developed by Miles et al.—rely essentially on the case study/expert assessment approach.

3
Evaluating the Performance of Environmental Institutions: What to Evaluate and How to Evaluate It?

Ronald B. Mitchell

Introduction

Questions of performance are central to both scholars and practitioners interested in institutions. Whereas the preceding chapter focused on the extent to which institutions "make a difference," this chapter focuses on the extent to which institutions achieve particular objectives. Shifting the focus to performance adds a normative aspect, in the sense of "standards to assess by," to the questions, discussed in the previous chapter, of whether an institution causes outputs, outcomes, or impacts. Assessing an institution's causal significance requires comparing the state of the world in the presence of an environmental institution to a best estimate of what that state would have been in the institution's absence (see Underdal, chapter 2 in this volume). This chapter shares Underdal's focus on institutions as the main independent variable of interest but adds an *actual-versus-aspiration* comparison to the *actual-versus-counterfactual* used in such causal analyses. The aspirations considered can be those held by creators of the institution, other interested parties, or the evaluator. In short, performance analysis seeks to identify how much an institution contributed to whatever progress was made toward a specified goal.

Questions of institutional performance highlight two issues that often go unremarked in analyses of institutional causality: in what dimensions should institutional performance be evaluated; and, for any given dimension, how should researchers go about evaluating performance? As the beginning of a response, discussion here reviews work on institutional performance to date and identifies new research frontiers. The focus is on international environmental institutions; however, the arguments presented may apply equally well to environmental institutions at other scales, from the local to the international and from the highly formalized to the completely informal.

Definitions and Terminology

It is useful to define several terms central to the still-young field of environmental institutional performance. The term *performance dimension* refers to the various criteria against which institutions can be evaluated. Institutions can be evaluated against either the primary or the subsidiary goals for which they were designed, but they can also be evaluated against the goals of actors outside an institution in question. Thus, nongovernmental advocates, scholars, or students may be as interested in evaluating an institution in terms of equity, social justice, or broad notions of sustainability as in terms of the environmental quality or environmentally related behaviors that motivated its creators. Evaluating institutional performance requires at least one *performance scale* or system of measurement for each dimension being evaluated. Often several scales are available for a given performance dimension. Each scale requires a *performance reference point* to which observed outcomes can be compared. Reference points facilitate the estimation of the counterfactual state of affairs along the chosen dimension—the likely scenario had there been no institution. Estimating the counterfactual situation is necessary because claims of causality underpin assessments of performance evaluation. But such scales also include *performance standards*, deviation from which the evaluator can use to categorize an institution as performing well or poorly, as with, for example, standards of compliance or collective optima. Finally, a *performance score* is the numeric or nonnumeric value that some scholars assign to observed institutional outcomes on a given scale relative to either a reference point or a standard. Table 3.1 summarizes these definitions.

Progress to Date

Over the past decade and a half, the IDGEC research program has made considerable progress in understanding—and identifying the sources of—institutional performance, often as part and parcel of work on institutional causality. IDGEC-related research on institutional causality (see Underdal, chapter 2 in this volume) has sometimes addressed performance, identifying not only how environmental institutions have made

Table 3.1
Performance-related terms

Performance dimension	A specific aspect of an institution under evaluation
Performance scale	System of measurement for a given performance dimension
Performance reference point	Counterfactual point to which observed outcomes can be compared to identify institutional influence
Performance standard	Normative point to which observed outcomes can be compared to assess the magnitude of institutional influence
Performance score	The numeric or nonnumeric value assigned to an institutional outcome on a given scale

the world "different," but also how they have made it "better." Institutions have the potential to induce a wide range of effects: intended and unintended, positive and negative, and direct and indirect (Young and Levy 1999). To date, researchers evaluating environmental institutions have tended to use the goals established by institutional creators and participants and have focused on behavior change, environmental improvement, or, less frequently, both as the performance dimensions of interest.

Counterfactual Reference Points: Behavior Change
Performance research has made considerable progress when focusing on environmentally related behavior as a performance dimension. Particularly in the international relations field, efforts in the 1980s and 1990s to refute then-dominant assumptions that international institutions do not have an independent effect on state behavior prompted an initial focus on the extent to which states complied with the specific behavioral requirements of formal legal agreements (Young 1979, 1989a, 1992; Fisher 1981; P. Haas 1989; Chayes and Chayes 1991, 1993; R. Mitchell 1994b; Brown Weiss 1997; Brown Weiss and Jacobson 1998; Underdal 1998). Defined as behaviors by actors to conform to the explicit institutional requirements governing those behaviors (Chayes and Chayes 1993; R. Mitchell 1993), compliance has some attractive analytic features. First, most institutions define compliance such that high levels of compliance correspond to desired levels of environmental quality, making it reasonable to assume that compliance contributes to, even if it

does not equate with, environmental improvement. Second, even when compliance offers few immediate and direct environmental benefits, it may be an important institutional objective because of the more diffuse and longer-term benefits that derive from fostering the legitimacy of international environmental institutions (R. Mitchell 2005). Third, many institutions (though not all) establish clear compliance standards, reducing the analytic assumptions required to identify a performance standard.

Compliance research fostered performance research in several ways. It contributed to a broader shift in the focus of international relations from regime formation to regime effectiveness. The challenge of realist scholars that institutionalists demonstrate the causal influence of international institutions (Strange 1983) prompted important intellectual developments. Not least this included highlighting the need to define no-institution counterfactuals explicitly to avoid misattributing particular behavioral outcomes to institutions (see Fearon 1991). Although institutional advocates and international lawyers have sometimes been more concerned with whether individuals comply with domestic laws, and states with international commitments, than with rigorously assessing the role that relevant institutions play in such behavior, even early scholars who discussed institutional performance in terms of compliance were usually careful to estimate what would have happened in the absence of the institution (Young 1989a; P. Haas, Keohane, and Levy 1993; R. Mitchell 1994b; Brown Weiss 1997; Brown Weiss and Jacobson 1998).

Yet over the course of the 1990s, several analytic shortcomings of compliance research became evident. First, compliance is dichotomous whereas institutional performance is better conceptualized as continuous. Second, compliance is distinct from—and not always coincidental with—institutional causality: compliance is often not institution induced (being either endogenous or coincidental), whereas noncompliance often can be (as when good-faith efforts to comply fail) (R. Mitchell 2007). Obviously, institutions cannot be considered to have performed well if compliance is largely coincidental and would have occurred anyway, as when fishing fleets come in under treaty-established quotas because of declining fish stocks rather than because of restraint in fishing effort. On the other hand, an institution may be considered to have performed well if it induces actors to make "good-faith" efforts, even if those efforts fall short of established compliance standards, as when countries make significant efforts to meet a treaty's requirements for a 30 percent reduction in pollution discharges or emissions but end up achieving only 10 or 15 percent reductions. Third, institutions may induce important behavioral changes not captured by the notion of compliance. Many institutions strive to induce behavior changes by actors who are not subject to the rules and hence cannot be defined as compliant or not. Both the Montreal Protocol and Convention on International Trade in Endangered Species of Wild Fauna and Flora (CITES) seek to influence nonmember countries by banning members from trading prohibited substances or species with nonmembers. And institutions may unintentionally induce changes by nonsubject actors. In a positive vein, the United Nations Framework Convention on Climate Change (UNFCCC) may induce member states to develop emission-reducing technologies that prove economically attractive and are adopted by all countries, regardless of treaty membership; or the convention may create norms that influence members and nonmembers, albeit to different degrees. On the negative side, both domestic and international fishery management institutions may reduce fishing pressure on regulated species in regulated regions while increasing the pressure on unregulated species and in unregulated regions. Fourth, many institutions that target particular behaviors lack, or have only vague, compliance standards. The international wetlands convention requires states to make "wise use" of their wetlands but does not define that phrase in ways that allow identification of compliance or noncompliance. With informal institutions and uncodified norms, it may be difficult to identify what (or even whether a) behavioral standard has been established, making identification of compliance impossible.

In response to these shortcomings, much research shifted during the 1990s to a focus on the broader concepts of behavior change and effectiveness (see, for example, Underdal 1992; Victor, Raustiala, and Skolnikoff 1998; Young 1999a; Miles et al. 2002; but note continuing progress in compliance-focused research, as in Reeve 2002; Breitmeier, Young, and Zürn 2006). Several large research collectives, and the edited volumes they produced, demonstrated the value of broadening the analytic focus from legal notions of compliance to social scientific notions of effectiveness—of whether environmental institutions contributed to positive environmental progress (P. Haas, Keohane, and Levy 1993; Keohane and Levy 1996; Victor, Raustiala, and Skolnikoff 1998; Young 1999a; Miles et al. 2002). Although theoretical conceptions of effectiveness have almost always included both behavior change and environmental quality, several factors (as elaborated below) led most scholars to examine behavioral indicators more frequently than environmental indicators. Indeed, the analytic shift from compliance to behavior change allowed scholars to engage a range of interesting, but previously obscured, questions. Scholars could now evaluate the performance of institutions that had important environmental effects but no clear compliance standards (Paarlberg 1993), induced positive effects on behavior that fell short of compliance or exceeded it (M. Levy 1993), and induced unintended or negative behaviors that made environmental matters worse (Connolly and List 1996; Barnett and Finnemore 1999).

Adopting behavior change as the performance dimension of interest has several advantages. Relative to a compliance focus, assessment of behavior change avoids the need to determine (or trust others' determinations of) whether particular behaviors were or were not compliant. It also avoids the analytic problems in assessing compliance that arise because of ambiguity about the compliance standard itself or its application to the behaviors involved. Additionally, focusing on behavior has advantages even for those committed to the view that environmental quality is the only valid metric of institutional performance. First, institutions can improve environmental quality only by causing changes in human behavior. For a variety of reasons, however, evidence that an institution induced dramatic positive behavioral changes need not imply that the institution also performed well in terms of environmental quality. By contrast, evidence that an institution did not change human behaviors undermines any claim of that institution's influence on environmental quality, even in the face of dramatic improvements. In short, good institutional performance in behavioral terms is a necessary, but not sufficient, condition for good environmentally related performance. Second, behaviors are closer in the causal chain to institutions than is environmental quality. This means there are simply fewer—even if not few—alternative explanations of why behavior changed than of why environmental quality changed. Thus, the analytic task of isolating institutional from noninstitutional influences is easier with behavior than with environmental quality. Third, more—and more consistent—evidence is often available about behaviors than about environmental quality. Because they are of concern for nonenvironmental reasons, data on many environmentally related human behaviors have been collected since long before environmental concern arose. Production and trade statistics and species harvest statistics, for example, are collected for economic reasons, using relatively consistent data-collection techniques over long periods of time. The question then arises whether those techniques facilitate subsequent evaluation. By contrast, knowledge of environmental quality requires explicit efforts to collect data that in many cases is simply harder to identify. For example, fish harvest data are both more abundant and more reliable than fish population data. Indicators of environmental quality are often difficult to use to evaluate institutional performance because they are collected by scientists studying a particular region, species, or pollutant using a methodology tailored to their specific research question and differing from those used by others studying an indicator that is nominally the same. Also, since such studies are often limited in temporal or spatial coverage, it can be impossible to find evidence that would support systematic evaluation of institutional performance.

For various types of institutions, behavioral change may be the most appropriate dimension in which to evaluate performance rather than a "second-best" alternative to environmental quality. Thus, evaluating institutional performance in terms of environmental quality seems particularly ill suited for institutions that, because of political constraints or by design, regulate only a small fraction of the anthropogenic sources of the problem. CITES regulates only trade in endangered species, even though species loss is driven by many larger anthropogenic pressures (including habitat loss and degradation, climate change, ambient pollutants, and domestic human predation). Even if such institutions induce dramatic behavioral changes, analysis is unlikely to reveal much variation in environmental quality. Likewise, focusing on environmental quality is inherently unlikely to identify any positive influence of institutions that have clear environmental quality objectives but require actions with attenuated links to those objectives. Thus, numerous institutions seek to promote scientific research and monitoring. The string of policy and behavioral changes that would be required to lead, in turn, to improved environmental quality is sufficiently extensive to make adopting an environmental quality standard unreasonable. Yet other institutions delineate clear behavioral prescriptions and proscriptions but identify vague or broad environmental objectives that would be difficult to operationalize as performance dimensions. Thus, it is unclear what pattern of environmental quality changes (as opposed to behavioral changes) would constitute movement toward the objectives of the many national and international institutions designed to ensure sustainable development, that strive to promote both conservation and "rational exploitation" of a species, or that seek to coordinate responses to (rather than avoidance of) oil spills, nuclear incidents, or other accidents. Finally, an institution's influences on behavior often occur, or are evident, long before their influences on environmental quality. For example, evidence of the Montreal Protocol's influence on chlorofluorocarbon (CFC) production rates has been available for decades, whereas corresponding improvements in the stratospheric ozone layer may not be evident for many years (Parson 2003).

Counterfactual Reference Points: Environmental Quality

Despite the virtues of using behavior change as the only dimension for evaluating institutional performance, the result often proves unsatisfying. Besides knowledge that institutions are altering human behavior, we seek assurance that the changed behavior produces improved environmental quality. Institutions can change behaviors without significantly improving environmental quality if they target the wrong behaviors, target too few of the right behaviors, target the right behaviors with the wrong tools or insufficient vigor, or target the right behaviors too late. In short, impressive behavioral performance may fail to produce visible environmental progress. A claim of high overall performance for an institution that induced dramatic behavioral changes without significant identifiable environmental improvement or prospects thereof would have little credibility.

Attention to the influence on environmental quality of the many local, national, and international institutions committed to mitigating or eliminating particular environmental problems makes particular sense. Many regulatory institutions specify environmental targets and timetables involving, for example, ambient levels of air pollutants, concentrations of river and marine pollutants, or population figures for threatened species. Even institutions that lack specific environmental quality targets often delineate goals in environmental quality terms, however vaguely. Thus, treaties exist that seek to protect the ozone layer and the climate system, to conserve and develop whale or fish stocks, or to improve marine and river water quality.

Evaluation of performance in environmental quality terms may even be appropriate for institutions that do not specify such terms. Environmental quality goals can be readily inferred for some institutions. Other institutions target behaviors with broad but diverse environmental benefits. And many economic, security, and social welfare institutions have potentially large environmental impacts, as is evident with respect to the

World Trade Organization (WTO), the International Monetary Fund, and arms control agreements addressing environmentally harmful substances. In such cases the analyst cannot rely on an institutional definition of the environmental quality goal but must elicit embedded environmental goals or potential environmental effects to use as performance dimensions.

Evaluating environmental quality performance makes particularly good sense when variation in environmental quality is dominated by anthropogenic drivers. Certain pollutants (e.g., nuclear waste, marine garbage, certain chemicals) and certain types of habitat destruction (e.g., deforestation, wetland drainage) have few, if any, natural causes. Accurate measurements of, say, ambient levels of pollutants or the extent of habitat destruction can provide strong evidence of whether such institutions have achieved their goals. Examples of research along these lines include the environmental Kuznets curve literature (see Shafik 1994; Grossman and Krueger 1995; Selden and Song 1995; Harbaugh, Levinson, and Wilson 2000) and the "free trade and environment" literature (see, e.g., Esty 1994; Antweiler, Copeland, and Taylor 2001). Here, researchers examine the intentional or unintentional influence of national and international economic institutions, respectively, on national levels of environmental degradation. Other scholars have examined the influence of democracy and other broad political institutions on environmental quality (Crepaz 1995; Lafferty and Meadowcroft 1996; Midlarsky 1998; Scruggs 1999; Bernauer and Koubi 2006). Deeper and more sustained interaction between social and natural scientists would ensure that such analyses account for both the anthropogenic and nonanthropogenic sources of environmental change.

But, as noted, environmental quality often proves an elusive dependent variable because of the strong influence of nonhuman factors. Natural fluctuations are often so large as to drown out the much smaller "signal" of institutional influence. Thus, air pollution and water pollution are influenced by wind patterns and river flows, respectively; these influences vary so dramatically over time in ways that often cannot be readily modeled that their influences on environmental quality cannot be easily distinguished from any institutional influences that may exist.

Goal Achievement, Problem Solving, and Collective Optima
The most recent developments in evaluating institutional performance have involved a shift from questions of "how far have we come?" to

"how far do we have to go?" Rather than relying exclusively on counter-factual performance reference points (whether behavioral or environmental), new research challenges us to evaluate performance against more normative standards. Three types of standards have been proposed: goal attainment, problem solving, and collective optima (Underdal 1992; Helm and Sprinz 2000; Hovi, Sprinz, and Underdal 2003a, 2003b; Young, 2003a; Breitmeier, Young, and Zürn 2006; Siegfried and Bernauer 2006). A counterfactual reference point asks simply whether an institution induced behavioral or environmental movement along a performance dimension. A "goal attainment" approach assesses progress toward the institution's formal goals. A "problem-solving" approach assesses progress toward resolving the problem as defined by the originators of the institution. A "collective optima" approach assesses progress toward an "ideal" or "perfect" solution of the problem as defined by a disinterested analyst (Sprinz et al. 2004; Siegfried and Bernauer 2006). These standards make progressively greater demands on the analyst.

These three approaches differ not in the performance dimension used but in the standard against which performance is measured. Goal attainment evaluates institutions *on their own terms*. Presumably, institution creators concerned about individual and institutional reputational effects of poor performance establish relatively unambitious goals that can be readily achieved. Institution creators may also adopt goals in light of the political, economic, and social constraints that they expect will later interfere with their achievement of those goals. All institutions will perform poorly if we adopt standards that are inattentive to the political will, economic resources, and other factors that inhibit the progress an institution even *attempts* to make. Therefore, it often will be a more compelling critique to show that an institution failed to achieve even the unambitious goals it set for itself than to show that it failed to meet standards that those creating it would have considered unrealistic at the time. An "institutional" goal also can be an "average" goal of institution participants that no individual participant actually holds. Institutional goals may consist simply of a mutually acceptable position reached by participants with competing, orthogonal, or hidden goals. In such cases it may be more appropriate to evaluate institutional performance in a disaggregated manner, examining the goals held by industrialized versus developing countries, indigenous versus nonindigenous cultures, or resource-rich versus resource-poor participants.

A problem-solving approach takes one step toward a more ambitious performance standard. This approach accepts the limitations implied by how institution creators defined the problem but not those implied by what ambitions they set for resolving it. A collective optimum standard goes yet further, highlighting that institutions may define a problem in narrow or limited ways that inhibit the more significant environmental progress that might be possible with a more expansive or holistic problem definition. Thus, the whaling convention can be assessed by how much progress it made in increasing whale populations by reducing harvests in line with annual quotas (the goal), in conserving whale stocks while promoting "the orderly development of the whaling industry" (the problem as institutionally defined), and in protecting various whale species from extinction (arguably, one element of an "optimal" environmental solution). Likewise, CITES can be evaluated in terms of its progress in reducing trade in endangered species (the goal), protecting threatened and endangered species (the problem as institutionally defined), and protecting the health and balance of ecosystems and biodiversity more generally (arguably, one element of an "optimal" environmental solution to endangered species protection).

The advantage of problem definition and collective optima standards is that they remove the constraint of evaluating institutions only on their own terms. This approach creates space to assess whether institutions that do well at achieving the goals they embody nonetheless perform poorly in a different sense because they include insufficiently ambitious goals or an inappropriate definition of the environmental problem. Both approaches also have the virtue of fostering cross-institutional comparisons, since they allow an analyst to apply a single performance standard to a range of institutions rather than having to adopt each institution's self-defined performance "yardstick." Thus, despite the obvious problems involved in defining the collective optima for many institutions, applying that standard across a wide range of institutions provides comparability that is impossible if institutional goals are adopted as the performance standard.

The concomitant problem of these approaches, however, revolves around who defines "resolution of a problem" or "the collective optimum." Institution creators have at least some standing in defining performance standards. Beyond them, however, it is unclear what standing academic analysts, scientists, policy makers, or nongovernmental organizations have in defining the "best solution available." For some institutions opinions may converge, making a particular performance standard the obvious choice. But for many more, opinions will vary widely, making any choice among them arbitrary or reflective of the analysts' biases.

The UNFCCC is a case in point, as identifying a collective optimum requires designating, at a minimum, the appropriate target level of greenhouse gas emissions and the year by which that level should be achieved, and, at a maximum, a year-by-year and gas-by-gas trajectory for achieving those results.

Recent efforts have laid both a theoretical and an empirical foundation for progress in using performance standards in addition to, if not instead of, counterfactual reference points. The debate over the Oslo-Potsdam approach has clarified that combining explicitly identified performance standards with explicitly identified counterfactual reference points generates institutional performance scores that may facilitate comparison across institutions and that often correspond to intuitive notions of institutional progress and performance (Sprinz and Helm 1999; Helm and Sprinz 2000; Hovi, Sprinz, and Underdal 2003a, 2003b; Young 2003a; Siegfried and Bernauer 2006). Despite the empirical difficulties faced in seeking to assess institutions using performance standards, at least two large collective projects have used goal attainment, problem-solving, and/or collective optimum standards to compare large numbers of international institutions (Miles et al. 2002; Breitmeier, Young, and Zürn 2006). Equally important, these studies have been able to make claims, albeit cautiously, that not only differentiate better- from worse-performing institutions but that also identify institutional features as well as contextual factors that foster or inhibit institutional performance.

Generating Performance Scores
Researchers have gone beyond debating the appropriate dimensions in which to evaluate performance and have begun defining performance scales and scores in ways that improve the ability to compare institutional performance. The Oslo-Potsdam team has proposed one model of performance scores (Hovi, Sprinz, and Underdal 2003a). They generate a performance scale for any performance dimension that runs from 0 at the counterfactual to 1 at the collective optimum. An institution's performance score corresponds to the level it reaches between these extremes, with a completely ineffective institution scoring 0 and a perfectly effective institution scoring a 1. This score defines performance as the fraction of the "distance" between a noninstitutional counterfactual and the collective optimum potentially induced by an institution. Although criticized on various grounds (Young 2003a), this approach has some attractive characteristics. It allows meaningful comparison of a wide range of insti-

tutions on a conceptual scale that provides an intuitive normalization of different institutional problems. It permits the statement "institution X moved 65 percent of the way toward the collective optimum relative to a noninstitutional outcome while institution Y moved only 30 percent of the way." Notably, the same approach could be applied using goal attainment or problem-solving standards instead of collective optima. In practice, as its proponents acknowledge, numerous obstacles undercut confidence in the estimates of the counterfactual and the collective optimum for any given institution (Sprinz et al. 2004), let alone consistency in such estimates across institutions. That said, the inclusion of a counterfactual and a collective optimum in performance scales begins to capture what we often mean by performance.

Another alternative performance standard involving "regime effort units" has been proposed that seeks to account for the difficulty of inducing behavioral change (as well as the amount induced) to foster more meaningful comparisons across institutions (R. Mitchell 2004). This model, for example, could take into account the fact that institutions that demand a 30 percent reduction in sulfur dioxide or CFC emissions require far less effort by participating actors than do those that demand 30 percent reductions in carbon dioxide or methane emissions.

Scholars have begun comparing institutions in the international arena using these or alternative means (Miles et al. 2002; Sprinz et al. 2004; Breitmeier, Young, and Zürn 2006; Siegfried and Bernauer 2006). Two large-scale projects have asked experts to generate individual institutional performance scores using quite rigorous (though different) research protocols (Miles et al. 2002; Breitmeier, Young, and Zürn 2006). The conceptual logic of both projects parallels that of the Oslo-Potsdam solution: researchers assess institutional performance relative to both no-institution counterfactuals and the stated goals (as opposed to collective optima) of the institutions. The divergence is far greater in their choice of performance scales. In contrast to the 0-to-1 point system of the Oslo-Potsdam approach, both projects adopted ordinal scales. This choice has the advantage that researchers can avoid promising more than they can deliver: they create (and place institutions into) a relatively few categories of performance that correspond to the researcher's inductive calculation of how accurately and precisely they can observe and evaluate institutional performance. This permits the ranking of institutional variation without creating a score the precision of which the underlying research cannot support.

Independent Variables

IDGEC research on performance has also made significant progress in identifying the sources of variation in institutional performance. We now have an extensive set of independent variables that explain observed differences in performance, regardless of the dimensions, standards, or scales used to describe that difference. A crucial insight has been that institutional performance depends on both institutional and noninstitutional factors (Young 1989a; Underdal 1998; Breitmeier, Young, and Zürn 2006). Early on, Ostrom (1990) identified eight design principles of successful commons-governing institutions as well as noninstitutional (exogenous) factors that influence both institutional design and implementation. Haas, Keohane, and Levy (1993) identify institutional effectiveness in terms of governmental concern, political and administrative capacity, and the contractual environment. Jacobson and Brown Weiss (1998) relate institutional performance to a more detailed set of twelve country characteristics, six international environment characteristics, eight institutional characteristics, and four characteristics of the activity involved. Victor, Raustiaula, and Skolnikoff (1998) identify systems of implementation review and other factors as important explanations of the effectiveness of international institutions. Young (1999a) and his colleagues focus on six different causal pathways and behavioral mechanisms by which institutions influence behavioral change. Various authors have delineated numerous other institutional and exogenous factors that explain institutional performance. Nor are exogenous factors always fully independent drivers of institutional performance, as they may interact with institutional features to condition performance. Thus, institutions often incorporate features designed only to influence actors that lack certain political, financial, or administrative capacities but not intended to influence others.[1] For example, financial assistance, technical training, and scientific exchange programs are designed primarily to foster environmental improvement in recipient, not donor, countries.

In short, the IDGEC community has an "embarrassment of riches" with respect to factors that explain institutional performance. Simply compiling the plethora of extant explanatory variables would produce a list of factors that lack overall coherence even though each item in it might have compelling logical and empirical support. Conceptually similar variables are often referred to using different terms, variables in different taxonomies often involve quite different levels of resolution and range, and variables in one taxonomy often do not map readily to those in others. Work is needed to systematize the sources of variation in institutional performance into a comprehensive and coherent explanatory framework that could foster the development of more cumulative knowledge on the subject.

Other Dimensions of Institutional Performance: What Should Be Evaluated Next?

The significant progress made in performance evaluation with respect to behavioral change and environmental quality has not been matched by corresponding progress with respect to other performance dimensions. Scholars could investigate a range of alternative dimensions to improve understanding of the performance of institutions that affect the environment.

In many policy realms, inputs to policy development or revision processes require that performance assessments be completed before the institution can be expected to have any influence on behavior, or at least before compelling evidence of such influence is available. Evaluating the influence of the Kyoto Protocol to the UNFCCC only after the end of the first commitment period might be analytically more relevant but would be necessarily policy irrelevant: negotiations of successor rules to the Kyoto Protocol will have already been completed. Policy relevance often demands that performance evaluation make use of projections of institutional influence from limited and/or poor-quality data on indicators other than behavior or environmental quality. Interest in assessing the performance of particular institutions seems, unfortunately, to wane over time with little scholarly attention paid to many institutions with decades-long track records (R. Mitchell 2003a).

Nor are all environmental institutions regulatory in nature. Many institutions do not address environmental problems by proscribing or prescribing particular behaviors but, instead, provide a forum for collective decision making (procedural institutions), encourage the pooling of resources for projects that would not be undertaken unilaterally (programmatic institutions), or promote certain norms and social practices (generative institutions) (Young 1999b, 28–31). These institutions are intended to set in motion social transitions that, it is believed, will eventually reduce environmentally malign behavior and improve environmental quality. For example, institutions often establish information or education campaigns, foster scientific research and environmental monitoring, and fund "portfolios" of capacity-building projects that may even contain components expected to fail. It is difficult to know where (or when) to look for the influences of such institutions on environmental quality, let alone convincingly to demonstrate causal links between such institutions' immediate influences and whatever eventual changes in environmental quality they may induce.

Changes in behavior and improvements in environmental quality are not always the sole institutional objectives, which further complicates performance analysis. Environmental institutions have a set of quite direct and predictable effects that are nonenvironmental but not unimportant. Indeed, opposition to environmental institutions as often stems from concerns about their nonenvironmental effects as their environmental ones. In such cases the focus is on whether an institution's large nonenvironmental effects have any, or sufficiently large, offsetting environmental benefits. Thus, environmental institutions are criticized for, inter alia, their direct economic costs, the drag they place on development, the equity of their distribution of costs and benefits, and their cultural and social impacts. In addition, concern can exist about how well institutions perform in functional and institutional terms.[2] Institutions that promote transparency, accountability, and stakeholder participation may be valued even when those institutional traits inhibit or delay behavioral change and environmental improvement. A set of performance dimensions that includes how institutions operate can serve, then, to provide a richer picture of institutional performance.

"Leading Indicators" of Institutional Performance

Even for institutions that directly target behaviors and environmental quality, there are situations where it is helpful to evaluate institutional performance in other terms. Existing theory suggests that the effects of most institutions are frequently indirect and rarely instantaneous. Thus, just as economists and policy makers evaluate and adjust economic policies using "leading economic indicators" considered to be good predictors of subsequent economic growth, institutional analysts could look more seriously and carefully at the processes that institutions entrain, that is, at the ways institutions may create, strengthen, or redirect social processes that will eventually generate (or enhance) institutional effects. Indeed, using proximate nonbehavioral and nonenvironmental indicators of institutional performance has some advantages. The ability convincingly to link institutions to their impacts declines as a function of the lag between institutional action and those impacts. The longer the lag, the more likely it is that other factors that also influence the behavioral or environmental indicator will have changed in ways that run counter to (and hence obscure) institutional effects or that coincide with (and cannot be readily discounted as causes of) institutional effects.

"Leading" institutional performance indicators involve direct and immediate institutional effects that, over time, can be reasonably assumed to generate the ultimate effects of interest. They are instrumental indicators that are reasonably good predictors of ultimate institutional performance but on which evidence becomes available long before it becomes available for the latter. Leading indicators discussed here include environmentally related behaviors and environmental quality in hopes of prompting consideration of other, similar, indicators. There is no reason, however, not to identify leading indicators for any of the other performance dimensions discussed below. For example, an absence of low-cost alternatives to an environmentally harmful behavior would likely be a good predictor of the cost-effectiveness of an institution designed to address that behavior. Similarly, in assessing the cultural impacts of an environmental institution, a valuable indicator might be youth emigration from traditional communities. In general, however, the best leading performance indicators will be those that can be observed soon after institutional action, that can be clearly and convincingly linked to institutional action, and that have strong logical and empirical bases for claims of successfully predicting institutional performance with respect to the indicators of ultimate interest.

Public Commitments and Changes in Policy Outputs and Economic Decisions For many institutions, certain public commitments and changes in public policies or economic decisions can be identified as necessary, though not sufficient, conditions for institutionally induced environmental improvement. Membership, as when governments ratify environmental treaties or companies accept environmental codes of conduct, is often taken as a near-term proxy for institutional performance. Although states, corporations, and individuals may ignore their public commitments, it seems reasonable to assume that important actors that assume public institutional commitments are, on average, more likely to contribute to the goals of those institutions than those that publicly reject such commitments. Even stronger predictors of subsequent behavioral change and environmental improvement exist in the form of internal

institutional changes, for instance, in the legislation, regulations, or policies of the actors targeted by an institution. Thus, changes in domestic
legislation, executive branch rule making, or corporate policy and planning documents constitute compelling evidence that more than lip service
is being paid to institutional goals and of another step in a trajectory toward environmental improvement.

Improved Scientific Understanding of a Problem and Potential Solutions
For many environmental institutions, evidence of improved scientific
understanding of the environmental problem and potential solutions
provides a useful leading performance indicator. Institutions may promote scientific understanding or research and development directly (as a
sole objective or as one part of a larger regulatory effort) or indirectly by
raising the salience of an environmental problem so that government,
private, and academic scientists (and their national, local, corporate, or
private funders) dedicate more resources to the problem. Better knowledge of the causes of an environmental problem and of technological
alternatives can increase the motivation to avert environmental change
while decreasing the countervailing pressures that inhibit changes to
existing behavior patterns. Initial research into how institutions designed to promote scientific understanding influence policy and behavior
(Andresen et al. 2000; R. Mitchell et al. 2006) suggest that improved
scientific understanding can, under certain circumstances, lead to environmental policy and behavior changes that ultimately lead to environmental quality improvement. Evidence of institutional influence on
scientific understanding might include funding of science related to the
problem or articles published on that problem.

Creating or Strengthening Environmental Norms   Norms are known to
influence behavior under at least some circumstances, and hence evidence of an institution creating or strengthening an environmental norm
is likely to presage corresponding changes in behavior and, ultimately,
environmental quality (Finnemore 1993; Katzenstein 1996; March and
Olsen 1998). Sorting out causality with respect to norms is particularly
challenging, since the causal links between norms and behavior are
bidirectional: norms at time period T influence behavior at $T + 1$, but
so too does behavior at $T + 1$ influence norms at $T + 2$ (R. Mitchell
2005). Yet norm creation and strengthening is an important potential
path of influence for many institutions. Norms usually involve deeply

held values that are the foundation for a wide range of behaviors. Therefore, if they can successfully alter norms, institutions can wield significant long-term influence over behavior. Yet precisely because altering actors' normative convictions takes time, unambiguously identifying an institution as the cause of such changes proves difficult. Evidence of institutional influence on norms might include the frequency with which particular phrases (such as "sustainable development") appear in speeches and news articles or the terms in which justification is given for actions that run counter to the norm the institution seeks to promote.

Economic Performance Dimensions

Beyond environmental impacts, policy makers and researchers are, for obvious reasons, interested in the direct and often relatively predictable economic influences of institutions. Yet the performance dimensions of costs, cost-effectiveness, and cost-efficiency have still to receive significant attention.

Economic Costs   In terms of the costs of establishing and maintaining an environmental institution, research could begin with the categorization of costs. Creating an environmental institution at the international, national, or local level almost always involves considerable time and resources, often as much or more by institutional opponents as by institutional proponents. The costs of negotiating the creation and operation of many institutions can be obvious. Costs are relatively easy to identify when institutional tasks use budgets funded by participating actors. Institutional costs are far harder to calculate for the many institutions that coordinate the behavior of various actors, whether those actors are the nation-states of international regimes or the individuals involved in local commons institutions (Ostrom 1990). Which behaviors should count as institutional tasks, and which of the associated costs should count as institutional costs? Efforts to implement institutional requirements, to monitor behavior and environmental quality, to reward or punish implementation efforts, to conduct scientific research, and to evaluate and negotiate new rules may all form part of the picture. In all these cases it may prove difficult to sort out which costs were incurred because of the institution and which would have been incurred in any event. Equally important, these tasks as often involve important nonmonetizable costs and significant opportunity costs in which resources used by one institution are unavailable for other, more effective, efforts. Little work has been done in this area, and considerable value would stem from developing typologies of institutional costs along with rigorous accounting methodologies.

Economic Benefits   In some cases environmental institutions may generate economic side benefits. Efforts to identify "no-regrets" policies, particularly with respect to efforts to address climate change, provide an example. Institutions created for environmental reasons may prompt actors to revisit various economic decisions, upsetting the inertia that often leads both individuals and government bureaucracies to continue behaviors adopted when they were economically beneficial but that have become economically costly. Likewise, institutions can induce actors to make investments that will generate a larger stream of benefits over the long term than current behaviors, investments they would not otherwise make because of high initial costs.

Cost-Effectiveness   Once researchers identify the types and magnitudes of costs incurred in developing and maintaining an institution, clearly the next question relates to cost-effectiveness. Do institutional benefits exceed institutional costs? Although cost-benefit analysis is a well-developed field of study and ad hoc calculations surely occur for decisions by countries or individuals to create, maintain, and support many social institutions, systematic efforts to evaluate the cost-effectiveness of environmental institutions have been rare. Given the possibility of determining that an institution is not cost-effective, institutions have some, often strong, incentives not to engage in rigorous self-assessment. This provides all the more reason for scholars to take on this task, one that would entail developing methods to identify cost-benefit ratios; distinguish cost-ineffective, somewhat cost-effective, and very cost-effective institutions; and, at a higher level of resolution, identify which institutional tasks are "cost centers" and which are "profit centers." The results of such research could provide a foundation for making existing institutions more cost-effective and for indicating when creating a new institution is not warranted. Although not usually couched in cost-effectiveness terms, initial research on institutional interplay (Stokke 2001a; Young 2002a; Gehring and Oberthür, chapter 6 in this volume) has examined overlapping regimes that are redundant or create ineffective "divisions of labor." To the extent that different existing institutions are fully or partially interchangeable, removing such redundancies would reduce costs and improve cost-effectiveness.

Cost-Efficiency   Finally, cost-effective institutions need not be cost-efficient. Cost-effective institutions are those whose benefits exceed their costs; cost-efficient institutions are those whose cost-benefit ratios are better than the corresponding ratios for other institutions addressing the same problem. For a given task, one institution may generate benefits 20 percent greater than the costs incurred to establish and maintain it. Although this may appear to be "great value," alternative institutions might produce benefits that exceed institutional costs by 40, 80, or 200 percent. Addressing such questions requires broad knowledge of "typical" institutional cost-benefit ratios and the range of such ratios. At present, little information exists on whether most institutions are cost-effective in an absolute sense, let alone relative to other institutions that are—or could be—established in response to the same problem.

Indirect Dimensions of Performance

Institutions have a range of indirect and unintended effects that merit attention and that are often central to the political and policy debates surrounding institutional formation and operation. Whether driven by sincere concerns about negative collateral impacts of environmental institutions or by more strategic efforts to "expand the scope of conflict" in order to build support for or opposition to an institution (Schattschneider 1960/1975), advocates often evaluate the performance of environmental institutions in many dimensions besides environmental or behavioral change. From a scholarly perspective the advocacy underlying the resulting evaluations causes them often (though not always) to lack the analytic rigor necessary for credibility. Nevertheless, the numerous advocacy documents generated in support of or opposition to various environmental institutions identify many potential institutional effects that are quite susceptible to the analytic tools used to evaluate environmental benefits and economic costs. Some of these effects are identified below in hopes of leading others to create a more comprehensive list.

Economic Growth and Development   Beyond direct economic costs and benefits, considerable concern exists regarding the influence of environmental institutions on economic growth and development. Some environmental institutions may impede, while others may foster, economic growth, and these effects may differ between developing and industrialized countries. For example, significant greenhouse gas emission reductions incorporated in any future climate change agreements almost certainly will reduce gross domestic product growth in some, if not all, countries, but some fisheries agreements have mitigated the economic problems of fleet overcapitalization and of stock crashes. Net economy-wide effects (whether positive or negative) are likely to reflect gains for some countries, actors, and sectors and losses for others. The need arises to identify the processes by which environmental institutions promote environmentally positive or benign economic growth, and also to design institutions to minimize the drag they place on environmentally positive or benign economic activities. More research is needed into how, under what conditions, and which environmental institutions move economies toward sustainable development by mitigating the ways in which economic development runs counter to environmental protection. Such analyses should examine not only the extent to which environmental institutions contribute to a "sustainability transition" but also how that contribution compares to that of other social forces (Board on Sustainable Development Policy Division 1999; Kates et al. 2001; Kates and Parris 2003). The contributions of environmental institutions may be dramatic or minor when compared to economic markets, social movements, transnational actor networks, environmental exigencies, or various other forces that influence this larger process of social change.

Economic Equity, Cost Incidence, and the Distribution of Costs   Equity seems a particularly fertile arena for future research, given that institution creators have increasingly designed environmental institutions with equity at least somewhat in mind. The differentiated obligations, flexibility mechanisms, and financial transfers in the UNFCCC and the Montreal Protocol reflect acknowledgment that variation in an actor's historical responsibility for a problem, ability to pay for a problem's resolution, level of economic development, or other attributes should influence the institutional obligations that actor is asked to assume and the costs that actor should bear. Institutions generate equity concerns in at least three ways. First, institutional rules dictate which actors must change their behaviors and by how much, and whether those actors must pay the associated costs or whether others are to pay or share those costs. Second, the benefits of improved environmental quality accrue unevenly across actors. Third, efforts to remedy environmental problems cause indirect impacts. Even an institution that is completely ineffective from an environmental perspective may influence equity by imposing costs on some actors and not others. The first two concerns usually influence those actors participating in the institution. The third more often involves actors affected by, but not participating in, an institution.

Many researchers have skirted issues of equity because of the normative judgments involved. But the distributional impacts of institutions can be submitted to rigorous empirical analysis that either remains agnostic about or sequesters normative aspects (see, for example, Parks and Roberts 2006). Analysis of equity performance could move forward by separating empirical identification of an institution's distribution of costs and benefits from normative and/or prescriptive judgments about that distribution. Such a separation might allow researchers to build a collective research program that provides a common, and descriptively accurate, empirical foundation on which competing normative claims could be made. As with other performance dimensions, careful analysis of an institution's influence on cost incidence requires counterfactuals: comparing the observed distribution of costs and benefits to a clearly specified and empirically supported finding regarding who would have borne the costs and received the benefits of an environmental problem had it not been addressed.

Careful counterfactual analysis could provide a more explicit and grounded discussion of why certain actors benefited from an institution while others were harmed, whether this was intentional and justified, and how undesirable institutionally induced inequities could be mitigated or eliminated. Researchers could also examine different definitions of equity more rigorously and systematically. In terms of climate change, some have analyzed alternative emissions source categories and the corresponding "implied responsibility for emissions if an agreement is based on some form of the polluter-pays principle" (Subak 1993, 68; Sebenius et al. 1992). But there exists a wide, and largely unexamined, variety of potential equity criteria, including not only responsibility for the problem but also ability to pay; inequities in economic, social, and other realms; and so on. For most such criteria there are numerous ways of both measuring and weighting the influence on the actors involved. Although it may be possible to provide an equity-based argument for almost any distribution of costs and benefits deriving from an institution, more rigorous empirical evaluations might provide the basis for more reasoned discussion about these issues.

Social Justice   Closely connected to questions of economic equity are those of institutional performance and social justice. Environmental institutions may alter the balance between rich and poor both within and across countries. Indeed, the impact of environmental problems or the environmental institutions that address them on already disadvantaged societal groups has become an increasing concern for many environmental advocates. The environmental justice literature has identified numerous examples in which large-scale social forces (i.e., informal institutions) displace the environmental problems of the rich onto the poor, domestically and internationally (Princen, Maniates, and Conca 2002; Lee 2006). But researchers have increasingly recognized that opposition to environmental institutions designed to protect endangered species and biodiversity can come from the already disadvantaged populations whose livelihoods are adversely affected when "charismatic megafauna" trample their farms, eat their livestock, or threaten their children (Biermann 2006; A. Gupta 2006). Equity also can be framed in broader, noneconomic terms related to how environmental institutions mitigate or exacerbate ethnic, racial, or other social conflicts. Nor can all institutional costs be monetized. Some economically disadvantaged cities, provinces, and countries, for example, have rejected imports of hazardous wastes despite large financial transfers, demonstrating that rights, sovereignty, and other norms may not be readily converted into economic terms. Prior informed consent rules, for example, are likely to lead some countries to import more hazardous chemicals and wastes and other countries to import fewer. Both the "absolute" patterns of postinstitutional imports and the change from preinstitutional patterns will influence political perceptions of whether these are "good" institutions or need revision. Some efforts are already being made to investigate such issues (Paavola and Adger 2006). At a deeper level, efforts to redefine universal human rights to include the right to a clean environment or the right to certain environmental amenities entail a corresponding claim that the preinstitutional distribution of costs from most environmental problems falls disproportionately on certain disadvantaged groups and that environmental institutions should be created precisely to remedy this situation (Shelton 1991). More rigorous evaluation of equity as an institutional performance dimension would allow scholars to contribute more usefully to legal and policy debates that inform the creation and revision of many environmental institutions.

Cultural Impacts   The impact of environmental institutions on cultures, particularly traditional cultures, constitutes another important but understudied dimension of performance. In some cases environmental preservation and cultural preservation can directly conflict, as in provisions in the whaling convention that allow indigenous whaling of endangered species. In other cases such conflicts are less obvious but no less real, as when lands sought for habitat preservation have been traditionally occupied by an indigenous group or the preservation of environmentally important lands requires imposing constraints on indigenous use. Yet other cases may exhibit positive synergies, as when institutional constraints on development protect both threatened habitats and indigenous cultures. An increasing number of institutions recognize that preserving natural ecosystems may also require preserving associated traditional knowledge and culture, or that preservation of traditional knowledge can help extend our knowledge of the Earth's natural systems farther back in time (Jackson 1997, 2001; Jackson et al. 2001) or, as with knowledge regarding medicinal properties of plants and animals, provide more instrumental benefits (Blum 1993; Zebich-Knos 1997). Environmental institutions may help preserve—or speed the demise of—traditional environmental knowledge and traditional cultures.

Although studying "cultural" impacts of institutions has usually meant traditional or indigenous cultures, environmental institutions may also have significant impacts on nonindigenous cultures. Environmental institutions, perhaps domestic ones especially, have significant implications for the structure of cities (e.g., the setting of urban growth boundaries), land use and the economic activities in which people engage (e.g., efforts to inhibit aquaculture or promote pesticide-free, organic, and/or small-scale agriculture), and how lives are led (e.g., environmentally driven efforts to promote telecommuting as an influence not only on daily travel patterns but also on the type and frequency of a person's daily interactions). Issues of how institutions influence cultures of all types, alter traditional knowledge, and address the balance between cultural and environmental preservation deserve greater attention.

Good Governance and Functional Performance

Finally, we often care how institutions act as institutions, that is, how well they perform certain functions or meet certain standards of governance. In many countries and internationally, institutions are increasingly judged not only by how well they achieve their goals but also by their degree of stakeholder participation, accountability, transparency, legitimacy, and other criteria of good governance (Wirth 1991; M. Stewart and Collett 1998; Grant and Keohane 2005; Hood and Heald 2006).[3] Institutions that achieve effective environmental improvement by violating human rights will generally be viewed unfavorably, whereas those that build the capacity of stakeholder groups to participate in environmental decision making will generally be viewed more favorably, independent of their environmental influence. Less starkly, there may be trade-offs in which institutions that are significantly more accountable or transparent are preferred even though they may take longer to produce environmental results.

In other contexts, institutions may be assessed on how—and how well—they perform certain functions, temporarily without consideration of whether the performance of those functions produces some set of subsequent benefits (Hovi, Sprinz, and Underdal 2003a, 74). Procedural, programmatic, or generative institutions may be judged in terms quite different from those applied to regulatory institutions (Young 1999b, 28–31). Environmental institutions differ significantly in how well they foster joint decision making among their members. Some institutions successfully address difficult problems quickly and proactively seek out new ones; others struggle to produce collective responses to even relatively benign problems (Miles et al. 2002). Institutions often seek to promote, inter alia, environmental monitoring, social capacity building, and project financing (Kanie and Haas 2004). Although institutions that induce such efforts seem likely, ultimately, to improve environmental quality, the fact that such efforts may be three, four, or more causal steps removed from such improvements may lead to acceptance of failure or success in inducing those efforts as a valid indicator of institutional performance.

In yet other contexts, creators of some institutions that address environmentally related problems show little concern with the institution's immediate influence on environmental behaviors or outcomes. An institution's major influence may be the alteration of social processes, leading actors to adopt new social roles, perform new functions, and engage in new social processes that affect environmental quality so indirectly that these direct influences become, essentially, valued in their own right. Efforts to change decision making, policy making, and regulation in certain ways may be ends in themselves, as they are assumed to generate a wide but diffuse range of environmental benefits. For regimes such as the 1998 United Nations Economic Commission for Europe (UNECE) Aarhus Convention on Access to Information, Public Participation in Decision-making and Access to Justice in Environmental Matters and the 1991 Espoo Convention on Environmental Impact Assessment in a

Transboundary Context, it is difficult to know what behavioral or environmental outcomes would make good indicators of performance. Nor is it easy convincingly to trace observed changes in those indicators back to institutional efforts. Indeed, it seems unlikely that rules promoting access to information, public participation, or the use of environmental impact assessments would be discarded even if they were definitively shown to make environmental problems worse. The focus of past work on the environmental or behavioral effectiveness of institutions has left considerable room for research into how well institutions perform their governance tasks.

Performance Scales: How to Measure Performance Dimensions

Despite their diversity, some claims can be made about how to develop performance scales that apply to many, if not all, performance dimensions. Rather than seeking the "best" performance scale, this section delineates several useful criteria that can help develop, and clarify the advantages and disadvantages of, different scales.

Construct Validity, Accuracy, and Reliability   The value of any scale lies in the degree to which researchers and practitioners accept the scores assigned to institutions as reasonable approximations of those institutions' performance in the given dimension. Such acceptance depends on the scale being construct valid, accurate, and reliable. *Construct-valid* scales are those that accurately capture the central elements of the concepts of interest claiming to be captured (DeVellis 2003). *Accurate* scales (and scores on those scales) are those that involve observing or measuring institutional variables in ways that maximize the chances that the scores assigned are the "true" values for the institution. *Reliable* scores are generated by devising and applying systematic procedures with sufficient consistency that other researchers evaluating the institutions with the same scales would be likely to produce the same score (Carmines and Zeller 1979; Neuendorf 2002).

Transparency   Ensuring confidence in a carefully constructed and systematically applied scoring system requires research transparency. Scoring systems are more likely to be used if users can see for themselves how the research generated the scores and that the system is construct valid, accurate, and reliable. This is best accomplished by documenting and making publicly available the rules used to code results that form the basis of scores and the evidence used in assigning scores to particular institutions.

Comparability   A scoring system should also allow meaningful comparison across institutions. Regardless of the performance dimension or performance scale used, scores should allow institutions to be classified as performing similarly to some institutions and differently from others (for nominal scales) or better than some and worse than others (for ordinal, integral, and ratio scales). This requires not only consistent scoring procedures but also the defining and designing of scales in ways that make sense when applied to a wide range of institutions. Notably, the "natural" units for measuring performance are not always meaningfully comparable across institutions. Comparing reductions in sulfur dioxide emissions under the 1979 UNECE Geneva Convention on Long-Range Transboundary Air Pollution to reductions in carbon dioxide emissions under the UNFCCC's in "tons of pollutant reduced" makes little sense since emission quantities of the two pollutants differ so markedly. Scales based on percentage changes take an initial step toward meaningful comparison (R. Mitchell 2002), while normalizing such changes (whether using a collective optimum or some other standard) goes yet further (Hovi, Sprinz, and Underdal 2003a). Finding scoring systems viewed widely by researchers as supporting meaningful comparisons may be challenging for some performance dimensions. Yet in other performance dimensions scales may readily allow such comparisons. The oft-used Gini index of inequality, for example, may provide a credible basis for comparing different institutions in terms of their impacts on equity.

Scales Appropriate to the Performance Dimension   Performance differences can be recorded using nominal, ordinal, interval, or ratio scales. Nominal scales differentiate performance without ordering it, fostering nuanced description of institutions as performing "differently" without requiring a judgment of which performed better. Theoretical and empirical constraints make it difficult to place variation in some performance dimensions along a single line. Especially when several aspects of a single performance dimension are interdependent or difficult to disentangle, a scale that builds on typological theory or factor analysis may be valuable. Typologies might distinguish, for example, among those that influence only resource-rich actors, those that influence only resource-poor actors, and those that influence both sets of actors; between those that drain a country's economy generally and those that also boost certain economic sectors; or between institutions that preserve smaller pristine ecological habitats by exclusion and those that preserve larger, less "pure" habitats by fostering ecotourism. Nominal scales can foster systematic comparison of complex, multidimensional institutional variation in ways that a single ordinal ranking system would obscure. For example, an "environmental justice" scale could characterize how well institutions achieve both social justice and environmental goals, even though those achievements could not be aggregated. Ordinal scales become appropriate for performance dimensions that seem sufficiently simple and independent of other performance dimensions that institutions can be placed on a single, unidimensional scale. Ordinal scales move beyond claims that institutions performed differently to claims that one institution did "better" than another on a given dimension. Comparative effectiveness research has adopted precisely this approach, comparing how institutions addressing fisheries, pollution, and other environmental problems have performed in terms of environmental improvement (see, for example, Miles et al. 2002). Interval and ratio scales go yet further and estimate the size of performance differentials. The intervals used should reflect the resolution with which the scoring system can detect variation, so that scales from 0 to 5 or 0 to 10 will often be more appropriate than scales from 0 to 100. For example, we might want not merely to identify whether each in a sequence of amendments to a treaty constituted an improvement but also to obtain a specific estimate of how much improvement each amendment made. The challenge in such cases, as is evident in the debate over the Oslo-Potsdam solution (Hovi, Sprinz, and Underdal 2003a, 2003b; Young 2003a), lies not so much in creating such a scale but in convincing users that scores correspond to real differences between institutions in a given performance dimension.

Methods   Finally, there are multiple methods for evaluating institutional performance. The methodology can involve an array of alternative qualitative or quantitative techniques; can base findings on expert or nonexpert assessments; can rely on process tracing, counterfactuals, or statistical algorithms; and can include control variables in analyses or control for exogenous factors through case selection. The best methodological choices are likely to be made when the researcher bases selection on how well different methods fit with the theoretical state of play and available empirical evidence, with an eye toward addressing as rigorously as possible the concerns of those most skeptical of institutional influence.

Ambitions for the Future

The foregoing discussion has proposed a foundation for future performance research. It is, nonetheless, more limited than it need be. Other directions exist, some particularly promising, for expanding and building on that foundation.

Broadening the Scope of Comparability

Building on prior efforts, scholars should seek to develop methodologies, scales, and scores that allow comparison of quite different types of institutions. Much important research remains to be done in comparing the performance of two governments, of two nongovernmental organizations (NGOs), or of two treaties. Because most, though not all, researchers focus on particular types of institutions, few have compared across institutional types. But there is considerable value in comparing across categories: for example, assessing whether fish harvests are best constrained (and fish stocks best protected) by a network of local institutions, NGO-corporate certification programs, NGO awareness-building efforts, private fishery management corporations, national ministries of fisheries, or international fishery commissions. Over the long term researchers could make a crucial contribution to knowing the conditions under which social, political, and economic resources are better invested in one type of institution versus another (R. Mitchell 2007, 920).

Developing a Multifaceted View of Performance

In the years ahead, significant improvement in understanding institutions will depend on developing richer and more nuanced pictures of their performance. Accomplishing this requires that scholars move toward evaluating institutions in multiple dimensions, with multiple scales, and against multiple standards.

Multiple Dimensions   Just as we derive a more complete picture of a figure skater by assessing both artistic merit and technical difficulty, so do we derive a fuller picture of institutional performance by evaluating a variety of performance dimensions. The choice of which dimension or dimensions to evaluate will—and should—reflect the different analytic goals and normative preferences of the researcher. For various reasons delineated above, behavior change is likely to remain a central focus of performance research, and improvements in that realm could foster more meaningful comparisons among institutions. That said, expanding the range of performance dimensions deemed "appropriate" for research would allow analysis of the many institutions for which behavior change is not a feasible or relevant performance dimension. Such an approach would also build a more nuanced picture of those for which behavior change is one, but only one, element of performance.

Multiple Scales Institutional performance research would also be improved by using multiple scales to measure a given aspect of a phenomenon. A single scale can rarely capture the observed variation in a given performance dimension. Concerns with equity might lead to research on how an international institution influenced the distribution of resources between industrialized and developing countries, between rich and poor citizens within each country, between different ethnic groups within each country, and between men and women within each country. Multiple subscales, as opposed to a single metric, would paint a more accurate picture of institutional influence. A summary performance score that aggregated across those subscales could still allow a single ranking across institutions. As long as the subscale scoring and aggregation methods were transparent, users could assess how much confidence to place in the summary ranking.

Multiple Standards Finally, although all performance research must use counterfactuals as a reference point, a fuller picture emerges when a range of standards is adopted (see, e.g., Siegfried and Bernauer 2006). We can place more confidence in claims (whether negative or positive) of institutional performance derived from convergent evidence of compliance, behavior change, goal achievement, problem resolution, and collective optima. And, equally important, inconsistencies among such evaluations shed light on the exact character of institutional strengths and weaknesses.

Measuring Dynamic Performance

Much progress also can be made by building on nascent efforts to evaluate institutional performance in dynamic terms (Gehring 1994; Siegfried and Bernauer 2006). The temporal profiles of institutional performance can vary considerably. Some institutions perform well initially by channeling the attention that led to their creation, while others perform poorly because of lack of knowledge, resources, experience, and support. Some institutions perform reasonably well initially, improve through maturation effects, and then decline because of institutional senescence. Others require large financial, institutional, and normative investments that provide "returns" only many years later. Because an institution's performance may vary over time for either institutional or exogenous reasons, developing methods to assess this dynamic aspect requires methods for generating dynamic estimates of counterfactuals. Changes in an environmental problem may make it more benign or more malign, that is, easier or harder to resolve. Exogenous changes in the broader context—for example, the end of the Cold War or the 9/11 attacks—may facilitate or impede institutional efforts. In terms of good governance criteria, there may be concern with how well institutions can adjust in response to operational experience and/or new or changing knowledge and with how resilient, flexible, and robust they are in response to exogenous changes. Having generated counterfactuals in light of these considerations, it is possible to imagine performance scores based on the "area" between a line of observed outcomes and a corresponding line of counterfactual points. Although the "area" between those lines may constitute a useful assessment of "life span" performance, including dynamic performance, assessments will require further methodological work since "life span" performance scores that are equal in area can be generated by quite different institutional profiles relative to their counterfactuals. Two institutions working on an identical problem might generate equivalent "areas" of influence but with one having a large influence for a short period of time and the other having a much smaller influence over a more sustained period. Equally important, institutional goals often change over time, whether becoming more aggressive, less aggressive, or simply altering the emphasis placed on competing goals, as seen in the histories of the International Whaling Commission or the World Bank.[4]

Being Open to the Negative Effects of Institutions
Most performance research to date has unself-consciously assumed that the influence of institutions is either absent or positive. But, among other negative influences, institutions can "take up space" and so inhibit the

development of more effective institutions; can channel limited societal resources toward addressing one problem by, however unintentionally, siphoning them away from other problems; or can squander the resources that are devoted to them. They can provide venues in which the effort to build collective action delays unilateral action by leaders without offsetting benefits in fostering actions by laggards. Far more attention could be devoted to identifying the "pathologies" particular to environmental institutions and identifying when they are likely to arise and how they can be avoided.

Evaluating Nonenvironmental Institutions

An area where scholars have made progress, although more remains to be done, is in evaluating the environmental performance of nonenvironmental institutions. As touched on above, interest often lies in the environmental effects of economic institutions ranging from domestic policy approaches to formal organizations such as World Trade Organization and the International Monetary Fund to the broader institutions of free trade and globalization more generally (Esty 1994; Shaw and Cosbey 1994; Kingsbury 1995). Corporations are increasingly held to account for their environmental impacts, and certification programs—such as the Forestry Stewardship Council and Marine Stewardship Council—seek both to evaluate and to influence the environmental performance of essentially economic institutions. These are all areas in which additional research could be undertaken to advantage.

Addressing Institutional Interaction and Policy Diffusion

Nascent research on institutional performance also has implications for performance evaluation. Institutional interplay may create redundancies and conflict but may also create healthy competition among institutions (Stokke 2001a; Young 2002b; P. Haas 2004; Oberthür and Gehring 2006c). Contexts involving institutional nesting, overlap, and interplay will require careful attention to parsing the influence of different institutions (see Sprinz et al. 2004; Gehring and Oberthür, chapter 6 in this volume). Thus, should behavioral changes related to the emission of ozone-depleting substances be attributed to the framework convention regulating such substances, the subsequent protocol, or subsequent amendments and adjustments? How should credit be allocated for dramatic improvements in a local environmental problem when those improvements reflect the direct and immediate influence of new local institutions, when those institutions would never have developed without earlier, structural political or economic changes? As institutional interplay increasingly reflects conscious policy coordination rather than the unintended consequence of unilateral institutional action, it may become difficult to attribute positive, or negative, outcomes to one institution, another, a combination, or the meta-institution constituted by policy coordination efforts. An institution may wield influence by propagating institutional metanorms or design principles—for example, emissions markets, the precautionary principle, or the framework-protocol approach to treaty writing—that are adopted by environmental institutions at the international, national, and local levels.

Attending to Problem Structure and Endogeneity
Finally, although latent in much of the foregoing, the issue of problem structure deserves explicit mention. Researchers often start by attributing improvements in the outcome of interest to an institution when a more likely source of such variation is problem structure. Both the theoretical and empirical foundations for taking problem structure seriously have been laid, but considerable work lies ahead (Young and Levy 1999; R. Mitchell and Keilbach 2001; Miles et al. 2002). The notion that problems vary from benign to malign provides a useful analytic starting point (Miles et al. 2002). More nuanced taxonomies, however, would allow us to estimate not only the ease or difficulty of addressing a given problem but also which functions an institution might be expected to perform well and which poorly, as noted in the literature addressing "institutional fit" and "institutional mismatch" (see Young, chapter 1 in this volume; Galaz et al., chapter 5 in this volume; as well as Young 2002b; R. Mitchell 2006).

Issues of institutional endogeneity also have yet to receive sustained analytic attention from researchers working on institutional performance (though see Ringquist and Kostadinova 2005). Environmental institutions are not designed independently of the social, economic, and political characteristics of the environmental problem they address. Some institutions may perform poorly because they face constraints that make them "designed to fail" or because they are intended to influence politics rather than policy and behavior. Other institutions may benefit from (or be harmed by) a context characterized by, say, political creativity, entrepreneurship, and political opportunities. To take one example, the ongoing debate over whether sanctions are crucial to institutional performance cannot be resolved by simply comparing institutions that include sanctions to those that do not, because, at least at the international level, perpetrating countries may accept sanctions as part of the institutional response to a "tragedy of the commons" problem but will reject them in response to an upstream/downstream problem (Chayes and Chayes 1995; Downs, Rocke, and Barsoom 1996; R. Mitchell and Keilbach 2001; R. Mitchell 2006). Endogeneity influences institutional member- ship as well as institutional design: good reasons exist to assume that those who join voluntary-membership institutions, including all interna- tional institutions, have systematically different incentives to alter their behavior and address the problem than those who do not join. Improv- ing our assessments of institutional performance requires that researchers take both design endogeneity and membership endogeneity far more seri- ously in the future than they have in the past.

Conclusion

Research into the performance of institutions that influence global envi- ronmental change has made significant progress over the past decade and a half. Scholars have developed careful methods for distinguishing insti- tutional effects from other factors, have identified a range of institutional and exogenous factors that explain variation in institutional perfor- mance, and have done considerable empirical work in evaluating—and in some cases comparing—institutional performance. This past progress provides a solid foundation on which to build future efforts to under- stand institutional performance and its sources better. To develop a rich and nuanced picture of institutional performance that is satisfying to researchers and useful to practitioners requires open-mindedness in terms of both the dimensions of institutional performance evaluated and the metrics used for evaluation. The diversity of interests and skills within the research community can be put to good advantage by encouraging those interested in institutional performance to evaluate performance in more than their preferred dimension and to do so employing as many metrics as are available and feasible to use. Following past practice, re- search should make careful use of behavioral and environmental counter- factuals but also use goals, problems, and optima as standards. Building on past practice, researchers should evaluate institutions in terms of leading indicators; economic, social, and cultural impacts; and criteria for good governance and institutional function. Methods should be developed and applied for comparing institutional performance, treating performance as multifaceted rather than unidimensional, evaluating performance dynamically, evaluating the environmental impacts of nonenvironmental institutions, and carefully accounting for problem structure and endogeneity. This represents a challenging research agenda but one that offers researchers the opportunity, over time, to discover why some environmental institutions perform differently than others, why some perform better than others, and what institutional and exogenous factors influence those outcomes. Such an understanding, in turn, will allow scholars to make more valuable contributions to the practitioners engaged in designing and operating environmental institutions to mitigate human impacts on the Earth.

Acknowledgments

This chapter has benefited greatly from comments and suggestions from Thomas Bernauer, Mark Halle, Norichika Kanie, Peter Sand, Arild Underdal, Oran Young, Maria Gordon, and the numerous scholars who commented on the presentation of an earlier draft of this work at the IDGEC Synthesis Conference in Bali, Indonesia, in December 2006. This chapter is based upon work supported by the National Science Foundation under Grant No. 0318374 entitled "Analysis of the Effects of Environmental Treaties" September 2003–August 2008. Any opinions, findings, and conclusions or recommendations expressed in this material are those of the author and do not necessarily reflect the views of the National Science Foundation.

Notes

1. I am indebted to Joyeeta Gupta for this insight.
2. I am indebted to Norichika Kanie for this insight.
3. I am indebted to Oran Young for this insight.
4. I am indebted to Liana Bratasida for this insight.

4

Building Regimes for Socioecological Systems: Institutional Diagnostics

Oran R. Young

Introduction

Practitioners and analysts alike take institutional design seriously. They regard institutions—treated as assemblages of rights, rules, decision-making procedures, and programmatic activities that guide or govern human activities—as key determinants of the course of human affairs, and they invest remarkable amounts of time and energy in efforts to shape the content and character of these arrangements (Chayes and Chayes 1995). Visions of constitutional conventions in which enlightened founders work hard to produce constitutive agreements covering a wide range of issues and expected to remain in place on an indefinite basis loom large in thinking about such matters (D. Stewart 2007). But unlike constitutional conventions, which are relatively rare occurrences, efforts to design regimes covering more specific activities occur all the time. Nowhere is this more apparent than in the case of resource and environmental regimes created to guide or manage human-environment interactions in a variety of settings.

The fact that institutions figure both as causes of environmental problems and as mechanisms for addressing them helps to explain the appeal of institutional design in this realm. There are numerous cases in which prevailing institutions emerge as sources of problems arising in human-environment interactions. To take a single prominent example, the essential argument associated with what we call the "tragedy of the commons" centers on the absence of rights and rules capable of controlling entry and, more generally, regulating the actions of users of common-pool resources (G. Hardin 1968; Baden and Noonan 1998). More generally, problems arise when institutions fail to adjust or evolve in ways needed to cope with the growth of human populations, increases in material consumption, and the introduction of new technologies. A collection of rules of the game that gives rise to perfectly acceptable results under some conditions may fail miserably when human pressures on natural resources or ecosystem services rise in response to population growth or the advent of powerful new technologies.

At the same time, adjusting existing institutions or creating new ones to fill gaps in prevailing arrangements can become a part of the efforts to solve or at least to manage a wide range of problems associated with the impact of human actions in socioecological systems. Newly created or restructured clusters of rights and rules can impose restrictions on the use of common-pool resources in the interests of preventing severe depletions, require those whose actions produce negative externalities to internalize the harm done to others, and compel users of "free" ecosystem services to incorporate the value of these services into their cost calculations. The effectiveness of institutions varies from one situation to another; institutions never account for all the variance in the anthropogenic drivers of human-environment interactions. Nonetheless, there is general agreement that institutions can and often do loom large as determinants of the trajectories of socioecological systems (Ostrom et al. 2002; Underdal and Young 2004; Breitmeier, Young, and Zürn 2006).

As we move deeper into the Anthropocene (Crutzen and Stoermer 2000), interest in creating effective institutions to guide human-environment relations has grown rapidly. Human actions have long played a significant role in shaping various features of the biophysical environments in which humans operate (Turner et al. 1990). But now we have embarked on an era of human-dominated ecosystems, a situation in which human actions—both deliberate and unintended—have emerged as major driving forces in the dynamics of biophysical systems (Vitousek et al. 1997; Schellnhuber et al. 2004; Steffen et al. 2004). The implications of this development for institutional design are profound. Whereas institutions have always been important as determinants of efficiency and equity in human affairs, they are now major factors in determining the prospects for sustainability or, in other words, the ability of humans to avoid disrupting socioecological systems on a planetary scale.

Because environmental and resource regimes are socially constructed and therefore assumed to be malleable, it is a short step from analyzing the roles they play to launching efforts to engage in institutional design. The difficulties associated with controlling population growth or guiding the path of technological innovation are well known. By contrast, we are inclined to believe that it is feasible to create or adjust institutions with comparative ease in order to address a wide range of problems arising from human-environment interactions. For reasons discussed in a later section of this chapter, many assumptions about the feasibility of (re)designing institutions to solve specific problems are naive. But this has not dampened enthusiasm for efforts to sharpen our ability to design institutions to meet all sorts of challenges.

One appealing strand of thinking about design directs attention to the formulation of design principles or, in other words, general propositions about conditions that determine whether institutions will prove effective in solving problems or, in some formulations, whether regimes will have the robustness and resilience needed to endure over long periods of time (Ostrom 1990). In their strongest form, design principles feature statements spelling out necessary or sufficient conditions for institutional effectiveness. Thus, they may assert that success will follow if some specified conditions are fulfilled. More modestly, these principles may assert that success cannot occur unless certain specified conditions are met.

Appealing as the resultant principles are, their application to real-world situations is fraught with difficulties. Actual problems generating a demand for governance differ from one another in many ways that are relevant to the character of the regimes needed to solve them. Equally important is the fact that there are alternative routes to solving similar problems arising in human-environment interactions. What is to be done, then, to help us realize the potential for altering existing regimes or creating new ones to come to terms with a variety of real-world problems? Our answer to this question centers on the idea of institutional diagnostics.

The goal of this chapter is to explore the nature of this approach to institutional design and to show how institutional diagnostics gives rise to a method for identifying features or elements of resource and environmental regimes best suited to addressing specific situations calling for the development of governance systems. What we call the diagnostic method does not yield simple recipes that anyone can use with a high probability of success. Rather, it provides a way forward that can increase the likelihood of success in the hands of skilled and experienced practitioners. The analysis to follow makes use of specific issues relating to environmental problems to illustrate the nature of institutional diagnostics. But the diagnostic method is useful in addressing the need to develop governance systems in every social setting.

A Cautionary Note

Lest readers harbor any illusions about the efficacy of this enterprise, it is worth taking a moment to note a number of constraints on what we can expect to achieve through exercises in institutional design. Efforts to design resource and environmental regimes can succeed only to the limit of the causal roles that institutions play in steering human-environment relations in the relevant issue areas. Neorealists and others inclined to regard institutions as epiphenomena reflecting the operation of underlying forces (e.g., the exercise of power or the grip of dominant discourses) will be skeptical at best about the investment of time and energy in efforts to solve problems through the (re)formation of institutions (Strange 1983). No general response to the arguments of such skeptics is possible. Work carried out under the auspices of our project and other long-term research programs leads to the conclusion that there is a lot of variation from one domain to another and even from one case to another in the same domain regarding the significance of institutions. Yet the appeal of institutional design and the willingness of practitioners and analysts alike to invest time and energy in such endeavors are notable.

No matter how clever designers are, agreement on the principal features of an institution offers no guarantee of success in solving any given problem. Major changes arising in the process of moving agreed-upon arrangements from paper to practice are normal. A variety of economic, social, and technological conditions can impede the operation of any institutional arrangement. The same design may prove highly successful in one setting but perform dismally in other settings. Under the circumstances we can expect to find ourselves devoting a good deal of attention to the specification of scope conditions. We always need to think carefully about features of the broader socioeconomic setting that are likely to promote or impede the operation of issue-specific institutional arrangements. Even then we can expect to be surprised on a regular basis by the outcomes resulting from the creation of institutional arrangements to address specific real-world problems.

What is more, the socioecological systems in which institutions operate are virtually always complex and dynamic. The conditions prevailing at the time of regime formation are subject to constant change. Many of these changes will be nonlinear in character, giving rise to substantial uncertainties and the prospect of rapid-change events. It follows that a design that was appropriate under conditions prevailing at the outset may prove dysfunctional as circumstances change. This puts a premium on what is commonly called adaptive management, or the ability of management systems to monitor changes in the relevant settings, provide early warning regarding the onset of major changes, and make suitable adjustments to maintain the effectiveness of a regime in changing circumstances. In dealing with human-environment relations, nothing is more likely to fail than a regime that lacks the flexibility to make—in some cases far-reaching—adjustments needed to address changes in the relevant problems and settings.

All this guarantees that efforts to design institutions to address environmental problems will frequently fail or produce suboptimal results. Should this lead us to lose faith in institutional design and to reach the conclusion that some other approach to problem solving is needed to address large-scale environmental problems? Research carried out under the auspices of our project does not warrant this inference. As in all human endeavors, failures as well as successes are common in the realm of institutional design. The fact that both market failures and government failures are common occurrences does not lead to any general rejection of private or public arrangements intended to guide or govern human actions. The important thing is to take steps to maximize our success rate or "batting average" in this field and to be alert at all times to opportunities to adjust our practices in ways that can lead to improved performance.

The Diagnostic Method

Nowhere are the concerns about relying on an approach featuring the development of propositions about conditions deemed necessary for success more apparent than in efforts to address the institutional dimensions of large-scale environmental changes (Young et al. 1999/2005; Young 2002b). Those who work in this area have learned two important things about the roles that institutions play. Although everyone understands that institutions never account for all of the variance in the behavior of socioecological systems, studies of terrestrial, marine, and atmospheric systems conducted in every part of the world reinforce the conclusion that institutions are among the important forces both in explaining what can go wrong in human-environment interactions and in responding effectively to problems arising in this domain (Miles et al. 2002; Ostrom et al. 2002; Young 2002b; Lambin, Geist, and Lepers 2003; Breitmeier, Young, and Zürn 2006; Lambin and Geist 2006).

The second—and equally important—lesson is that there is little prospect of developing simple generalizations capable of explaining the roles that institutional arrangements play in a variety of settings and, in the process, providing a sound basis for designing new institutions to deal with an array of problems arising in conjunction with human-environment interactions. What works perfectly well in addressing one issue may fail miserably in addressing another. It follows that there are compelling reasons to pay close attention to scope conditions in any effort to apply general knowledge about institutions to the design of new regimes or governance systems.

This is a rich domain for analysis.[1] Because the problems we must address differ from one another in many ways that have implications for the sorts of institutional arrangements needed to alleviate them, it is essential to avoid jumping to simplistic conclusions based on a superficial assessment of isolated features of complex problems or on facile analogies to other, more familiar problems. What is needed instead is a systematic process in which diagnostic queries probe the nature of specific issues and the institutional arrangements needed to guide the behavior of the key actors. The use of the term *diagnosis* in this connection brings to mind the efforts of physicians to ascertain the nature and causes of ill health before arriving at prescriptive conclusions. But similar practices occur in other fields of applied science. Architects must assess the relevant features of a site and the expected uses of a building before formulating specific plans for new structures; engineers use similar diagnostic queries before devising designs for bridges or dams that are appropriate to specific sites and conditions.

For best results it is advisable to consider a sizable set of diagnostic queries before formulating recommendations regarding matters of design in specific situations. It is impossible to maximize across a number of dimensions simultaneously. But a multidimensional picture of the problem and the range of systems of rights and rules available to solve it will provide policy makers with the sophistication needed to confront the inevitable trade-offs involved in institutional design. Yet those using the diagnostic method must confront several practical matters. The range of potential queries is vast, but the time and resources available to policy makers responsible for dealing with specific issues are always limited. Those with modest resources and overcrowded agendas will find these constraints particularly severe. Even those whose resources are extensive will find it necessary to draw the line somewhere. This makes it essential to establish priorities, focusing attention on those queries that seem most important and proceeding down the list to the extent that available time and human resources permit. The following account emphasizes queries that members of our research community have identified as matters of high priority. But readers may well be able to draw on their own experience to augment or restructure this set of queries for their own use.

As our experience with institutional diagnostics grows, we may well be able to streamline this process by identifying syndromes or combinations of conditions that typically go together. In cases where there are significant incentives to cheat, for example, it is likely that monitoring systems, enforcement mechanisms, and dispute-resolution procedures will all be important. As in other fields (e.g., medicine), knowledge of common syndromes can make efforts to use the diagnostic approach to institutional design simpler and more efficient. At this stage, however, we lack the knowledge needed to identify and understand the implications of such syndromes relating to environmental governance. This may well emerge as a research priority for those who will carry this research program forward in the coming years.

Diagnostic Queries: The Four Ps

Concrete examples will serve both to clarify the nature of the diagnostic method and to demonstrate how diagnostic queries can provide information needed to design effective regimes in the real world of environmental policy making. Our research points to a number of queries that are particularly important for those whose major concern is to craft institutional arrangements that will prove effective in terms of the criterion of problem solving. For convenience we group these queries into what we call the Four Ps: Problems, Politics, Players, and Practices.

Problems
Problems arising in human-environment interactions can take a variety of forms. Although our research has not confirmed the existence of a single dimension (e.g., from benign to malign) that can be used to

characterize environmental problems, we can identify a variety of features of problems that have major implications for design (Young 1999b, 2002b; Miles et al. 2002). Thus, the diagnostic method starts with an assessment of the major characteristics of the problem at hand and proceeds to analyze the implications of these characteristics for the nature of the regime needed to solve the problem or, failing that, to manage it. We emphasize the following queries in this category.

Is the problem well understood, and are the parties in agreement about the basic character of the problem and appropriate procedures for solving it?

While there is substantial consensus regarding the nature of some problems, others are subject to widely different interpretations on the part of major players. Some treat climate change, for instance, as a problem of controlling concentrations of carbon dioxide and other greenhouse gases in the Earth's atmosphere. Others approach it as a matter of decarbonizing industrial societies. Differences of this sort can lead actors to internalize international commitments quite differently when it comes to domestic implementation. A cap-and-trade system is likely to appeal to those thinking in terms of decarbonization. Those who see the problem as a matter of controlling concentrations of greenhouse gases, on the other hand, are more likely to be attracted to arrangements featuring carbon capture and storage or carbon sequestration schemes. These differences suggest focusing on the bottom line (e.g., success in meeting targets calling for reductions in net emissions by an agreed-upon date) while giving individual regime members leeway in meeting their obligations. More generally, the greater the differences among the parties regarding the character of the problem and procedures for solving it, the more important it is to set general goals that leave ample scope for the members of the resultant regime to choose their own methods for fulfilling their commitments or obligations under the terms of constitutive agreements.

Does the problem take the form of a coordination problem or a collaboration problem?

Coordination problems, such as the need to establish air-traffic control systems and designated shipping lanes to minimize dangers associated with air and marine transport, lend themselves to solutions that do not generate incentives to cheat on the part of individual subjects (e.g., ship captains or pilots). By contrast, collaboration problems, such as the need to devise rules governing the behavior of individual harvesters in a fishery, call for the development of mechanisms to deter and to sanction violations on the part of individual actors (e.g., fishers). The creation and implementation of effective compliance mechanisms are critical to the success of efforts to solve collaboration problems. By contrast, there is no need to devote time, energy, and resources to eliciting compliance when the problem at hand takes the form of a coordination problem.

Is a one-off solution possible, or is it necessary to find ways to address the problem on an ongoing or long-term basis?

Sometimes it is possible to take steps that solve a problem once and for all. Once an oil tanker is designed and built to comply with equipment standards requiring the installation of segregated ballast tanks, for example, the problem is solved in the sense that there is no way for operators to eliminate or disable these tanks (R. Mitchell 1994a). Rules calling on tanker captains to comply with detailed regulations governing the discharge of oily wastes or bilgewater at sea, on the other hand, require constant monitoring to ensure compliance. In some cases there is a fuzzy middle ground regarding such matters. Even after catalytic converters became standard equipment on automobiles, for instance, regulators worried about the possibility that individual owners would take steps to disable them. But to the extent that solutions involve irreversible actions, the need to create institutions capable of monitoring behavior and enhancing compliance on an ongoing basis will decline or even disappear.

Is the problem self-contained or will efforts to solve it impact preexisting institutional arrangements?

Some resource or environmental regimes are able to operate successfully with few if any consequences for other institutional arrangements in place at the time of their creation. The regime created in 1959 to govern human activities in Antarctica, for instance, is largely self-contained in this sense. But interplay between or among individual regimes is both common and becoming more widespread as interdependencies among human activities rise and the density of institutional arrangements governing these activities increases. The regimes dealing with ozone depletion and climate change interact because chlorofluorocarbons (CFCs) and some of the more attractive substitutes for them (or by-products of these substitutes like trifluoromethane, HFC-23) are greenhouse gases.

Many environmental regimes interact with the world trade regime because trade measures used to maximize compliance with the provisions of individual environmental regimes can conflict with key rules of the trade regime. International regimes dealing with issues like climate change can cause problems at the domestic level when they call on the governments of member states to act in ways that exceed their authority or to adopt policy instruments (e.g., cap-and-trade systems) that are alien to their political and legal cultures. It follows that efforts to design regimes to address specific problems need to consider the likelihood of institutional interplay and take steps needed to deal with these interactions whenever they will have major consequences for the effectiveness of one or more of the affected regimes.

Do the actions of government agencies, private corporations, individuals, or some combination of the three lie at the heart of the problem?

Problems vary with regard to the identity of the actors or principals whose behavior is at issue. In the case of a ban on nuclear testing, for instance, the problem is to control the actions of government agencies. Efforts to clean up the Rhine River, by contrast, have focused on the actions of corporations. Regimes dealing with the conservation of endangered plants and animals seek to eliminate or drastically curtail the activities of individual poachers. Some cases are complex in these terms. Arrangements designed to reduce long-range air pollution, for example, commonly apply both to municipal or publicly owned power plants and to privately owned utilities. But the message is clear. Government agencies, private corporations, and individuals differ from one another in behavioral terms; successful regimes must be designed with these differences in mind.

Is the problem cumulative or systemic?

Some problems, such as the loss of biological diversity, are cumulative in the sense that they are place based, recur in many different settings, and must be tackled piecemeal or, in other words, in one setting at a time. Others, such as climate change, are systemic in the sense that they are global in character and must be dealt with in holistic terms. The implications of this distinction for effectiveness and compliance are somewhat counterintuitive. The incentive to shirk and become a free rider is apt to be particularly strong with regard to global problems; individual actors may hope that others will take steps to solve the common problem. But any level of compliance on the part of any actor whose behavior is a source of such a problem will make a contribution toward solving the whole problem. With regard to cumulative problems like the loss of biological diversity, on the other hand, success requires behavioral change on the part of particular actors in a position to protect specific species. Although any level of compliance on the part of members of the relevant group is helpful in addressing systemic problems, high levels of compliance on the part of specific actors operating in "hot spots" is required to solve cumulative problems.

Is the problem likely to give rise to changes that are abrupt, nasty, and irreversible?

There is an understandable tendency to assume that socioecological systems are relatively stable and to focus on problems that develop in a gradual, linear, and readily observable fashion. But not all problems fit this mold. Fisheries that cross critical thresholds can experience sudden collapses. The Earth's climate system is known to have experienced abrupt changes or system flips in the course of our planet's history. The need in such cases is to learn about the mechanisms that can cause abrupt changes and to monitor the relevant systems closely enough to detect their onset when it is still possible to take steps to change course to avoid overshoot (Meadows, Meadows, and Randers 1992). What is required here is an ability to encourage adherence to existing rules while at the same time cultivating a capacity to adjust or replace existing institutional arrangements in order to maintain a proper fit between the character of the problem and the nature of the governance system created to address it.

Politics

The (re)formation and implementation of institutions is a political process all the way down. Whether the problem arises at the local level, the national level, or the international level, there will be multiple stakeholders seeking to promote their own causes in processes of institutional design; the capacity of such players to exercise influence invariably looms large as a determinant of the course of regime (re)formation. The pervasive influence of politics is not necessarily a bad thing. There is no substitute for the exercise of political will in processes calling upon members of a social system to make collective decisions. It is important to draw a distinction, however, between analyzing the overall political landscape as part of the diagnostic process and focusing on the ins and outs of institutional bargaining in specific settings. There is no need to become preoccupied with the dynamics of institutional bargaining at the design stage, but institutional design needs to take into account the basic features of the political setting. Those who advocate designs that are clearly incompatible with prominent features of the broader political setting and that are therefore utopian in nature are unlikely to be able to contribute to making significant progress in efforts to solve or manage a variety of problems arising in human-environmental relations. From a design perspective, the trick is to devise innovative and even visionary arrangements to solve complex problems while staying within the bounds of political feasibility. Our work draws attention to the following queries in this category.

Is power or influence among the stakeholders concentrated or dispersed?

The diversity of political settings in these terms is great, ranging from cases in which there is an issue-area hegemon (i.e., a single dominant actor) at one extreme to situations featuring a symmetrical distribution of power among a sizable number of stakeholders at the other. The consequences of this factor for institutional design are often far-reaching. Where there is a hegemon, institutional design must reflect the preferences of that actor. In extreme cases the set of stakeholders may constitute an Olsonian privileged group in the sense that a single dominant actor values the establishment of a regime more than the cost of supplying it, regardless of the contributions of others (Olson 1965). The result is apt to be an institutional arrangement spelled out in relatively precise terms, even though some of the lesser actors may be unhappy about specific provisions of the arrangement. In cases where power is highly dispersed, on the other hand, the trick is to reach agreement on matters of institutional design without incurring transaction costs that are prohibitively high. Avoiding precision in specifying a regime's provisions is likely to be necessary in such cases. In reality, most cases fall somewhere between the two extremes in terms of the concentration of power, a fact that explains why some treaties, statutes, and other constitutive agreements are more precise regarding their provisions than others.

Are there negotiating blocs or coalitions whose interests in the relevant issue area clash or diverge sharply?

In cases of regime (re)formation where no single actor is in a position to dominate the process, a small number of blocs—typically two to four—commonly become the major players both in forming and in implementing institutional arrangements. The case of climate change in which blocs representing the European Union; the United States, Japan, and several other like-minded states; the Group of 77 plus China; and the small-island developing states are the principal protagonists exemplifies this situation. The implications of the existence of such blocs for institutional design are twofold. More often than not, the resultant arrangements involve amalgams of the design features preferred by the major blocs, and negotiators frequently resort to ambiguous language as a necessity for eliciting agreement from the principal players. This is one reason why efforts to deal with a variety of problems (e.g., climate change, the loss of biological diversity) frequently commence with framework agreements that have little substantive content but that are intended to launch processes leading to the acceptance of more substantive supplements over time.

Does the problem fit comfortably into some established and widely accepted discourse and lend itself to the use of well-known policy instruments?

Although every problem has features that make it unique, it is often possible to frame individual problems as exemplars of broader types or classes of concerns, and the act of doing so will have far-reaching consequences for institutional design. Prominent examples include discourses centering on limiting entry to common-pool resources, finding ways to charge for the use of ecosystem services, and devising cost-sharing mechanisms to fund the provision of public goods. Discourses—like institutions themselves—are socially constructed; they change over time. This means both that the preferences of key players regarding appropriate discourses may vary and that the choice of a discourse to use in thinking about any specific problem can be contentious. But once a dominant discourse emerges, it will shape the process of institutional design (Litfin 1994). All regimes that approach problems as matters of devising mechanisms to limit entry on the part of users (e.g., Individual Transferable Quotas), for instance, have a number of things in common. Under the circumstances, design can focus on more detailed provisions than is the case when there is no prior agreement on the basic character of the problem under consideration or the appropriateness of specific policy instruments. Similar remarks are in order regarding these more detailed provisions or, as they are often called, policy instruments. Once players become familiar with specific instruments (e.g., cap-and-trade systems applied to pollutants), it becomes relatively easy to apply them to new issue areas.

How pervasive are corrupt practices or manipulative activities intended to promote the interests of individual players in the issue area a regime addresses?

In thinking about institutional design, there is an understandable tendency to assume that regimes will work as envisioned in their constitutive agreements. But there is always a gap—sometimes a large gap—between the ideal and the actual regarding such matters. Partly this is a consequence of normal complications arising from moving institutional arrangements from paper to practice. But in part the gap arises from the existence of opportunities for shrewd and unscrupulous players to manipulate regimes to their own advantage or to violate their provisions with impunity. When corruption in this sense (e.g., illegal logging; illegal, unregulated, and unreported or IUU fishing) is a major concern, institutional design must focus on matters of compliance and enforcement. Transparency is one important factor in this context. It is harder to violate standards requiring actors to demonstrate compliance with key rules prior to engaging in a given activity, for example, than to violate prescriptions calling for compliance on an ongoing basis. In the latter case the establishment of systems of implementation review capable of monitoring the behavior of addressees in a systematic and sustained manner becomes important (Victor, Raustiala, and Skolnikoff 1998).

Players

Regardless of the character of the problem at hand and the major features of the political landscape, the diagnostic method directs attention to a number of matters relating to the principal actors or players—individuals as well as collective entities like states and corporations—responsible for causing the problem, likely to experience harm arising from the problem, or located in a position to play a role in solving it. Insights regarding such matters may have far-reaching implications for the nature of the regime required to address a problem effectively. Our research suggests the importance of paying particular attention to the following queries regarding the principal actors or players.

Do the principals behave as rational actors—in the sense of being self-interested utility maximizers—or are their actions influenced significantly by other sources of behavior, including a sense of legitimacy or the force of habit?

Regardless of the identity of the principals, there is a strong tendency —at least among those desiring to develop tractable models—to assume that those subject to rules are rational utility maximizers. The result is a preoccupation with incentives and an emphasis on finding ways to minimize the cost of compliance or drive up the cost of noncompliance in designing specific arrangements. But most actors are influenced, at least in part, by normative commitments, a sense of propriety, or the force of habit (Hart 1961; March and Olsen 1998). In such cases finding ways to internalize the rules and, in the process, to turn compliance into an automatic response emerges as a central concern. With regard to individuals or small groups, this suggests the importance of taking steps to enhance socialization and opportunities to participate in the rule-making process in contrast to relying on the use of conventional sanctions. In dealing with governments or, more likely, individual government agencies, it leads us to think in terms of routinization. An agency that incorporates the maintenance of a system of rights, rules, and decision-making procedures into its own mission may even adopt the provisions of a regime as elements of its organizational identity.

Are the subjects unitary actors or is their behavior a product of internal dynamics?

It is conventional in studies of decision making to assume that actors are not only rational but also unitary in the sense that they have well-defined preferences or utility functions. But as the literature on two-level games makes clear, the appropriateness of this assumption is questionable in a variety of settings (Putnam 1988). Individuals can experience contradictory motivations that trigger intense inner struggles when it comes to matters of compliance with the requirements of regulatory regimes. Although they are subject in principle to the discipline of maximizing profit in relevant markets, corporations often experience internal contests, as exemplified by the case of DuPont with regard to efforts to phase out CFCs and other ozone-depleting substances starting in the 1980s. The scope for the impact of two-level games is even greater with regard to the actions of states. When key actors are affected significantly by the occurrence of these internal dynamics, compliance may depend as much on appealing to domestic constituencies as on establishing more conventional international compliance mechanisms (Dai 2005).

How large is the group of subjects?

Efforts to enforce rules dealing with illegal fishing or illegal harvesting of wildlife on the part of large numbers of small operators are apt to cause nightmares, even in cases where regulators have ample authority to penalize violators. Bringing pressure to bear on a few actors who cannot easily hide their actions and who are concerned about their reputations is considerably easier than controlling the actions of thousands or even millions of individual subjects. Given the accuracy with which modern technology can detect the signals of nuclear explosions, for example, the effect of numbers explains why it is easier to deter violations of test-ban agreements than to stamp out the activities of illegal fishers or poachers of wildlife. It also accounts for the pronounced tendency to focus on the comparatively small number of producers of airborne pollutants (e.g., sulfur dioxide) in contrast to the millions who consume the relevant products (e.g., electricity), despite the fact that the problem lies in the final analysis with the actions of the consumers.

Is the group of subjects homogeneous or heterogeneous?

In some cases it is reasonable to treat the group of subjects as homogeneous, a situation that makes it feasible to create compliance mechanisms that treat them in a uniform manner. A group of industrial fishers using similar gear or a group of ranchers with similar needs for grazing lands exemplifies this case. But more often than not, a regime's subjects will differ from one another in important ways. There are, for instance, coastal and flag states, upstream and downstream users of water, and owners and nonowners of land. A particularly difficult asymmetry arises when there is little or no overlap between those whose actions cause a problem and those who are victims in the sense that they are likely to bear the brunt of the problem's impact. This is one reason why the problem of climate change is so hard to solve. When conditions of this sort prevail, rules must differentiate among groups or categories of subjects, and those responsible for creating or implementing regimes must frame the relevant rules in a manner that takes these differences in roles into account. A critical step in forming and implementing a regime, therefore, is to assess the extent and nature of heterogeneity among the anticipated subjects and to frame the rights, rules, and decision-making procedures accordingly.

How transparent are the actions of the regime's addressees?

Even in the absence of formal compliance mechanisms, the degree to which subjects can violate rights and rules clandestinely or without their actions becoming known to the public can make a big difference in behavioral terms. Although formal sanctions may be needed to deal with hard-core violators, the prospect of negative publicity will suffice to deter others from breaking or skirting the rules. Under the circumstances it is worth noting that it is much easier to keep tabs on the actions of a handful of large producers of ozone-depleting substances than to monitor the actions of millions of consumers of products containing these substances. It makes sense in this case to create a regime that focuses on the actions of producers, even though it is the behavior of consumers that is the ultimate concern. Similarly, with regard to the disposal of oily wastes at sea, it is virtually impossible to engage in clandestine violations of equipment standards requiring the inclusion of segregated ballast tanks, whereas discharge standards are easier to flout. Despite the fact that equipment standards seem less efficient than discharge standards in principle, therefore, it is easy to see why those engaged during the 1970s in efforts to reform the regime governing intentional oil pollution at sea shifted their attention from discharge standards to equipment standards.

Practices

Efforts to (re)form resource and environmental regimes take place in broader or overarching settings featuring well-established social practices or metapractices that those endeavoring to (re)form issue-specific regimes must normally accept as given. Some of these practices (e.g., the legal procedures applicable to negotiating, signing, and bringing into force legally binding treaties and conventions) are formal in nature. Others (e.g., practices involving such matters as the principle of common but differentiated responsibilities at the international level) are much less formal but widely acknowledged to be important nonetheless. Because practices are socially constructed, they are subject to change over time. Still, from the point of view of those seeking to design regimes to deal with specific problems, these overarching or metapractices are facts of life much like the distribution of power or the character of relevant actors. It does little good to advocate the creation of regimes that cannot work in the relevant setting. As a result, those seeking to design effective regimes to solve or manage specific problems must pay attention to the nature of the relevant metapractices and craft arrangements that are compatible with these practices. Our work has led us to pay particular attention to the following queries in this category.

Are the parties free to make choices regarding the types of constitutive agreements to employ in addressing specific problems?

Some but not all settings allow those engaged in regime formation to choose among constitutive agreements varying from formal, legally binding documents (e.g., statutes or treaties) to relatively formal instruments that are not legally binding (e.g., ministerial declarations) and on to informal agreements of various kinds (e.g., an exchange of notes) (Lipson 1991; Abbott and Snidal 2000). Legally binding agreements are apt to exert a greater compliance pull than less formal agreements, but they normally take longer to negotiate, end up with less substantive content, and are harder to amend or adjust than less formal agreements. In this connection, institutional design typically involves significant trade-offs. If a problem is both urgent and subject to changes that are hard to anticipate, an informal agreement may constitute the preferred strategy. When issues feature incentives to cheat and violation tolerance is low, on the other hand, there is much to be said for holding out for provisions that have the force of law behind them (Zaelke, Kaniaru, and Kruzikova 2005).

Do prevailing practices permit starting with a core group of committed and like-minded players and expanding the membership of the resultant regime over time?

Because institutions ordinarily exhibit the characteristics of public goods—nonexcludability and nonrivalness—there is no incentive to form minimum winning coalitions in efforts to (re)form regimes. On the contrary, it is generally beneficial to maximize the proportion of the members of the relevant community bound by the terms of constitutive agreements. Yet it may be time-consuming and costly to wait until (almost) all the members of a group sign on to such an agreement before launching a new governance system. In such cases key players may decide to form a regime based on the participation of a relatively small core group, assuming that others will be motivated to join or can be persuaded to join at a later time (Schelling 1978). The Montreal Protocol on ozone-depleting substances is a case in point. The challenge to those engaged in institutional design in such cases is to decide on the composition of the group best suited to become first movers in the creation of a regime to address a specific problem.

Is the principle of common but differentiated roles and responsibilities both acknowledged and in use in the relevant setting?

The practices prevailing in some settings exhibit a strong tendency to treat all relevant players equally. This is especially true in international society where the idea of sovereignty is often construed as giving all states equal legal status or standing. Yet actors seeking to solve particular problems often differ in ways that are relevant to devising and implementing solutions. Flag states, port states, and coastal states play different roles regarding maritime shipping. Advanced industrial states and small-island developing states have strikingly different concerns regarding the consequences of greenhouse gas emissions. Where prevailing practices are inflexible, it is hard to factor these considerations into diagnostic assessments. Where there is flexibility, by contrast, providing for heterogeneity regarding roles and responsibilities becomes an option. The move in recent years to accept the principle of common but differentiated responsibilities at the international level is a highly significant development in these terms (Young 2001a).

Is it permissible to opt for a framework agreement at the outset with the intention of adding substantive amendments or protocols over time as the regime develops?

Another consideration has to do with the extent to which prevailing practices call for the negotiation of a comprehensive agreement (e.g., the 1967 Nuclear Non-Proliferation Treaty or the 1982 Law of the Sea convention) at the outset. Where issues are relatively stable and transaction costs are not exorbitant, there is much to be said for this approach. But many issues pertaining to socioecological systems do not meet these requirements. In cases like ozone depletion, climate change, and biological diversity, the parties have adopted what has become known as the framework-protocol approach, starting with a modest framework agreement and working to add one or more substantive protocols over time. The two strategies are fundamentally different from the perspective of institutional design. Where prevailing practices allow for choice in these terms, therefore, it is critical to consider this matter with care in the diagnostic process.

Is the regime expected to operate as a stand-alone governance system, or will it be embedded in some larger system of institutional and organizational arrangements?

In domestic society new regimes are typically assigned to an existing agency (e.g., the Environmental Protection Agency or the Forest Service in the United States) for purposes of implementation. Stand-alone arrangements (e.g., the whaling regime, the regime for Antarctica) are more common in international settings. But we should not exaggerate this difference between national and international settings. The California Coastal Commission operates in a relatively autonomous fashion, several arrangements dealing with plant genetic resources at the international level are merging into a recognized cluster or complex, and a number of arrangements dealing with maritime shipping and pollution are administered in a coordinated fashion by the International Maritime Organization (R. Mitchell 1994a; Raustiala and Victor 2004). Where the participants have options in these terms, the choice between stand-alone and embedded regimes becomes an important design issue. If association with an existing organization is likely to lead to crippling biases, the case for creating stand-alone arrangements will be strong. When problems relating to administrative capacity and the availability of funds are more prominent, on the other hand, there will be a stronger case for lodging institutional arrangements in an existing organization that has the capacity to implement these arrangements effectively.

Are there practices in place in the overarching setting that address matters of implementation review, reauthorization, and amendment?

Where procedures dealing with evaluation and implementation review are well defined in the overarching setting, there is no need to incorporate separate procedures dealing with such matters on an issue-specific basis. In the United States, for instance, continuing oversight, regular reauthorization, and occasional amendment are all standard practices. As a result, it would be superfluous to invest energy in working out separate arrangements regarding such matters for inclusion in specific regimes, such as the regime covering sulfur dioxide emissions established under the Clean Air Act Amendments of 1990. But such procedures are not standard practices in many other settings. In the case of the various protocols to the European Long-Range Transboundary Air Pollution regime, for instance, no such procedures are in place. Under the circumstances it is important to pay attention in the design of specific regimes to the extent to which overarching practices addressing these concerns are in place and to act accordingly.

Institutional Bargaining and Operating Principles

The diagnostic method is a design tool; it is intended to help those seeking to work out the contents of the provisions of regimes or governance systems needed to solve problems ranging from the consumptive use of living resources in small-scale settings to global issues like anthropogenic interference in the Earth's climate system. The idea of political feasibility is inherent in the diagnostic method in the sense that it includes queries intended to probe the compatibility of proposed arrangements with the defining features of the political setting in which they are expected to operate. This prevents the expenditure of time and energy on utopian exercises that have little or no relevance to real-world problems, whatever their attractions in purely hypothetical terms.

Yet the fact that an arrangement is politically feasible in some general sense does not mean that the parties will succeed in reaching agreement on its principal features in the process of institutional bargaining associated with regime (re)formation in specific situations. The knowledge required to design the major elements of effective regimes will be of little help unless we can master the art of forming and implementing agreements including these elements, and of adjusting them over time to maintain their effectiveness in the face of a steady stream of changes that are both endogenous and exogenous in nature. It is easy to find examples of elaborate designs for institutional arrangements that have little or no prospect of being adopted in the first place or that end up as dead letters or purely paper arrangements, even when they are enshrined in formal agreements. No matter how well designed they are, moreover, regimes cannot produce lasting solutions to problems associated with socioecological systems if they lack robustness in the sense of a capacity to withstand day-to-day stresses and resilience or the capacity to adapt to shifting demands without experiencing breakdown or undergoing fundamental change. Problems develop and evolve, sometimes quite rapidly. Institutions that are unable to adjust to changing circumstances are unlikely to produce lasting solutions to the problems they address.

From this perspective it is essential to recognize that all processes of regime (re)formation are political in character. Issues of governance or management always give rise to what analysts call competitive-cooperative or mixed-motive interactions (Schelling 1960). Participants in such situations have incentives to cooperate, but there is an understandable tendency as well for actors to focus on promoting their own interests, even while recognizing the need for cooperation to reap joint gains or avoid joint losses. Put another way, although coordinated behavior is needed to achieve outcomes lying on the welfare or Pareto frontier, participants will strive to reach agreements at specific points on the frontier that are most favorable to themselves. Much of the literature on strategic interaction—especially those works produced by political scientists—is just as concerned with distribution or who gets what as with solving collective-action problems (Rapoport 1960; Schelling 1960; Young 1975). But behavioral research regarding such matters points to a much more complex reality. Faced with tough choices in situations resembling the prisoner's dilemma, for instance, some actors manage to cooperate even in the absence of communication (R. Axelrod 1984). When allowed to communicate and especially when encouraged to pay attention to the shadow of the future, individuals succeed in arriving at cooperative—and even Pareto optimal—solutions with some regularity (Oye 1986; Kahneman 2003).

How should those desiring to maximize social welfare or to contribute to the common good by creating regimes and helping to maintain their effectiveness and resilience over time proceed in such a world? Usable knowledge in this context takes the form of guidelines based on best practices in contrast to the diagnostic queries emphasized in the preceding section (Kahneman and Tversky 2000). Because institutional dynamics are governed by a multiplicity of factors, and our understanding of the processes involved does not allow for the identification of necessary or sufficient conditions for the (re)creation of effective regimes, we cannot offer simple directions that those responsible for creating regimes and adjusting them in the face of changing circumstances can employ as surefire recipes. Even so, we know enough about these processes to articulate a set of guidelines that can help those dealing with institutional dynamics to arrive at socially desirable outcomes. To illustrate the use of guidelines in processes of regime (re)formation and to make the take-home messages memorable, the following account focuses on operating principles relating first to regime formation and then to institutional re-formation featuring adaptive changes.

Forming Institutions

The absence of propositions stating invariant relationships makes it impossible to offer advice that says you must meet certain conditions if you hope to succeed in creating resource or environmental regimes, much less that you can be sure of succeeding if you do meet certain conditions. Yet we know a lot about the processes involved in regime formation, and this knowledge is sufficient to support a number of operating principles reflecting best practices in this realm. Our research directs attention to a number of guidelines of this nature.

Formulate the problem in a manner that highlights opportunities to generate mutually beneficial results.

Regimes are normally created to solve or alleviate problems. But there is no objectively correct way to define or frame specific problems, and the formulation ultimately accepted can have far-reaching consequences for efforts to create an effective regime. Is fisheries management, for instance, basically a problem of allocating total allowable catches among competing users of specific stocks, or is it better approached as a matter of applying principles of ecosystem-based management to secure the long-term health of marine systems? Is climate change essentially a problem of decarbonization of industrial societies or a matter of limiting concentrations of greenhouse gases in the Earth's atmosphere? Decisions about such matters are not only important determinants of the prospects for success in regime formation; they also can have powerful impacts on who gets what as a result of such efforts. This suggests the importance of proceeding with care in framing issues for consideration in efforts to form regimes. As a rule of thumb, it makes sense to highlight opportunities to move toward the Pareto frontier in contrast to dwelling on movements from one location to another along a well-defined welfare frontier.

Recognize that regime formation involves distinct stages differing from one another in important respects.

Processes of regime formation encompass several distinct phases or stages. Although thresholds separating individual phases from one another are often fuzzy, it makes sense to differentiate among the phases of agenda formation, in which the problem is framed and moves to a high enough place on the political agenda to become actionable; negotiation, in which the terms of a constitutive agreement are hammered out and articulated in a law, a treaty, or some other formal document; and operationalization, in which the mechanisms needed for implementation (e.g., commissions, secretariats, monitoring systems, financial mechanisms) are put in place. The significance of this distinction lies in the fact that the determinants of success differ from one stage to another.

Agenda formation requires cognitive innovation, negotiation calls for bargaining skills, and operationalization depends on administrative capacity (Young 1998). Failure in any of these areas will lead either to stalemate or to the creation of arrangements that are consigned to the status of dead letters.

Focus on producing results that participants see as equitable and legitimate rather than on meeting ill-defined standards of efficiency.

We are accustomed to thinking about regime formation from the vantage point of efficiency, and there are good reasons for doing so. If we can increase the size of the pie, there will be more gains to be shared among all the participants. But it is easy to overemphasize the usefulness of thinking in these terms. The true welfare frontier is often hard to locate empirically, and many discussions of efficiency take the view that there is nothing wrong with ending up with losers as well as winners so long as this increases some measure of social welfare. In real-world situations, all parties are concerned acutely with the degree to which outcomes are equitable in the sense that the distribution of benefits and burdens conforms to some reasonable standard of fairness, and processes are legitimate in the sense that they meet reasonable standards of due process. Regimes that leave a lot to be desired in terms of efficiency form all the time. By contrast, regimes that fail to meet basic standards of equity and legitimacy seldom form. And when they do, compliance is almost always an ongoing problem. Thus, efficiency is desirable, but some general sense of equity is essential.

Bear in mind that institution building thrives on efforts to form maximum winning coalitions rather than minimum winning coalitions.

Those who analyze legislative politics often emphasize the importance of creating minimum winning coalitions. So long as a coalition is large enough to ensure victory in such a setting, it makes sense to minimize the number of parties in a position to claim a share of the gains (Riker 1962). Whatever the merits of this argument in the context of legislative politics, it does not apply to bargaining over the constitutive provisions of social institutions. These arrangements generally exhibit the qualities of public goods—nonexcludability and nonrivalness. This suggests that there will be gains to be had in terms of burden sharing from bringing more members of the affected group into a regime in contrast to losses arising from the need to apportion the gains among more claimants. It is not essential to work with the grand coalition from the outset. Efforts to establish regimes are often more successful when a subset of the members of the group affected by a problem or concerned with an issue take the lead or become first movers in launching an institutional arrangement (Schelling 1978). Nonetheless, the logic of expanding the size of the group over time is compelling.

Be aware of the implications of the theory of the second best.

There are many situations in which institutional options that seem preferable on paper turn out to be inferior to other options in practice. Assuming perfect compliance, for example, a regime employing discharge standards to regulate intentional oil pollution at sea would be more efficient and hence more desirable than a more rigid arrangement requiring all new tankers to meet the same equipment standard. In practice, however, it turns out that discharge standards are prohibitively expensive to monitor and enforce in political as well as economic terms, whereas it is comparatively easy to ensure compliance with rules requiring all new tankers to be built with segregated ballast tanks. In effect, a regime that succeeds in practice turns out to be better than an alternative arrangement that looks more attractive on paper but is ineffective in practice (R. Mitchell 1994a). This phenomenon—often described in terms of the theory of the second best—occurs frequently in the realm of resource and environmental regimes.

Be prepared to take advantage of windows of opportunity in processes of institutional bargaining.

Opportunities to make progress in forming regimes are fleeting; they come and go over the course of time. Elections produce new governments that are more or less interested in particular problems. Economic upturns or downturns leave parties feeling more or less concerned about the probable costs of implementing commitments made under the provisions of specific regimes. Technological developments make it easier or harder to monitor the actions of those subject to a regime's rules. Under the circumstances, agreements on specific issues reached during the course of institutional bargaining cannot be counted on to remain in play indefinitely. Upon entering office in 1981, for example, the Reagan administration in the United States repudiated tentative agreements made by its predecessor regarding key provisions of the negotiating text for the Law of the Sea convention. Taking office in 2001, the Bush administration promptly renounced agreements regarding climate change made by its predecessor. There is much to be said, therefore, for striking while the iron is hot in efforts to form regimes to deal with specific issues.

Reforming Institutions

Change is endemic to social institutions. Specific changes may be endogenous or exogenous in origin, gradual or abrupt in character, and systemic or compartmentalized in scope. Knowledge regarding the nature of change in the institutional dimensions of socioecological systems remains relatively underdeveloped. Yet we already know enough to articulate some operating principles of interest to those desiring to improve the performance of specific regimes, strengthen the resilience of existing arrangements, or replace arrangements that have become dysfunctional. Our research highlights the following operating principles or guidelines relating to institutional reform.

Differentiate between developmental change and change that calls into question the robustness or the resilience of a regime.

Some changes are developmental in character in the sense that they feature the natural evolution of a regime over time. Arrangements starting with the articulation of framework agreements that are meant to be supplemented with the addition of substantive protocols are familiar cases in point. So also are arrangements that have well-defined procedures allowing for amendments to existing provisions to take account of ongoing developments (e.g., new harvesting technologies in fisheries, new knowledge about the side effects of chemicals like CFCs). Other changes, by contrast, reflect shifts in the power, interests, or identities of important members of resource or environmental regimes. Whereas developmental changes normally enhance the resilience of institutional arrangements, changes to accommodate shifts in configurations of power or interests can undermine the resilience of the affected regimes. Both types of change are common, but it is clear that they raise different problems and call for different responses on the part of those dedicated to the maintenance of effective institutions.

Remember that ineffective regimes can gain strength with the passage of time and that the performance of high achievers can deteriorate.

Like interpersonal relationships, institutions require continuous attention to ensure that they remain effective in fulfilling their goals and solving the problems that led to their creation. Institutions that are frustratingly slow to take hold sometimes cross a threshold or pass a tipping point that precipitates a comparatively rapid rise in their effectiveness. The regimes created to clean up the Rhine River and to regulate pollution in the North Sea illustrate this phenomenon. Conversely, arrangements that are highly successful at first can encounter problems at later stages that they are ill equipped to handle and that undermine their effectiveness. Arrangements capable of managing fisheries effectively given the use of one set of harvesting technologies, for example, can deteriorate quickly following the introduction of new harvesting technologies (e.g., the high-endurance stern trawler). The pervasiveness of change calls for constant vigilance on the part of those concerned with the maintenance of effective regimes.

Strike a balance between excessive institutional flexibility and arrangements that make intentional change overly difficult to achieve.

Regimes that are too rigid cannot remain resilient as providers of governance in settings featuring dynamic socioecological systems. Conversely, arrangements that give way at the first indication of pressure will become epiphenomena that have little or no significance as determinants of collective outcomes in human-environment interactions. This is a familiar issue in the creation of constitutive arrangements. How stringent should the rules governing amendments to the constitutive agreement be? There is no simple formula allowing us to achieve an ideal or optimal balance regarding such matters in specific cases. We can say that the more dynamic socioecological systems are, the more important institutional flexibility becomes. But there is no substitute for addressing this issue in case-specific terms.

Use adaptive management where possible, but bear in mind that the requirements for success in such endeavors are difficult to fulfill.

The idea of adaptive management has gained in popularity but lost in precision in recent discussions of the importance of maintaining a close fit between problems arising in socioecological systems and the regimes created to cope with them. Adaptive management calls on those responsible for implementing regimes to approach their initiatives in experimental terms in the interests of acquiring the knowledge needed to achieve better results with the passage of time. But in many cases this is easier said than done. Adaptive management is not just a fancy term for learning by trial and error; it requires a sustained effort to simulate the characteristics of successful scientific experiments through measures like establishing control groups, avoiding the pitfalls of selection bias, and being alert to the danger of spurious correlations (Lee 1993a). Adhering to these methodological standards is never easy; it will normally be harder under real-world conditions in which the welfare of numerous human subjects is at stake. Nonetheless, in a world in which institutional change is pervasive, there is no substitute for taking advantage of all available means to enhance the adaptability of institutions to avoid a loss of effectiveness.

Encourage broader processes of social learning.

No matter how good processes of adaptive management become, they cannot guarantee success in efforts to overcome the influence of standard operating procedures or the impacts of agency cultures that make it hard for international secretariats or domestic agencies charged with implementing the provisions of regimes to make the continual adjustments needed to stay abreast of various forms of institutional change. It is common to encounter social traps in this realm that have the effect of making administrative bodies more rigid with the passage of time, while institutional change gives rise to a growing demand for increased flexibility and adaptation to maintain the effectiveness of regimes (Cross and Guyer 1980). There is no sure cure for this problem. But it is important to recognize the prevalence of such occurrences and to find ways (e.g., the use of simulation exercises and the development of scenarios) to help those responsible for institutional implementation to spot changing circumstances and to take steps to address emerging problems before they get out of hand (Social Learning Group 2001).

Take advantage of opportunities associated with the onset of changes that are nonlinear and abrupt.

Practitioners and analysts alike regularly fail to anticipate nonlinear and abrupt institutional changes and to take full advantage of the opportunities they present. A striking case in point is the collapse of the former Soviet Union and the effort to restructure the economic systems of the successor states in the wake of this development. It is now clear that many of the measures adopted, including so-called shock therapy and ill-considered schemes for transferring state-owned assets into private hands, were little more than ad hoc measures advocated by individuals and groups who were quite inexperienced with such situations or championed by those expecting to benefit from the resultant transformations. Institutions are normally sticky, and it can be frustrating to wait for rare opportunities to introduce substantial changes. Yet the costs of failing to prepare carefully for these occasional opportunities are steep. Russia, for instance, is still paying the price for institutional mistakes made ten or fifteen years back. From a social welfare perspective, or for that matter any reasonable perspective on the common good, there is a compelling case to be made for committing the modest resources needed to engage in rigorous efforts to prepare in advance for opportunities created by occasional, fleeting, and often unpredictable interludes of nonlinear and abrupt change.

Conclusion

Institutions account for only a part of the variance in the behavior of socioecological systems. Numerous other factors—both biophysical (e.g., climate change) and socioeconomic (e.g., technological innovation)— play important roles in such systems as well. This means that it never makes sense to attribute problems such as severe depletions of living resources or improper treatment of various types of wastes solely to institutional deficiencies. Conversely, it is naive to suppose that restructuring existing institutional arrangements or replacing them with new ones can solve all our problems. Not only is the role of institutions a variable, but institutions also interact with other factors to determine the behavior of complex socioecological systems. This is bad news for those hoping to develop institutional design principles stated as necessary or even sufficient to create effective regimes or to ensure their resilience in a variety of settings. But it presents a fascinating challenge for those willing to work with colleagues in a broad range of disciplines to make use of the diagnostic method in dealing with the dynamics of institutional arrangements.

Are there transportable principles that are nontrivial in character and that apply across a wide range of cases of institutional (re)formation (Miles 2006)? Research carried out under the auspices of our project suggests that there are no principles that are nontrivial in content and that apply to all cases of regime (re)formation. Even such familiar propositions as the need for monitoring and graduated sanctions to ensure compliance with the rules of the games do not apply in cases where the regime does not emphasize behavioral prescriptions, there are no incentives to cheat, or the regime becomes superannuated once the problem is solved. What is portable is the diagnostic method itself. This method provides a step-by-step procedure that is applicable to specific issues in every issue area. What we need in every case is a set of recommendations crafted to take into account the constellation of conditions brought to light through the use of the diagnostic method.

Can we train policy makers, administrators, and analysts alike in the use of the diagnostic method? The answer to this question is surely yes. We can codify a relatively large collection of design queries, teach users how to employ them by introducing a range of practical examples and simulated cases, and establish priorities among them for those whose time and resources are limited. The process here resembles that of any applied discipline, such as medicine, architecture, or engineering. As in all such disciplines, the use of diagnostic queries as a method for crafting the provisions of resource and environmental regimes involves talent as well as skill. Some practitioners in each of these fields have an unusual gift for diagnostic work. They are widely recognized as exceptional diagnosticians, and their services are both sought after and well rewarded by those expecting to benefit from the insights they produce. Even so, any intelligent individual who is prepared to make a concerted effort can acquire usable diagnostic skills. By studying specific queries along with examples of their application to a wide range of situations, they can learn how to make use of this method to address new problems arising in their domains of responsibility. The development of teaching tools relating to institutional design will become more and more important as we strive to meet the growing challenges of life in the Anthropocene.

Note

1. For a parallel account of diagnostics in the field of economic development and growth, see Rodrik (2006).

III

Analytic Themes

5

The Problem of Fit among Biophysical Systems, Environmental and Resource Regimes, and Broader Governance Systems: Insights and Emerging Challenges

Victor Galaz, Per Olsson, Thomas Hahn, Carl Folke, and Uno Svedin

Introduction

Human and biophysical systems are closely interconnected. Yet not only have scientists and practitioners largely failed to recognize the tight coupling between these systems, but the stakes of failing also to harness the dynamic behavior of socioecological systems are getting higher. Two clear signals of this failure are the loss of vital ecosystem services at a global scale (Millennium Ecosystem Assessment 2005) and the far-reaching societal challenges posed by global environmental change (Steffen et al. 2004). Although analysts can project some of the future impacts on ecosystems and livelihoods, other effects will surface completely unexpectedly because of limited understanding of the strong interconnectedness of social and biophysical systems. Impacts will occur across many scales, with effects measured across time and space and at different levels of social organization and administration where humans and the environment intersect (Holling 1986; S. Schneider and Root 1995; S. Schneider 2004). Hence the need arises to consider how well the attributes of institutions and wider governance systems at local to global levels match the dynamics of biophysical systems. This is what Institutional Dimensions of Global Environmental Change (IDGEC) research denotes as "the problem of fit" (Folke et al. 1998; Young et al. 1999/2005; Brown 2003; Young 2003b).

Our discussion reviews this problem from particular perspectives. Reference to governance in addition to institutions places a strong, appropriate emphasis on the multilevel patterns of interaction among actors, their sometimes conflicting objectives, and instruments besides institutions that are chosen to steer social and environmental processes within a particular policy area (see Stoker 1998; Pierre 1999; Pierre and Peters 2005; Stoker 1998; Jordan, Wurzel, and Zito 2005). The focus of this review of fit is through a "resilience lens," concentrating on the capacity of institutions and broader governance mechanisms to deal with environmental change as linked to societal dynamics and to reorganize after unforeseen impacts. In this sense the governance challenge lies not only in developing multilevel institutions and organizations for multiscale ecosystem management, but also in aligning with the dynamics of biophysical systems while taking social systems into full account. Governance needs to meet the demands both of incremental change when things move forward in roughly continuous and predictable ways and of abrupt change when experience is often insufficient for understanding, consequences of actions are ambiguous, and the future of system dynamics is often uncertain (e.g., Adger et al. 2005). This discussion looks particularly at how to avoid the pathways of socioecological misfit institutions and wider governance that lead to constrained options for societal development and future capacity for adaptation (Gunderson and Holling 2002; Berkes, Colding, and Folke 2003).

Carl Folke and colleagues (1998) and Young (2003b) have elaborated the problem of fit in detail. Our intention here is to provide a transdisciplinary update, linking insights from research on socioecological systems with advances in the social sciences related to governance theory, which encompasses research on institutions. The resilience literature generally uses the term *social-ecological systems* to highlight the strong interconnectedness and coevolution of human-environmental systems (Berkes and Folke 1998; Berkes, Colding, and Folke 2003). In this chapter, however, we use the term *socioecological* to contribute to the compatible and uniform use of key terms and concepts in the book.

We aim to outline the "anatomy of misfits," illustrate their underlying mechanisms, and present strategies derived from research to cope with the identified mismatches. We explore the tight connection between social and ecological systems. Human dependence on the capacity of ecosystems to generate essential services and the vast importance of ecological feedbacks for societal development show that social and ecological systems are not merely linked but rather *interconnected*. In line with Berkes and Folke (1998), the need arises to address the interplay and fit between social and ecological systems by relating management practices based on ecological understanding to the social mechanisms behind these practices in a variety of geographical settings, cultures, and ecosystems.

We also present insights concerning the social processes and institutional structures that seem to build resilience in socioecological systems, that is, a capacity for living with and learning from change, expected or unexpected. We examine worldwide changes in the sociopolitical landscape, such as decentralization, public-private partnerships, and the emergence of network-based governance. Here we highlight the need to recognize the dynamic nature of not only socioecological but also governance systems, as well as the notion and features of adaptive governance.

The combined dynamics of social and ecological systems leads to a number of emerging governance challenges that will become important as a consequence of the increased interconnectedness of social, economic, technical, and ecological systems (Held 2000; Young, Berkhout, Gallopin, et al. 2006); the nonlinear nature of interconnected socioecological systems; and global environmental change (Steffen et al. 2004). The problem of fit in this context leads to discussion also of the importance of innovations in knowledge production, to understand better the behavior of interconnected systems, and the need to create stronger linkages to policy.

The Anatomy of Misfits between Biophysical and Environmental and Resource Regimes

How do we identify a "misfit"? The answer has important policy and scientific implications. Policy makers who are aware of a mismatch between an institution and a biophysical system see the real-world social and ecological implications. Identification of poor institutional fit forces researchers to specify the underlying and often interacting biophysical and social mechanisms (Hedström and Swedberg 1998) that explain the lack or loss of resilience in the institutional arrangements. In table 5.1 we elaborate different kinds of misfits between governance and biophysical systems and their underlying mechanisms.

The table shows how institutional solutions differ considerably for different sorts of misfits. The aim here is not to provide a complete or all-encompassing list of solutions, but rather to highlight the need for a range of solutions. It is of particular interest also that the identification

Table 5.1
Types of misfits between ecosystem dynamics and governance systems

Type of misfit	Definition and mechanism	Examples	Solution(s) suggested in the literature
Spatial	Institutional jurisdiction too small or too large to cover or affect the areal extent of the ecosystem(s) subject to the institution.	I. Administrative boundaries do not match hydrological boundaries, which creates collective-action problems, misallocation of responsibility, and hydrological and ecological degradation (Lundqvist 2004).	I. River basin/integrated water resources management (Global Water Partnership 2000) Bioregionalism (McGinnis, Woolley, and Gamman 1999)
	Institutional jurisdiction unable to cope with actors or drivers external or internal and important for maintaining the ecosystem(s) or process(es) affected by the institution; e.g., institutional arrangements can be "too large" when providing centrally defined "blueprints" that ignore existing local biophysical circumstances (Scott 1995).	II. Local institutions for management of sea urchin are unable to cope with the development of global markets and highly mobile "roving bandits" (Berkes et al. 2006). III. Central managers design rules and implement "one size fits all" institutions that are inappropriate to the local social or ecological context (Ostrom 1999).	II. Multiple-scale restraining institutions (Berkes et al. 2006) III. Collaborative, decentralized natural resource management (Wondolleck and Yaffee 2000) Adaptive comanagement (Olsson et al. 2004)

| Temporal | Institution formed too early or too late to cause desired ecosystem effect(s). | IV. In the 1950s and 1960s, governments in the West African Sahel promoted agricultural and population development in areas with only temporary productivity due to above-average rainfall. As the area returns to its low-productive state, erosion, migration, and livelihood collapse result (Glantz 1976). | Early-warning systems and national preparedness plans (Wilhite 1996) |
| | Institution (and possibly the actor interaction it entails) produces decisions that assume a shorter or longer time span than those embedded in the biophysical system(s) affected; and/or social response is too fast, too slow, too short, or too long compared to the time taken for biophysical processes involved (Holling and Meffe 1996; Scheffer, Westley, and Brock 2003). | V. The speed of impacts of invasive species is not matched by the speed of response of institutions, resulting in possible severe ecological and health implications (Meyerson and Reaser 2003; Miller and Gunderson 2004). | Adaptive management (Walters 1986) Adaptive comanagement (Olsson et al. 2004) Scenario planning (Peterson et al. 2003) |

Table 5.1
(continued)

Type of misfit	Definition and mechanism	Examples	Solution(s) suggested in the literature
Threshold behavior	Institution does not recognize, leads to, or is unable to avoid abrupt shift(s) in biophysical systems. Institution provides for inadequate response to contingencies (e.g., lack of rules for action in extreme conditions) or reduces variation in biophysical systems (e.g., by removing response diversity, whole functional groups of species, or trophic levels; and/or by adding anthropogenic stress such as pollution). Institutions fails to respond adequately or at all to disturbances that could have been buffered or that helped to revitalize the system before. Leads to practically irreversible biophysical shifts (Folke et al. 2004).	VI. Application of single species "maximum sustainable yield" triggers fish stock collapse due to overharvesting of key functional species (Pauly et al. 1998; Worm et al. 2006). VII. Food production is increased through monocultures at the expense of other ecosystem services (Rockström et al. 1999). Result is an increase in the risk of biophysical shifts and hence also rapid yield decline (e.g., Gordon, Dunlop, and Foran 2003).	Variable quotas, market-based incentives (Roughgarden and Smith 1996) Multiple-scale restraining institutions (Berkes et al. 2006) Adaptive management (Walters 1986) Adaptive comanagement (Olsson, Folke, and Berkes 2004) Adaptive governance (Folke et al. 2005) Scenario planning (Peterson et al. 2003)

Cascading effects	Institution is unable to buffer, or trigger further effects between or among biophysical and/or social and economic systems.	VIII. El Niño climate anomaly in 1972–73 led to excessive rainfall in usually arid regions while regions that usually receive abundant rainfall were plagued by drought. Sharp decline in commercial fish landings triggered sharp increase in prices of substitutes and shifts by U.S. farmers and Brazilian entrepreneurs to growing soybeans (Glantz 1990).	Adaptive governance (Folke et al. 2005) Steering of "networks of networks" (this chapter)
	Institutional response is misdirected, nonexistent, inadequate, or wrongly timed so as to propagate or allow the propagation of biophysical change(s) that entail(s) further causative changes along temporal and/or spatial scales (Kinzig et al. 2006).	IX. Western Australia: Abrupt shifts from sufficient soil humidity to saline soil and from freshwater to saline ecosystems might make agriculture a nonviable activity at a regional scale and trigger migration, unemployment, and weakening of social capital (Kinzig et al. 2006).	

of misfit mechanisms can serve as an "early warning signal" upon which institutional actors can act.

Discovery of the threshold misfit mechanism of loss or active removal of biological diversity in ecological systems could serve as an important signal of this kind. As observed by Folke and colleagues (2004), the loss of response diversity (i.e., species that can carry out the same ecosystem function[s] but that respond differently to disturbances [Elmqvist et al. 2003]) leads to more fragile ecological systems. This means that disturbances that were buffered and that may have helped revitalize a system before diversity loss can instead spark practically irreversible shifts in biophysical systems. The result in turn can be states with less capacity to support social welfare. This applies to both small- and large-scale ecological systems, including shallow lakes, coral reefs, landscapes, and even the global climate system (Scheffer et al. 2001; Folke et al. 2004; S. Schneider 2004).

While research shows that maintaining biophysical diversity helps prevent threshold effects, some researchers argue that institutional diversity is also important. As discussed by Bobbi Low and colleagues (2003), redundancy and diversity in environmental and resource regimes can become a major source of stability and strength, as they can provide multiple ways of coping with or reorganizing after change and unexpected events. The argument is that redundant systems can compensate for human errors and for unpredictable changes in circumstances. One simple example of this is technical redundancy in engineered systems such as the Boeing 777. Even though this redundancy is costly, multiple components that assume the same function can work as backup in case of partial technical failure or provide redundant strength, hence allowing for a higher margin of error. Both these types of redundancy can provide robust performance despite changing and uncertain environments (Low et al. 2003).

The inability of institutions, such as local resource regimes or national governments, to respond to rapidly changing circumstances—a *temporal misfit*—can also signal institutional failure. Examples include difficulty experienced by institutional actors at various administrative scales in monitoring and buffering the impacts of invasive species (Miller and Gunderson 2004) and the inability of international institutions to monitor and respond to the sequential depletion of key species in marine food webs (Berkes et al. 2006).

Interactions can occur among different sorts of misfits, as seen in spatial- and temporal-scale mismatches of institutions designed for water management where arrangements fail both to match the catchment area and to adapt to changing circumstances. Threshold and cascading mechanisms also occur in water management institutions, creating vulnerability to climate change due to an inability to avoid irreversible shifts and/or possible contribution to such shifts. The social situation is exacerbated when institutional arrangements fail to cope with resulting indirect social, ecological, or economic effects. Berkes and colleagues' analysis (2006) of "roving bandits" illegally harvesting sea urchins, for example, illustrates spatial (locally rooted institutions versus highly mobile fleets), temporal (relatively fast rate of ecological and market-driven change versus slow evolution of international and local institutions), and probable threshold misfits (risk of collapse due to inadequate institutional response). Interactions among misfit institutions and among misfit mechanisms have received little study; hence the examples and mechanisms presented here should be viewed as "ideal type" categories developed for heuristic reasons (see Doty and Glick 1994).

Coping with Misfit Regimes

A number of national and international policy initiatives have been strongly promoted to deal with some institutional misfits. Far too often, however, these initiatives have targeted only the first two categories of misfits: spatial and temporal levels of biophysical systems. Examples are river basin management, collaborative natural resource management, and participatory natural resources planning. These initiatives, however, *do not automatically create a better fit in preventing or dealing with abrupt threshold behavior or cascade effects in socioecological systems.* The importance of this observation should not be underestimated in the face of the multilevel and nonlinear character of interconnected biophysical systems (Gunderson and Holling 2002; Folke et al. 2004). In the same way the promotion of adaptive management to "manage around thresholds" (e.g., Rogers and Biggs 1999) does not automatically lead to a better fit in terms of a regime's capacity to avoid or not to trigger large-scale cascading effects with the potential to spill over into a diverse set of domains and policy fields.

As the type and number of misfits increase (e.g., from local spatial misfits to cross-national cascade effect misfits), so does the governance challenge. This results from the enlargement in the number of actors, spatial scales, and interactions across systems introduced by environmental and resource regimes operating on the multiscale and cross-system nature of global environmental change.

The Fundamental Importance of Time

Multilevel governance systems have to cope with both incremental change and fast and sometimes irreversible shifts in biophysical systems. Although certain regimes may be highly efficient in times of slow or small, often predictable changes, they might fail in times of fast, uncertain change (Duit and Galaz, forthcoming).

The behavior of complex adaptive biophysical systems sometimes requires institutional actors to respond "quickly," although this becomes a relative term as applied to a specific system and the level of change to be governed. In the case of threshold effects, governance must be able not only to coordinate relevant actors, but also to achieve coordination before critical and irreversible thresholds are crossed. Studies of management of biophysical systems indicate that the capacity to promote necessary mobilization tends to be either too slow to engage or even nonexistent compared to the speed and scope of change. This misfit regime behavior has had major consequences in several cases: collapsed fisheries at various spatial scales ranging from local to global (Berkes et al. 2006; Worm et al. 2006); drastic changes in the function and feedback in global biophysical systems (Steffen et al. 2004), as in the case of irreversible shifts in freshwater systems, coral reefs, and productivity of soils (Scheffer et al. 2001; Folke et al. 2004); and the often irreversible loss of ecosystem services, such as water purification, food production, mitigation of environmental hazards, carbon sequestration, and cultural values (Millennium Ecosystem Assessment 2005).

A similar argument applies to cascading effects. Not only must the proper response be achieved by individual or collective actors, but it must also be done within such a time frame that measures are implemented to buffer the ecological, social, or economic effects of the cascade. The question of time and regime fit, then, concerns how well institutional arrangements allow for biophysical system change that occurs gradually, the potential of a system to shift suddenly and irreversibly, and the possibility in a system of fast or slow unfolding of cascading effects. With both threshold and cascade effects, the issue of time brings high uncertainty as a factor to accommodate in regime design.

From Linked to Interconnected Biophysical Systems

Why do governance systems continually fail to protect vital ecosystem functions and resources? Important external factors can include a lack of alternative livelihoods, corruption, administrative fragmentation and inefficiency, and the presence of rent-seeking behavior—profiting from manipulation of the economic environment rather than through trade or production—at different levels and on a number of scales. However, a lack of acknowledgment of the dynamics of strongly interconnected socioecological systems appears to be a fundamental but seldom elaborated endogenous factor in institutional failure.

Socioecological systems are not just social and ecological systems, with some temporal and weak links in between (Westley et al. 2002). Nonetheless, this is the simplistic conventional understanding that sees the socioeconomic system extracting natural resources from the ecological system, which in turn receives disturbances (such as pollution and resource extraction) from the socioeconomic system. A number of recent syntheses point to the strong feedback and coevolution between social and ecological systems (illustrated in figure 5.1). Jianguo Liu and colleagues (2007), for example, elaborate how ecological change and decision making alternate in periods of time, creating reciprocal interactions between human and natural systems (see also Costanza, Graumlich, and Steffen 2005). At worst these interactions can push socioecological systems toward increased vulnerability, as elaborated for the Goulburn Broken Catchment in southeastern Australia. Loss of socioecological resilience in this case can be traced to ecologically uninformed, crisis-induced policy making (Anderies, Ryan, and Walker 2006). The need to understand fully the true, highly interconnected character of socioecological systems hence should not be underestimated.

Although certainly illuminating in a number of senses, conventional natural resource management studies tend strongly to investigate processes *within* the social domain only, treating the ecosystem largely as a "black box." Research makes the bold, implicit assumption that if the social system somehow performs adaptively, it will also manage the environmental resource base in a sustainable fashion. This assumption entails a view that environmental and resource regimes and other institutions need only to be well organized.

The flaw of this assumption shows up, for instance, in the collective action among coastal fishermen in Belize at the end of the 1960s. Signs

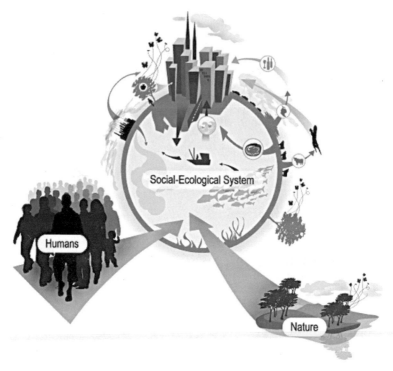

Figure 5.1
Interconnected socioecological system. (Illustration by Christine Clifstock)

of declining catches, and concerns about profits being lost to other actors in the market able to process, market, and export the resource, triggered the creation of fishers' cooperatives. This labor institution seemed to lead to a number of socially desired outcomes, such as increased revenue for the fishers. Although the strategy was initially economically successful, the increased collective action combined with technological development (i.e., fuel-based technology) led ultimately to excessive harvesting of stocks of lobster and conch, which in turn resulted in worse economic conditions (Huitric 2005). This example shows institutional interplay between the labor institution and the distant institutions governing the oceans. Ocean rules in this case plainly amounted to a misfit through the threshold mechanism of allowing depletion of biological diversity. The inadequacy of the ocean regime led to institutional interplay with the labor institution in the form of collective action where the effects of the latter were allowed to run unchecked into resource depletion. In a similar vein, the research of Allison and Hobbs (2004) shows how institutions created by political decision makers in response to environmental degradation in agricultural systems in Western Australia result in a "lock-in" to the response of natural resource users. The result is an institutional misfit characterized by the creation of a pathological trap (Holling and Meffe 1996) of continued erosion of the resource base and concomitant social decline in the region. A third example arises in the field of biodiversity conservation. The loss of biodiversity is often argued to be strongly interrelated with endemic corruption in developing countries (Laurance 2004). The data, however, show that even countries with transparent, otherwise effective, and noncorrupt governance systems have declining levels of species richness (Katzner 2005). Evidently the problem of fit plagues institutions designed to conserve biodiversity whether or not they appear well organized and no matter the place or type of governance system under which they operate.

Human society may show a great ability to design institutions, mobilize collective action, and respond to changing circumstances, but the institutional and other societal responses may occur at the expense of changes in the capacity of ecosystems. Recent reports highlight that human attempts to adapt to social or environmental change have caused a loss of ecosystem resilience, pushing many biophysical systems close to thresholds or into changed states with a lower capacity to generate ecosystem services (e.g., Scheffer et al. 2001; Folke et al. 2004). Thus, the result of poor fit is increasingly seen as important since it can lead ultimately to a failure of the resource to sustain societal development.

A focus on the poor fit of environmental and resource regimes alone to understand failures to manage environmental change cannot provide a full analysis. Nor is it sufficient to rely on ecological data to inform the design of environmental and resource regimes fully. Berkes and colleagues (2006) bring to light the societal and market processes that generate changes in large-scale ecological systems by showing how the sequential exploitation of marine resources is triggered by highly mobile "roving bandits" and rapidly developing world markets. Basing institutional design on ecological knowledge alone, without recognizing the fundamental impact of other institutions and social actors on ecological systems, is a simplistic approach that fails to appreciate the complexity of governance processes, mental models (Adams et al. 2003), and the social features that enable management of dynamic ecosystems (Folke et al. 2005). The result of such an approach will always be an environmental or resource regime misfit.

These examples illustrate why institutions formed to manage biophysical systems or their elements need to recognize that the separation of social and ecological systems is artificial and arbitrary. The intersection between social and ecological systems must be addressed in its full complexity, a coevolution that justifies the term *interconnected* rather than *linked*. Regime design needs to recognize this interconnection to form a successful fit with the biophysical system that it addresses.

Interconnected Socioecological Systems and the Problem of Fit

Lack of an integrative perspective on socioecological systems is only part of the story. This problem is exacerbated by the mismatch between not only temporal scales (as above) but also spatial scales of management and ecosystem change. Management worsens, of course, when scale mismatches contribute to rules and decision making that cause threshold and cascade effects. A number of studies show how blueprint, command-and-control approaches for managing natural resources can do just this, as they often fail to match the geographic range and therefore often the diversity of different local settings and the complexity of ecosystems (Holling and Meffe 1996; Wilson 2006). As a consequence, this management approach has pushed many ecosystems into degraded vulnerable states (Scheffer et al. 2001; Folke et al. 2004).

An institution set at too large or too small a level on a spatial scale will entrain management failure based on rules and procedures that address an insufficient number of ecosystem variables in their efforts to deliver efficiency, reliability, and optimality of ecosystem goods and services (Holling and Meffe 1996). Stabilizing production of a set of desirable goods and services can lead to an increased vulnerability of the system to unexpected change (Gunderson and Holling 2002; Folke et al. 2004). Wilson (2006) argues, for example, that the mismatch of ecological and management scales makes it difficult to manage the fine-scale aspects of ocean ecosystems and leads to fishing rights and strategies that tend to erode the underlying structure of populations and the systems themselves.

The shift from treating social and ecological systems separately to regarding them as truly interconnected complex socioecological systems, characterized by nonlinear relations, multiple stable states, and the potential for threshold behavior and qualitative shifts in system dynamics (Jervis 1997; Levin 1998), has triggered the emergence of analytical frameworks like *socioecological resilience, adaptive comanagement*, and *adaptive governance*, all of which can be related to matters of institutional function.

Enhancing Institutional Fit through Adaptive Comanagement

Adaptive comanagement refers to the multilevel and cross-organizational management of ecosystems. Such multilevel governance systems of institutional interplay often emerge to deal with crises and can develop within a decade (e.g., Olsson, Folke, and Berkes 2004). They combine the dynamic learning characteristic of adaptive management with the linkage characteristic of collaborative management (Gadgil et al. 2000; Wollenberg, Edmunds, and Buck 2000; Ruitenbeek and Cartier 2001; Folke et al. 2005). The combination aims to address the analytical and managerial shortcomings of both adaptive management and comanagement. Adaptive management addresses the humans-in-nature perspective and learning by doing (Holling 1978), but the approach has been criticized for not incorporating other knowledge systems (McLain and Lee 1996). Comanagement, on the other hand, addresses institutional and epistemological aspects, multistakeholder processes, and the sharing of power in natural resource management; but it often neglects fundamental ecosystem feedback and dynamics as well as larger governance dimensions.

Olsson, Folke, and Berkes (2004) discuss the role of adaptive comanagement in building *resilience* in socioecological systems. It has been almost three decades since the ecologist C. S. Holling introduced the term *resilience*. Since then, multiple meanings of the concept have appeared (table 5.2), all with different management and policy implications (Gunderson 2000). One such meaning considers return times as a measure of stability ("engineering resilience"). This definition arises from traditions of engineering, where the motive is to design systems with a single operating objective and to accommodate an engineer's goal of developing optimal designs. As argued by Lance Gunderson (2000), there is an implicit assumption of only one equilibrium or steady state; or, if other states exist, they should be avoided by applying safety measures.

For *ecosystem resilience* the challenge is to sustain the capacity of an ecosystem to generate valuable ecosystem services. *Social-ecological*

Table 5.2
Sequence of resilience concepts from the more narrow interpretation to the broader socioecological context

Resilience concepts	Characteristics	Focus on	Context
Engineering resilience	Return time, efficiency	Recovery, constancy	Vicinity of a stable equilibrium
Ecological/ ecosystem resilience Social resilience	Buffer capacity, withstand shock, maintain function	Persistence, robustness	Multiple equilibria, stability landscapes
Social-ecological resilience	Interplay disturbance and reorganization, sustaining and developing	Adaptive capacity, transformability, learning, innovation	Integrated system feedback, cross-scale dynamic interactions

Source: From Folke 2006

resilience (as defined by Folke 2006), on the other hand, emphasizes the reorganization, learning, and adaptive capacity of actors in response to ecosystem change, rather than attempts to design optimal strategies with one single objective in mind. Obviously, the ability to enhance resilience depends on the dynamics of the biophysical system as well as factors stemming from an institution created to manage these dynamics adaptively and with the capacity to handle surprise. The notion of *social-ecological resilience* endorses this challenge but explores further the institutional arrangements and the organizational and wider governance processes that enable adaptive comanagement of ecosystems (Folke et al. 2005).

Adaptive comanagement recognizes the fact that ecosystem management is an information-intensive endeavor and requires institutional design that facilitates and accommodates knowledge of complex socio-ecological interactions in order to create a very good fit with the biophysical system it addresses. Knowledge is applied and built on through monitoring, interpreting, and responding to ecosystem feedback at multiple scales (Folke et al. 2005). Because of the complexity involved it is usually difficult if not impossible for one or a few people to possess the range of knowledge needed for effective ecosystem management (Berkes 2002; Brown 2003; Gadgil et al. 2003; Olsson, Folke, and Berkes 2004).

Instead, knowledge for dealing with socioecological system dynamics becomes dispersed among individuals and organizations in society and requires social networks that span multiple levels in order for actors to draw on dispersed sources of information (Imperial 1999; Olsson et al. 2006).

Crisis, perceived or real, can trigger learning and knowledge generation (Westley 1995) and can open up space for new interactions and combinations of knowledge and experiences, as well as new management trajectories of resources and ecosystems (Gunderson 2003). For example, mobilization of different knowledge systems may take place in a *social learning process* (Lee 1993b), meaning "learning that occurs when people engage one another, sharing diverse perspectives and experiences to develop a common framework of understanding and basis for joint action" (Schusler, Decker, and Pfeffer 2003). In this way social learning integrates issues of knowledge generation, working out objectives, solving conflicts, and action. To achieve sufficient fit with a biophysical system, rights, rules, and decision-making procedures need to be premised on these kinds of knowledge-sharing and knowledge-generative processes.

The Social Foundations of Institutions for Adaptive Comanagement

Coordinating the required institutional and organizational landscape to enhance the fit between biophysical systems and governance is a far from simple task. Three related issues stand out as critical for success in this context: the first is the need to link organizations across levels, initiating interplay among their respective institutions; the second is the role of bridging organizations; and last is the importance of leadership.

Organizing linkages among institutions with relatively autonomous but interdependent actors and actor groups becomes crucial for avoiding fragmented and sectoral approaches to the management of ecosystem services and for enhancing the fit between governance systems and biophysical systems. Researchers have observed the active role of a few key individuals or organizations in linking institutions at different administrative levels as, for example, in connecting local communities to outside markets (Bebbington 1997; Ribot 2004; Pomeroy et al. 2006). Crona (2006) refers to individuals who act as middlemen to link fishers to markets in coastal communities of eastern Africa. As pointed out by González and Nigh (2005), intermediaries are no guarantee of more democratic decision making and can play a role in the implementation of hierarchical command-and-control institutions where policies are applied in a top-down fashion. Nongovernmental organizations (NGOs) also frequently play the role of coordinators and facilitators of the institutional interplay needed for comanagement processes (e.g., Halls et al. 2005) that can often improve or create good institutional fit.

Boundary organizations and *bridging organizations* are two forms of intermediaries tasked with establishing the institutional interplay typically necessary to achieve successful fit through adaptive comanagement. Boundary organizations can provide an array of important functions for linking researchers and decision makers (Guston 1999; Cash and Moser 2000). Although similar in some aspects, bridging organizations have a broader scope and address resilience in socioecological systems. A bridging organization provides an arena for trust building, social learning, sense making, identification of common interests, vertical and/or horizontal collaboration, and conflict resolution (Folke et al. 2005). The bridging organization is crucial for maintaining new collaboration among different stakeholder groups in order to foster innovation, generate new knowledge, and identify new opportunities for solving problems.

Malayang and colleagues (2006), for example, show how bridging organizations perform essential functions in crafting effective responses to change in socioecological systems. Bridging organizations create the space for institutional innovations and the capacity to deal with abrupt change and surprise. In Kristianstads Vattenrike, Sweden, most environmental governance activities are coordinated, but not controlled, by Ecomuseum Kristianstads Vattenrike, a small municipal organization acting as a bridging organization (Hahn et al. 2006). Its institution has led to the development of an explicit approach to conflict resolution and disturbances. Bridging organizations, like the one in Kristianstads Vattenrike, seem to play a central role in stimulating, facilitating, and sustaining adaptive comanagement and adaptive governance (Folke et al. 2005) and, by doing so, in avoiding the creation of misfit regimes. They can play a key role in collective learning processes that build experience with ecosystem change, enfolding it as "social memory"—the arena in which captured experience with change and successful adaptations embedded in a deeper level of values are actualized through community debate and decision-making processes into appropriate strategies for dealing with ongoing change (McIntosh 2000)—in an evolving institutional and organizational setting. Social learning contributes to the ability of actors to respond to feedback from a biophysical system and to direct the coupled social-ecological system into sustainable trajectories (Berkes, Colding, and Folke 2003). Seen to be essential in fostering sources of resilience in socioecological systems, bridging organizations and their institutions deserve more investigation. They serve as prominent examples (interestingly with the use of institutional interplay) of how to develop social practices, assign roles to participants, and guide interactions that facilitate environmental and resource regimes that achieve a successful biophysical system fit.

Leadership is another critical feature for increasing institutional fit through adaptive comanagement (compare Young 2001c). Key individuals can provide visions of ecosystem management and sustainable development that frame self-organization, that is, self-monitored collective action assumed without being guided or managed by an outside source (Agranoff and McGuire 2001; Westley 2002). Key individuals are important in establishing functional links within and between organizational levels, thereby facilitating the flow of information and knowledge from multiple sources to be applied in the local context of ecosystem management. Leadership has been shown to be of great significance for public network management. Network leadership and guidance differ greatly from the command-and-control style of hierarchical management (Agranoff and McGuire 2001). Steering is required to hold a network together (Bardach 1998), and the social forces and interests must be balanced to enable self-organization (Kooiman 1993). Socioecological systems that rely on only one or a few principal stewards, however, might not have the institutional capacity to prevent a misfit, as seen, for example, in the institutional response to change in the case of longleaf pine forest ecosystems in Florida (Peterson 2002).

Research reveals an important lesson in that it is not enough for institutions to create arenas for dialogue and collaboration or to develop networks that match the spatial scale of socioecological systems. Underlying social structures and processes for ecosystem management need to be understood and actively managed. Environmental and resource regimes must support social mechanisms and arrangements for accessing and combining knowledge to respond to ecosystem feedback at critical times (Olsson et al. 2006). However comprehensive the combined knowledge might be, complex socioecological dynamics always brings an element of surprise (Gunderson 1999, 2003). For institutional fit, the development of networks of actors and opportunities for interaction turns out to be essential, as it helps produce integrated adaptive responses to uncertainty and change (Stubbs and Lemon 2001; Hahn et al. 2006).

From Institutions to Dynamic Governance Systems

Institutional theory has made substantial advances in clarifying the importance of and the social mechanisms behind the emergence of self-organized institutions for natural resource management (Ostrom et al. 2002; Ostrom 2005). But the last decade of IDGEC research has brought other significant insights that highlight the dynamic and multilevel nature of governance systems with noteworthy implications for understanding the problem of fit.

From Government to Governance

Research advances in the field of institutions and natural resource management over the past two decades have occurred simultaneously with a number of worldwide shifts in the organization of society and politics. The trend has been toward less centralized styles of state governance (Stoker 1998; Pierre 2000) with several driving factors in play.

As argued by Bardhan (2002), decentralization as a fundamental and global policy experiment has proved the prime causative influence. Part of the logic driving this shift relates to the alleged failure and loss of legitimacy of the centralized state (Mayntz 1993; Bardhan 2002). Motive also lies in the expectation that a fragmentation of central authority will make government more receptive and efficient in its attempts to solve complex societal problems, such as chronic poverty (Datta and Varalakshmi 1999) and overextraction of natural resources (Ostrom 2005).

The growth of public-private partnership arrangements (i.e., cooperative ventures between the state and private business) is another trend in the same decentralized direction (Evans 1996; Osborne 2000). The motive in this case stems from the belief that collaborative interagency partnerships can achieve public policy goals and provide a more attractive alternative to full privatization or large-scale bureaucratic public-service organizations (Lowndes and Skelcher 1998). This shift is highly visible in the field of natural resource management (Ostrom 1999), ranging from water governance (e.g., Global Water Partnership 2000) and biodiversity conservation (Stoll-Kleemann and O'Riordan 2002) to capacity building for ecosystem management (Berkes 2002; Olsson, Folke, and Berkes

2004; Folke et al. 2005) and biotechnological research (Rausser, Simon, and Ameden 2000).

Governance scholars also note the augmented influence of NGOs and epistemic communities on policy processes at a number of political levels. Climate change policy (Gough and Shackley 2001), biodiversity policy (Fairbrass and Jordan 2001), and decision making in the European Union provide interesting cases in point. The existence of numerous access points into the institutional process and the large number of officials and organizations that have a role in the process all support the increased influence of nonstate political "entrepreneurs" such as NGOs and epistemic communities (Sabatier 1998; Zito 2001).

Last, the increased impact of multilateral agreements on domestic policy (Cortell and Davies 1996) and the spread of policy innovations across different nations (Busch and Jörgens 2005) also lead away from command-and-control state governance by central governments, increasing the influence of actors and policy makers beyond the state.

The Dynamics of Governance Systems and the Problem of Fit
Recent and ongoing shifts in governance have fundamental implications for understanding the problem of fit. Natural resource users trying to preserve ecosystem services and build resilience find themselves facing not only potential collective-action problems with other users (Ostrom 1990) but also a plethora of interlinked local, national, and international institutions and a diversity of actors and decision makers.

The case of property rights provides a good example. Much attention has been devoted to common-property regimes as alternatives to government-property or private-property regimes (e.g., Ostrom 1990; Bromley 1992). In common-property rights regimes, use rights, capital rights (rights to sell), management authority, and excludability may be distributed differently for different ecosystem services. Yet as the ecological level of management concerns increases, for example, to catchment or landscape level, generally a mix of property rights regimes exists, along with the need for coordination to reduce spillover effects in the form of external costs (e.g., the pollution from one harms all) and free riding (e.g., those who do not invest in biodiversity may still benefit from others' investments) among stakeholders—private landowners, communal land representatives, governmental agencies at different levels, and various NGOs. Because of their interdependence, stakeholders cannot fulfill their objectives in isolation from the actions of other stakeholders (Imperial 2005). At the larger ecological scale, the challenges are shifting from designing property rights per se to agreeing on goals and strategies for responding to environmental change and hence to developing a more dynamic governance system that achieves a good fit.

Although common-pool resources and institutional interplay undoubtedly play a fundamental role in the sustainable management of ecosystem services, they sit increasingly in the context of a highly dynamic, multisectoral, and multilevel governance landscape with a variety of actors and interests. This in turn increases the potential not only for misfits between institutions and biophysical systems (Folke et al. 1998; Cumming, Cumming, and Redman 2006) but also for a lack of fit between biophysical systems and *governance systems* of which institutions are a part.

By governance systems we mean the interaction patterns of actors, their sometimes conflicting objectives, and the instruments chosen to steer social and environmental processes within a particular policy area (Pierre 1999; Pierre and Peters 2005; Stoker 1998; Jordan, Wurzel, and Zito 2005). Although institutions certainly are a central component in governance (Pierre 2000), our ambition is to put a stronger emphasis on both the patterns of interaction between actors and the multilevel institutional setting under which they interact repeatedly, creating complex relations between structure and agency (Klijn and Teisman 1997; Rolén, Sjöberg, and Svedin 1997; Svedin, O'Riordan, and Jordan 2001; for applications see Bodin, Crona, and Ernstsson 2006; de la Torre Castro 2006).

One fundamental assumption is that differing multilevel institutional settings, combined with different interaction patterns (Scharpf 1997), will produce a diversity of outcomes related to the problem of fit. To be more precise, different institutional settings (not necessary related to natural resource management alone) and differing constellations of actors (i.e., differing by number, type, and bargaining resources) lead to different outcomes in social processes vital for managing the behavior of complex adaptive systems such as socioecological systems (e.g., high or low adaptive capacity; proficient, nonproficient, or nonexistent leadership; trust building or conflict propagating). All these in turn contribute to the degree of fit of any one institution, set of collaborating institutions, or overall governance system.

Harnessing Complexity through Adaptive Governance

Adaptive comanagement seems to be a step in the right direction for analyzing and coping with socioecological dynamics. On the other hand, it also faces analytical limitations associated with the multilevel character of both social and ecological change (Folke et al. 2005). How to create governance that is able to "navigate" the dynamic nature of multilevel and interconnected socioecological systems becomes a crucial issue in this context.

The notion of "adaptive governance" discussed by Dietz, Ostrom, and Stern (2003) and Folke et al. (2005) is interesting since it can address the possibilities and the need to draw on the multilevel changing nature of governance systems. Whereas "management" implies bringing together knowledge from diverse sources into new perspectives for practice, a focus on "governance" conveys the difficulty of control, the need to proceed in the face of substantial uncertainty, and the importance of dealing with diversity and reconciling conflict among people and groups who differ in values, interests, perspectives, power, and the kinds of information they bring to situations (Dietz, Ostrom, and Stern 2003). Such governance fosters social coordination that enables adaptive comanagement of ecosystems and landscapes. For such governance to be effective, joint understanding of ecosystems and socioecological interactions is required. This approach also recognizes the need both to govern social and ecological components of socioecological systems as well as to build a capacity to harness exogenous institutional and ecological drivers that might pose possibilities or challenges to social actors (Folke et al. 2005; see also Dietz, Ostrom, and Stern 2003). Folke and colleagues (2005) highlight the following four interacting aspects of importance in adaptive governance of complex socioecological systems:

1. Build knowledge and understanding of resource and ecosystem dynamics to be able to respond to environmental feedback.
2. Feed ecological knowledge into adaptive management practices to create conditions for learning.
3. Support flexible institutions and multilevel governance systems that allow for adaptive management.
4. Deal with external perturbations, uncertainty, and surprise.

Polycentric institutional structures—institutions with multiple and overlapping centers (M. D. McGinnis 2000)—are crucial in this notion. It has been proposed that these sorts of institutions can address environmental problems at multiple scales and nurture diversity for dynamic responses in the face of change and uncertainty. The argument is that large-scale, centralized governance units do not, and cannot, have the variety of response capabilities that can derive from complex, polycentric, multilevel governance systems (Ostrom 1998). Similarly, Imperial (1999, 459) argues that polycentric governance creates an institutionally rich environment that can "encourage innovation and experimentation by allowing individuals and organizations to explore different ideas about solving [complex] problems." Such arrangements allow for self-organization, and if efficiently linked across scales, they can increase the complexity of governance systems and therefore the variety of possible responses to change (Ostrom 1998).

A number of critical questions remain nonetheless. One concerns the elucidation of the type of institutional structures that enable and facilitate people to self-organize, collaborate, learn, innovate, reorganize, and adapt in response to threats or opportunities posed by environmental change. Accumulation of socioecological understanding and experience in a social memory seems critical for dealing with change. Furthermore, social networks can store social memories for ecosystem management, memories that can be revived and revitalized in the regeneration and reorganization phase following change (Folke, Colding, and Berkes 2003). There is also a need to understand the governance attributes that support and build social memory and hence resilience in the face of disturbance.

In connection with the need for social memory and the increase of public-private partnerships discussed earlier, Evans (1996) links public-private synergies to building the social capital important for economic development. He argues that social capital is often built in the intermediate organizations and informal policy networks, in the interstices between state and society. In the same issue of the journal *World Development*, Ostrom (1996) explores the constructability of such synergies between governments and groups of engaged citizens. Research is needed to discover the conditions that most easily facilitate such synergistic relations.

Furthermore, how people and societies respond to periods of abrupt change and reorganize in the aftermath is not well understood in relation to the problem of fit (Gunderson and Holling 2002). Governance research could certainly take on this challenge to a greater degree (Duit and Galaz, forthcoming). Explorative work, based on several case studies, suggests that four critical factors, interacting across temporal and spatial scales, seem to be required for resilience of socioecological systems during periods of rapid change and reorganization (Folke, Colding, and Berkes 2003):

• Learning to live with change and uncertainty
• Combining different types of knowledge for learning
• Creating opportunity for self-organization toward socioecological resilience
• Nurturing sources of resilience for renewal and reorganization

Lost Opportunities? Two Alternative Stories
Broadening the scope beyond institutional dimensions to study governance forces more inquiry into how the shifts in the larger sociopolitical landscape discussed earlier affect misfits between governance and ecosystems. At least two alternative perspectives arise. The first sees the changes in the organization of society as providing fertile ground for enhanced fit. The shift away from command-and-control methods of governing creates greater diversity of institutions, increased involvement of actors with complementary knowledge at a number of political levels, and polycentric institutions with noticeable multiscale linkages. Such changes could lead to increased overall diversity and redundancy as benefits that improve the resilience of social-ecological systems (Dietz, Ostrom, and Stern 2003; Folke et al. 2005). This is the view of those researchers who argue that institutional redundancy increases system reliability in the face of operational or environmental uncertainty (Streeter 1992). Independent planning teams, for example, may develop alternative management plans based on complementary observations and knowledge, enhancing the diversity of response options. Low and colleagues (2003) suggest that diversity and redundancy of institutions and their overlapping functions across administrative levels may play a central role in absorbing disturbance and in spreading risks. For vital components and functions, redundancy can prove economically efficient; the costs of redundancy should be weighed against the costs of designing components and functions that "never" fail, the costs of failure, and the costs of correcting failures when these occur anyway. Streeter (1992, 99) has referred to the backup function of redundancy as "failure absorption rather than failure correction."

A second, less optimistic picture highlights the risks of the decreased controllability of complex modern societies. Put bluntly, increased diversity and complexity in governance systems could result in decreasing levels of compliance among social actors (Mayntz 1993), higher un-certainty in outcomes of policy intervention due to more complex cause-effect relationships (Kooiman 2003), and decreased efficiency and legitimacy of central institutions and decision making owing to the enhanced autonomy of societal actors at diverse scales (Hirst 2000, 20–21). In addition, arguments against redundancy focus on avoiding policy inconsistencies—fragmentation, duplication, and overlaps—as well as the potential conflicts and high operational and transaction costs that may result when more people are involved in decision making (Imperial 1999).

It is of course impossible to know which of the two stories best describes the impacts of current sociopolitical shifts. There is a need, though, to highlight the risk of lost opportunities. The number of global initiatives pushing for greater diversity and complexity in governance systems seems to be increasing, as seen, for instance, in the following: political and expert-driven processes that promote integrated water re-sources management (IWRM); a move toward more participatory inter-national development strategies (H. Schneider 1999; Ellis and Biggs 2001); the acknowledgment of stakeholder participation, traditional knowledge, and innovations in ecosystem management by the Conven-tion on Biological Diversity; and approaches that promote sustainable development by building partnerships across scales and between stake-holder groups (e.g., the UN's Partnerships for Sustainable Development).

The networks and emerging cross-scale social interactions seen in such institutional arrangements may not promote a systematic understanding of the nonlinear behavior of socioecological systems. Such networks and interactions may not intentionally build a capacity to cope with abrupt as well as incremental change. These possibilities show the obvious risk that only more "messiness" rather than fit will be added to governance.

The role of social networks illuminates this point. Social networks can play a crucial role in the dynamic relationship between key indi-viduals and organizations as the groups responsible for implementing institutional arrangements (Westley 2002). It is also argued that social networks can enhance socioecological resilience as they lead to improved fit between biophysical systems and institutions (Olsson, Folke, and Berkes 2004; Folke et al. 2005; Hahn et al. 2006).

On the other hand, social networks can provide a conservative force that benefits from the existing misfit and therefore tries to block needed changes. The structure of social networks (i.e., the patterns of actor interactions in governance) is fundamental to governance whether the patterns promote adaptation and learning or vulnerable and maladaptive collaboration (Bodin and Norberg 2005; Janssen et al. 2006). As stated by L. Newman and Dale (2005), not all social networks are created equal; those composed of "bridging" links to a diverse web of resources strengthen a community's ability to adapt to change, but networks composed only of local "bonding" links, which impose constraining social norms and foster group homogeneity, can reduce adaptability.

Social networks also rely rather heavily on voluntary coordination and control (Jones, Hesterly, and Borgatti 1997). This implies that social networks are created and become robust only when they promote the joint interests of all parties. As a result, networks might impede institutional fit in several ways: use by central actors of their joint capacity to veto needed governance; failure to reach needed agreements (Mayntz 1993, 19; Pierre and Peters 2005); falling short in coping with ecosystem change because of embedded power relations (Galaz 2006); lack of incentives to deal with the direct and indirect effects of their actions on actors outside the network (Kooiman 2003).

The need arises for increased understanding of the role of networks in managing cross-scale interactions, dealing with uncertainty and change, and enhancing ecosystem management (Bodin and Norberg 2005; Bodin, Crona, and Ernstsson 2006). There is also a need to investigate further the potential of social networks and their cross-scale linkages to generate resilience through flexibility and the provision of response options in times of socioecological change. It is important also to understand how cross-scale dynamics can widen the scope of socioecological stability, helping to make systems more adaptable to change.

There is a risk that ongoing global sustainable development policy initiatives are missing the opportunity to explore the resilience-building potentials of decentralization, partnership arrangements, and the evolution of network-based governance. The addition of diversity and complexity to existing governance systems is often successful in the short term but may cumulatively become unfavorable to sustainability because of increased societal costs and diminishing returns from new institutions (Tainter 2004).

Emerging Challenges

Governance systems are just as dynamic as biophysical systems: the patterns of interactions among state and nonstate actors tend to change over time (Pierre 2000), existing social or policy networks oscillate between latency and activism (Mayntz 1993), societal actors adapt to and sometimes divert external governing attempts (Kooiman 2003), institutional development might embed path-dependent or positive feedback (Pierson 2003), and political support might erode and/or recover after extreme events (Dalton 2004). The combined dynamics of governance and socioecological systems poses a number of unexplored yet basic questions related to the problem of fit. The first of three such challenges is the recognition that large-scale crises can trigger political backlash that pushes governance systems toward more rigidity and hence greater vulnerability to change and surprises. The second challenge stems from the drastic tests of governance posed by cascading effects in socioecological systems. The third entails the possibility of tackling cascade effects by promoting governance that builds on managing "networks of networks" within existing yet diverse subpolicy fields.

Recognizing the Possibility of Backlash

Perceived crises often open a "window of opportunity" for learning and change, helping to overcome inertia and social dynamics, which often inhibit learning under "normal" conditions. This is an important insight from studies of organizations (Kim 1998), socioecological systems (Westley 1995; Gunderson and Holling 2002), and political decision making (E. Stern 1997). Crises may be caused by factors such as external markets, tourism pressure, floods and flood management, shifts in property rights, threats of acidification, resource failures, rigid paradigms of resource management, or new legislation or government policies that do not take local contexts into account (Berkes, Colding, and Folke 2003). Even crises that result in irreversible biophysical shifts that affect the economy and livelihoods of communities might trigger learning and the possibility of enhanced institutional and wider governance fit.

Backlash can also arise from crisis, however. The major U.S. governmental reorganization following September 11, 2001, provides a good illustration of political system response to large-scale crisis. Researcher James Mitchell describes how public policy had increasingly favored a broad engagement of civil society in hazard management in the latter part of the twentieth century. But in the wake of 9/11 there was a sudden return to governance that favored trained experts, centralized decision making, and secrecy over transparent, participatory, and decentralized approaches (J. Mitchell 2006; see also Gill 2004). The intense political debate in the United States after Hurricane Katrina (Waugh 2006) and in Sweden after the slow, ineffective response of the Swedish government to the devastating impacts of the Asian tsunami in December 2004 (Swedish Government Inquiry 2005, 14) showed strong advocacy of centralized "man-on-horseback" organizations. This concept entails institutional design that assigns centrally appointed leaders who, in times of crisis, clear the way for centrally controlled rapid-response teams of experts from the military and other action-oriented institutions (from J. Mitchell 2006, 230; compare Boin and t'Hart 2003).

The general trends toward more flexible, participatory, redundant, and polycentric institutions (and hence a possible better fit between biophysical systems and institutions) might be reversed by political processes triggered by large-scale crises that result from, for example, fast, abrupt changes in vital ecosystems (compare Scheffer, Westley, and Brock 2003); extreme, unexpected climatic or ecological cascade effects that in turn propagate social and economic effects that may also cascade because of the increased interconnectedness of systems (compare Rosenthal and Kouzmin 1997; Kinzig et al. 2006; Young, Berkhout, et al. 2006) and social and political responses to crises that create more rigid, more centralized, less fit, and therefore more vulnerable governance systems. How to avoid such crisis-triggered destructive feedback among ecological, economic, and political systems should be a major and urgent research area concerning the challenges of global environmental change.

Governance Challenges Posed by Cascading Effects

Suggestions to promote network-based governance (Kickert, Klijn, and Koppenjan 1997) or adaptive governance (Dietz, Ostrom, and Stern 2003; Folke et al. 2005; Lebel et al. 2006; Olsson et al. 2006) provide fruitful starting points in dealing with the complex behavior of socioecological systems. Yet relying too heavily on the powers of self-organized social networks to cope with the dynamic behavior of interlinked socioecological systems might, under certain circumstances, lead to serious governance failure.

Cases of catchment/river basin management and related proposals for governance based on "bioregionalism" illustrate this point. Natural resource management scholars (M. McGinnis, Woolley, and Gamman 1999; Lundqvist 2004) have widely acknowledged the success in overcoming the misfit between "natural" hydrological boundaries and institutions resulting from promoting catchment-based management and planning. Yet the usefulness of the institutions and networks involved in catchment- or region-based governance is nonetheless likely to be drastically reduced by shocks to the water system resulting, for example, from extreme events induced by global environmental change (Steffen et al. 2004; Munich Re 2006). In such cases the risk of triggering crisis, cascade effects, and nonlinear behavior in social, ecological, or economic domains stems from conditions beyond the spatial and time scales of the river basin or bioregion.

Kinzig and others (2006) provide a number of illustrations of the governance challenges posed by cascade effects or the possibility of causing such effects. Projections of social, economic, and ecological conditions in the Australian wheat belt reveal a number of interacting thresholds. Abrupt shifts from sufficient soil humidity to saline soils and from freshwater to saline ecosystems could render agriculture nonviable at a regional scale. This in turn might trigger migration, unemployment, and weakened social capital. Additional examples of the tight coupling and unexpected large-scale changes and cascade effects embedded in socioecological systems are detailed in two studies: Michael H. Glantz's research on the El Niño–Southern Oscillation shows how this phenomenon triggered droughts and floods that cascaded through a number of domains and indirectly led to the massive soybean plantations in Brazil (Glantz 1990); Pascual and colleagues (2000) describe cholera outbreaks related to El Niño–Southern Oscillation in Latin America and southern Asia that had serious health and livelihood implications.

This sort of misfit between existing governance and biophysical systems produces effects on coupled social, ecological, and economic systems that can be devastating for ecosystems and livelihoods. It becomes necessary to uncover ways to overcome limitations that may be embedded in network-based governance such as adaptive governance. As discussed, it should be recognized that network-based governance relies heavily on social coordination and control, collective sanctions, and reputations rather than on recourse to law and authority. The "complicated dance of mutual adjustment and communication" (Jones, Hesterly, and Borgatti 1997, 916) among social actors is based on the possibility of a number of things: repeated interactions (such as those provided by geographical proximity); restricting the exchange of actors in the network (to reduce coordination costs); and the development of shared understandings, routines, and conventions (to be able to cope with change and resolve complex tasks) (Larson 1992; Jones, Hesterly, and Borgatti 1997; Ostrom 2005).

But although they underlie network-based governance, these social mechanisms also highlight its limitations. As social and ecological processes propagate across scales, the problem-solving capacity of network governance will be highly limited when a quick response requires collective action and institution building at scales and in policy arenas other than those targeted by participants. Time to form shared understandings among actors and a "history of play" would be critically lacking (Ahn et al. 2001) because of limited earlier encounters. The possibility of applying collective sanctions in this sort of situation would be limited.

The implications should not be underestimated. Major drivers of change (e.g., climate change, continued decline in ecosystem services, changes in the dynamics of the Earth system) will trigger unexpected effects at spatial and timescales that could extend considerably beyond the problem-solving capacity of existing governance systems, network based or not. The social and ecological effects of events like Hurricane Katrina, the spread of pandemic diseases, the cross-national, cross-system challenges identified by Steffen and others (2004), and the type of large-scale unpredictable effects that could come out of major interconnected climatic and biophysical systems (S. Schneider 2004) clearly surpass the collective-action capacity of institutions and social actors, including institutions, at a wide range of levels on at least time, spatial, and administrative scales.

This does not imply a recommendation of "man-on-horseback" solutions based on the reemergence of centralized one-size-fits-all or command-and-control steering. It does, however, point to the need to discover whether it is possible, and if so how, to maneuver "networks of networks" of societal actors, including institutions, in a way that ensures avoidance of thresholds, prevention or mitigation of cascade effects, and a response and postcrisis reorganization capacity.

There are obvious normative implications. The gradual shift from hierarchically organized systems that govern by means of law and order to more fragmented systems that govern through self-regulated networks gives reason to explore this change in the context of democracy theory. As stated by Sorensen (2002), there is a need to reinterpret and reformulate the basic concepts of liberal democracy, such as "the people," "representation," and "politics," to make them more useful as guidelines for democracy in political systems characterized by network governance (Hirst 2000; Held 2004).

Steering Networks of Networks?

Can social and policy networks really be steered and coordinated temporarily and swiftly enough to cope with the nonlinear behavior of biophysical systems? What is meant by steering or directly influencing networks of networks is not conventional approaches to cross-sectoral (e.g., Lundqvist 2004; Krott and Hasanagas 2006) or transnational policy coordination (e.g., M. Hoel 1997). This sort of coordination seldom acknowledges the dynamic nonlinear behavior of complex socioecological systems but instead occurs in order to implement defined targets, say, a percentage reduction of some pollutant or the application of voluntary agreements or ecolabels (compare Jordan, Wurzel, and Zito 2005). Nor does such coordination refer to the creation of global monitoring or assessment programs (Young 2002d) or a "World Environmental Organization" (Biermann 2002c).

Instead, what is proposed is the temporary coordination of institutional interplay among existing social and policy networks in various policy arenas, such as water, security, land, health, or environment, to provide fast joint response to abrupt changes in biophysical systems that cascade through socioecological systems as well as time and spatial scales. The aim here is not the creation of new bureaucratic organizations, but rather the development of a capacity to utilize existing, or to compensate for nonexistent or maladaptive, social networks and institutions in diverse policy fields.

Although this might seem an impossible task, researchers analyzing the features of network-based governance have identified a number of network management strategies (Kickert, Klijn, and Koppenjan 1997). The strategies range from promoting mutual adjustment by negotiation and consultation to more direct interventions such as restructuring relations or the "selective activation" of networks (Kickert and Koppenjan 1997; Klinj and Koppenjan 2004). These management approaches are worth exploring in trying to match institutions and wider governance systems with biophysical systems containing the risk of devastating cascading effects.

As discussed earlier, leadership and bridging organizations also hold the potential to help address the possibility and occurrence of cascading effects, which bring particular exigencies: the coordinating challenges posed by fast processes on a temporal scale and large ones on spatial scales, the difficulties of managing a multinetwork landscape in terms of legitimacy and availability of resources, and the long-lasting ecological and social impacts of this management. As a result, response to cascade effects is likely to require heavy involvement of central state actors. State actors in stable democracies are likely to be the only actors in governance with the authority, legitimacy, and resources required to coordinate networks of networks in the interest of ensuring a good fit to biophysical systems.

First, as argued by Hirst (2000, 31), the state is the only actor able to distribute powers and responsibilities among itself, regional and local governments, and civil society. Second, the nation-state remains the main institution of democratic legitimacy that most citizens understand and are willing to accept. Effective democratic states can thus represent their population more credibly than any other body. Third, national governments in stable democracies have strong legitimacy with other states and political entities that take the decisions and commitments of stable, democratic governments as reliable. Thus, the external commitments of such governments can provide legitimacy for supranational majorities, quasi polities, and interstate agreements (Hirst 2000, 31; see also Lundqvist 2001; Pierre and Peters 2005). It follows that the state in stable democracies stands out as the only actor potentially capable of steering networks into maintaining high adaptability to changing circumstances and the capacity to promote collective action through binding agreements regarding long-term change (see March and Olsen 2006).

Research on adaptive governance of biophysical systems shows that the management of ecosystems and landscapes is often difficult to design and implement and therefore difficult to subject to planning and control by a central organization, such as a national government (Folke et al. 2005). The state has an important role to play in the governance of biophysical systems (Hirst and Thompson 1995; Lundqvist 2001), but its role may change from authoritative allocation "from above" to the role of "activator" (Eising and Kohler-Koch 2000). A challenge in this context is defining the boundary of participation. Different types of misfits in table 5.1 might require a plethora of organizational options and different patterns of interaction among actors at multiple levels. This means that the "boundary" has to be defined and actors mobilized in relation to the misfit type to be addressed. The activator has to have the capacity to facilitate the emergence of such policy networks. An example is the Mediterranean Action Plan, which was produced by a group of scientists, government experts, and NGO representatives (P. Haas 1992b).

Instead of superimposing ready-to-use plans for ecosystem management on local contexts, the role of central authorities and agencies could hence be to legislate to enable self-organization processes, provide funding, and create arenas for collaborative learning (Berkes 2002; Olsson, Folke, and Berkes 2004; Hahn et al. 2006). Folke, Colding, and Berkes (2003) refer to such an activator role as "framed creativity" of self-organization processes. Such learning processes require mechanisms for aggregating knowledge claims and interests among multiple actors. For ecosystem management there are several tools that can fill this function, for example, stakeholder dialogue and collaboration (Wondolleck and Yaffee 2000; Stubbs and Lemon 2001) and companion modeling (Trebuil et al. 2002). Other examples involve more ad hoc initiatives like the Great Barrier Reef Marine Park Authority in Australia, which held hundreds of community information sessions in regional and local community centers along the northeast coast to get stakeholders' input on a new zoning plan for the Great Barrier Reef (Thompson et al. 2004).

A similar argument has been raised for the emergence of different emissions-trading credits for carbon dioxide (CO_2) under the framework provided by the Kyoto Protocol. As argued by David Victor and colleagues, the fact that six parallel trading systems have emerged from the "bottom up" as the result of collaboration between state and private actors provides an effective way not only to decrease emissions but also to promote innovation and flexibility for changing circumstances. Self-organizing and diverse schemes each provide a "laboratory" with its own procedures, stringency, and prices. This makes it possible for policy makers to learn from successes and unworkability, with the contingency capacity to tap into alternative schemes when needed (Victor and House 2004).

Can There Ever Be a Fit?

Scientific consensus now holds that Earth systems have moved well outside the range of natural variability exhibited over at least the past half-million years. As stated clearly by Will Steffen and colleagues (2004), the nature of changes now occurring simultaneously, their magnitude, and their rates of change are extraordinary. But the sociopolitical landscape also displays a number of radical shifts to more decentralized governance more coupled to multilevel and multisectoral institutional arrangements with more complex decision-making structures. How well do these trends match? Is the fit between biophysical systems and institutions increasing, and for what type of environmental and resource problems? The answers depend on whether the increasing diversity and complexity in governance correlate with improved learning processes and an increased understanding of the dynamic behavior of socioecological systems, the encouragement of diversity and experimentation, and a capacity to mobilize collective action before critical thresholds are reached and/or in a way that does not trigger cascade effects or that can mitigate cascades already in motion. Good fit between governance and biophysical systems requires a government structure nested across levels of administration and with an adaptive capacity as suggested in research on multilevel environmental governance (e.g., Winter 2006). Essential also is a thorough understanding of the relevant ecological processes that operate across temporal and spatial scales (Gunderson and Holling 2002; Gunderson and Pritchard 2002).

Whether there can ever be a perfect fit between governance and biophysical systems is an important and difficult question. Does the uncertain, complex, multilevel, and interconnected nature of biophysical and social systems actually make it impossible for decision makers to design fully effective institutional arrangements and wider governance for the environment and natural resources?

The limits of institutional design are well known (see Young, chapter 4 in this volume). They stem partly from "institutional stickiness," path dependence, the lack of incentives for political actors to think long-term, and the ambiguity and unpredictability of institutional effects (Knight and North 1997; Pierson 2000). As noted by Paul Pierson (2000, 483), "... social processes involving large numbers of actors in densely institutionalized societies will almost always generate elaborate feedback loops and significant interaction effects which decision makers cannot hope to fully anticipate." In addition, even if the collective benefits of institutions are common knowledge, the most fundamental observed results of "rational choice" have given good and sufficient reason to expect dysfunctional results from rational individual choices (Sandler 2004).

As an example, Joyeeta Gupta (chapter 7 in this volume) provides an illuminating discussion of the politics behind decisions about the "appropriate" scale of institutional solutions to environmental problems. It seems clear that even if the ecological research community can come to consensus on the scale of the biophysical processes that maintain the functions and resilience of socioecological systems, the actual design of institutional solutions remains up for political grabs (see Young 1989b, 1994a).

The possibility this raises of destructive strategic behavior among social actors triggered by uncertainty and the complex interactions in socioecological systems adds another layer of challenges. Although some argue that uncertainty facilitates efforts to reach agreements in international regimes (Young 1989b, 361–62), distributive conflicts—conflicts over the allocation of resources—can arise and intensify over issues marked by considerable scientific, technical, economic, or environmental uncertainty. High levels of uncertainty can make benefit calculations associated with agreements difficult, which in turn can lead to disagreement about implementation based on such calculations. More importantly, actors uncertain about the future may simply discount it and focus on short-term gains or resist establishing any agreement that could potentially disfavor them, which usually increases conflict (Galaz 2006).

There is an additional puzzle worth highlighting: do some strategies aiming to cope with one type of misfit counteract efforts to cope with another? Do, for example, attempts to overcome spatial misfits add other misfits related to time, cascades, or thresholds to existing governance structures? Unfortunately there is no systematic research on this issue to allow for an informed answer. Yet it should be clear that simple blueprint solutions to misfits are difficult to design. Socioecological systems are highly dynamic multilevel systems that embed periods of both incremental and abrupt change, considerable uncertainty, and changes at multiple levels and speeds. Hence there is not one solution for one misfit. The challenge for governance is to allow for a diversity of solutions to all sorts of socioecological change.

Knowledge Production and the Problem of Fit

The conceptual issues that emerge from the problem of fit evidently hold implications for the evolution of knowledge-producing systems capable of informing and shaping well-matched solutions to difficulties arising in biophysical systems.

The shift in perspective from viewing ecosystems and sociocultural systems separately to consideration of one largely integrated system brings about a need for institutional reforms. The dominant logic and standpoint in natural science regarding means and goals of research contrast with those in the social sciences and humanities. The difference in stance has consequences for a shared socioecological perspective that may relate to the issue of context-dependent research objects (Svedin 1991) and to the tension between generality and partiality, or micro-macro relations. These issues present yet another institutional challenge concerning not only conceptual internal interdisciplinary challenges, but also the way research activities are organized to serve the new broader perspectives. This issue affects university organization, extramural platform building, interactions among universities and research institutes (with or without connections to industry), and industrial endeavors.

A second implication for knowledge production relating to institutional fit involves the concept of systems as objects of research. Comparatively new understanding sees phenomena as generated by the feeds backward and forward in complex interactions at various levels of uncertainty and predictability. This requires a new perception of the match needed in the interaction and integration of knowledge traditionally treated as specific to the different systems involved. In addition to structures to accommodate newly integrated research perspectives, new institutional arrangements are needed for knowledge production systems to support the strong integration (Rosen 1986) of thinking that used to consider systems as separate objects for knowledge production.

The process of knowledge production raises a third implication. As called for also in other chapters, knowledge stakeholders charged with informing the design of institutions need to come from both academia and practice. The differences between these groups require deliberately created processes that allocate time to certain tasks: a problem definition phase; a phase to devise and consolidate a strategy to gather and generate the knowledge needed; implementation of the resulting research program; a research consolidation phase, including consultations over results; and integration of feedback to produce a new round of knowledge production. Intricate new types of arrangements of the research process are needed for this approach.

A fourth consequence of fit-related concepts concerns the relationship between knowledge production per se and its connection to policy. Knowledge resulting from an iterative process among different types of actors working in different frameworks of logic and traditions requires connections to be established deliberately. This is sometimes referred to as the "bridging-the-gap issue." Major institutional challenges seem to arise in terms of suitably fitting the production of knowledge to legislative and administrative processes.

The increased recognition of the complexity of interconnected dynamics related to the problem of fit calls for policy making that connects to the knowledge production system in ways that make the normative aspects transparent. Other aspects involved in the need to address the process more than the product of knowledge include trust, democracy, and a broader cultural perspective, all of which have to be mobilized.

Possible Ways Forward: Concluding Remarks

Although the limitations of institutional design might seem to present overwhelming barriers to overcoming misfits between biophysical systems and institutions, it should be noted that windows of opportunity for change do open. Rigidity, veto points, and path dependence appear to be general characteristics of institutions, as do change and "punctuated equilibria" (Baumgartner and Jones 1991; True, Jones, and Baumgartner 1999). As described by Olsson and colleagues (2006), sometimes windows open because of exogenous shocks that can be used to enhance "fit" with biophysical system problems in specific regions with particular socioecological systems; Young points to the emergence of international regimes, such as that created for nuclear accidents after the 1986 Chernobyl disaster (Young 1989b, 372). The analytical and political test lies in identifying what circumstances, involving which exogenous shocks, will produce a "window of opportunity" in highly dense multilevel governance systems with multiple interacting actors.

During the preparation of this chapter in January of 2007, global environmental change issues such as climate change, extreme weather events, and the large-scale collapse of ecosystems were leading to media coverage in Sweden that was impossible to grasp because of its intensity. Not all amounts to the doom and gloom often portrayed in the public debate, however. We believe that there are indeed ways to cope with detrimental misfits between biophysical systems and governance, and that important insights, as outlined in this chapter and summarized below, have been reached in the past two decades that will prove critical in attempts to match institutions to both incremental and fast, often unpredicted, changes in socioecological systems:

1. Social and biophysical systems are not merely linked but interconnected. Institutions and policy prescriptions that fail to acknowledge this tight interconnection are likely not only to provide ill-founded advice

but also to steer societies onto undesirable pathways. An adaptive social system cannot fully compensate for ecological illiteracy, nor can an environmental policy or regime be effective without an understanding of the larger and dynamic social, economic, and political context.

2. Possible consequences of the problem of fit should not be underestimated. Changes in biophysical and social systems interact in poorly understood ways, creating the potential for major unexpected phenomena and "tipping points" in both small- and large-scale biophysical systems. Examples include practically irreversible shifts to degraded states in ecosystems such as coral reefs, freshwater resources, coastal seas, forest systems, savanna and grasslands, and the climate system.

3. Time is a fundamental aspect of the problem of fit. The question is not only how well governance can cope with incremental change and uncertainty, but also whether collective action can be achieved fast enough to avoid abrupt, irreversible shifts (threshold behavior) or to buffer cascading effects under high scientific and social uncertainty.

4. Governance systems are just as dynamic as socioecological systems. Turbulent times and perceived or real crises may justify a temporary deviation from adaptive governance approaches to more top-down, centralized, and vulnerable governance models. This contingency will become more likely if present global trends toward denser and "messier" multilevel governance systems result in actual or perceived reduction in governability of turbulent biophysical situations.

5. The promotion of multilevel governance and participatory approaches in environmental regimes does not guarantee an enhanced fit between ecosystem dynamics and governance. It is the quality of interaction that matters—how learning about ecosystem processes is stimulated; how different interests are bridged and common goals worked out; and how polycentric institutions are used to ensure political, legal, and financial support.

6. Once triggered, cascading effects pose a serious governance challenge because of the critical lack of time to respond and because of their spatial and cross-system character. Whether and how "networks of networks," using refined and deliberate institutional interplay and other interaction among other social actors, can be steered to buffer the impacts of cascades is a critical issue for the future.

7. The need to adapt knowledge production systems in accordance with the preceding observations is of great importance.

The fit between biophysical systems and environmental and resource regimes can be enhanced, but not without attention to the larger governance context and the dynamics of socioecological systems. It is essential to achieve a better grasp of the mechanisms behind different types of institutional misfits and to find governance solutions that build the capacity to harness these mechanisms in a highly dynamic and interconnected social, political, and ecological world in order to prepare for the challenges of an uncertain future.

Acknowledgments

The authors wish to acknowledge the comments and feedback from members of the IDGEC science community; workshop participants at the synthesis conference in Bali, Indonesia, December 2006; and three anonymous reviewers. We would also like to thank Gary Kofinas, Gail Osherenko, Will Steffen, Oran Young, Fikret Berkes, Andreas Duit, and members of the Natural Resource Management group at the Department of Systems Ecology (Stockholm University) for very helpful comments on an earlier draft of the chapter. Support from the Swedish Research Council for Environment, Agricultural Sciences and Spatial Planning (Formas) and the Stockholm Resilience Centre at Stockholm University is acknowledged. We are grateful in addition to Christine Clifstock for figure 5.1, "Interconnected socioecological system."

6

Interplay: Exploring Institutional Interaction

Thomas Gehring and Sebastian Oberthür

Introduction

Since the development of the Institutional Dimensions of Global Environmental Change (IDGEC) Science Plan in 1998 (Young et al. 1999/2005), institutional interaction has become an important subject of inquiry. The Science Plan put institutional interaction on the agenda of global change research when only a handful of scholars had raised the general issue. Their work drew attention to the risk of "treaty congestion" (Brown Weiss 1993, 679) and to an increasing "regime density" (Young 1996, 1) in the international system. Today it is widely recognized that "the effectiveness of specific institutions often depends not only on their own features but also on their interactions with other institutions" (Young et al. 1999/2005, 60). Many environmental issue areas are cogoverned by several international institutions with governance also involving institutions at lower levels of societal and administrative organization (regional, national, local) (Young 2002b, 83–138).[1]

Although research on institutional interaction is closely related to the study of the effectiveness of international institutions, it takes a distinct perspective and transcends the focus on individual institutions. Institutional interaction is part of the broader consequences of international institutions occurring beyond their own domains (Underdal and Young 2004). Exploration of such interaction supplements the traditional inquiry into the establishment, development, and effectiveness of individual international institutions. Focus turns to the relationship among institutions, however, whereas traditional institutional research addresses the relationship between actors and institutions.

We have made important headway in knowledge about institutional interaction since the inception of IDGEC. The IDGEC Science Plan identified three areas particularly worthy of research: the role of politics and political decision making and their relationship to functional linkages among different issue areas; specific types of interaction especially with respect to their significance for the performance of the institutions involved; and the exploration and characteristics of interaction as they create synergy or disruption among the institutions involved (Young et al. 1999/2005, 64–65). We show that through a huge expansion of both conceptual and empirical research, understanding especially of the second and third research areas has improved considerably. Although quantifying IDGEC's contribution to progress would prove elusive, IDGEC has without doubt provided an important focal point and inspiration for research on institutional interaction. Not least, it has provided an important forum for the coordination of research efforts and for the exchange of research results.

Our discussion of institutional interaction starts with a review of the empirical progress made as a result of the study of horizontal interaction among international institutions. Subsequently we examine the theoretical development and argue that we have made significant progress toward developing a theory of institutional interaction through the identification of a limited number of relevant causal mechanisms and ideal types. Next we introduce four principal strategies that have been employed in the exploration of institutional interaction. An analysis of the implications of institutional interaction for our understanding of international institutions and global environmental governance follows. The penultimate section explores the progress made in the specific research area of vertical interaction, which has largely developed separately from that of horizontal interaction. Finally, attention is turned to identifying a number of promising avenues for research on institutional interaction.

The Growth of Empirical Analyses

The number of empirical analyses of institutional interaction by both social scientists and lawyers has grown tremendously over the past decade. This work has confirmed the importance, ubiquity, and diversity of institutional interaction. Interinstitutional influence significantly affects the development and performance of virtually all institutions. Generally, the empirical research has focused on a limited number of "hot spots." A large potential exists for broadening the overall empirical coverage. Here we review progress in the most prominent areas of research.

The World Trade Organization and Multilateral Environmental Agreements

Trade-environment interactions are one of the "oldest" areas of relevant scientific inquiry. A number of trade-related multilateral environmental agreements (MEAs) have been found to interact with the World Trade Organization (WTO). MEAs concern, on the one hand, the regulation of international trade, such as the Convention on International Trade in Endangered Species of Wild Fauna and Flora (CITES), the Basel Convention on the Control of Transboundary Movement of Hazardous Wastes and Their Disposal, the Rotterdam Convention on the Prior Informed Consent Procedure for Certain Hazardous Chemicals and Pesticides in International Trade, and the Cartagena Protocol on Biosafety. On the other hand, MEAs, such as various fisheries agreements and the Montreal Protocol on Substances That Deplete the Ozone Layer, employ trade restrictions as an enforcement measure (e.g., Brack 2002; Eckersley 2004; Palmer, Chaytor, and Werksman 2006). Driven by the expansion of the world trade regime to cover, among other things, intellectual property rights and sanitary and phytosanitary measures, and by the emergence of further MEAs, the scope of trade-environment interactions has also expanded (e.g., Rosendal 2001a, 2006; Andersen 2002; Oberthür and Gehring 2006c; Chambers, Kim, and Young 2007).

Studies by social scientists and lawyers alike have highlighted the potential for conflict between the WTO and trade-related MEAs and have identified potential solutions. Contributions have especially drawn attention to the ways in which the WTO, backed by its comparatively strong dispute settlement mechanism, works against effective global environmental governance. The existing obligations under the WTO "chill" negotiations on MEAs because they constitute obstacles to agreement on environmental trade restrictions or limit the effectiveness of such restrictions (Brack 2002; Eckersley 2004). WTO obligations also undermine the effective implementation of MEAs by protecting free trade in goods irrespective of the environmental consequences of the underlying production processes. The identification of the conflicting areas has led to the analysis of various potential solutions, including mechanisms available in international law (Pauwelyn 2003) and options for institutional reform of the WTO (Tarasofsky 1997; Biermann 2001b).

More recent studies have investigated in more detail the response of MEAs to the influence of the WTO. This has led to the insight that MEAs are not as weak in this conflict as they might appear at first glance. Trade-environment interactions are not a one-way street because MEAs have proved surprisingly robust in influencing the WTO. Despite the chilling effect of the WTO, more than twenty MEAs comprise trade measures to date. Their proponents have found, and used, the room for maneuver to adapt to the WTO requirements while still pursuing their objectives with trade measures. Among other things this has led to specific efforts to avoid discrimination against nonparties (Palmer, Chaytor, and Werksman 2006). The introduction of trade-restrictive measures adapted in this way has in turn restricted the WTO's regulatory scope and authority (e.g., Oberthür and Gehring 2006b) and has triggered adaptations on the side of the WTO to allow for resulting multilateral trade measures. This has produced increasing acceptance of appropriately designed MEA trade measures as reflected in the interpretation of the WTO regulations by the WTO Appellate Body and in the proceedings of the WTO Committee on Trade and Environment. As a result, no dispute concerning the implementation of an MEA has yet been brought before the dispute-settlement mechanism of the WTO (Charnovitz 1998; Palmer, Chaytor, and Werksman 2006, 187).

Overall these results indicate that the interaction between the WTO and MEAs is more balanced than some early analyses might have suggested. An increasing number of studies during the past decade have highlighted the achievements of MEAs in shaping the balance between trade and environment. The emerging picture is one of an increasingly institutionalized (and thus recognized) division of competences and labor between MEAs and the WTO (Gehring 2007). Certainly the current balance may not be sufficient or satisfactory, and tensions may worsen in the future based on the persisting societal conflict between free trade and environmental objectives. However, the latent interinstitutional conflict between the WTO and MEAs highlighted in many early analyses appears to have been managed relatively successfully so far, as the conflict has not become acute. If this observation can be further confirmed, it would provide an indication that the current decentralized management of institutional interaction has been more successful than traditionally assumed (see "Implications for Policy Making," below).

Climate Governance

The growing literature on institutional interaction in climate governance illustrates the particular multi-institutional nature of this governance area. The international climate change regime that is based on the UN

Framework Convention on Climate Change and its Kyoto Protocol has an enormous scope. As a result, it overlaps and interacts with a multitude of other issue areas and institutions in a variety of ways. In addition to the multifaceted and multi-institutional nature of international climate governance, the paramount importance of climate change on the international (environmental) agenda has contributed to the emergence of a rich literature on the wide-ranging interactions with various other environmental institutions and with institutions not primarily environment oriented.

A number of studies that have explored interactions among the international climate change regime and other MEAs have in particular highlighted the potential hegemony of climate governance over other environmental concerns. The objective of maximizing carbon uptake by monocultural forest plantations may, reinforced by the economic incentives built into the Kyoto Protocol, defeat the competing objective of preserving natural biodiversity-rich ecosystems under the Convention on Biological Diversity (Pontecorvo 1999; Jacquemont and Caparrós 2002). The climate change regime drove the adoption, in 2006, of an amendment of the London dumping convention that allows carbon sequestration in deep-sea deposits (International Maritime Organization 2006). Similarly, activities under the Kyoto Protocol's Clean Development Mechanism (CDM), which helps fund climate protection projects in developing countries, have been found potentially to clash with efforts to phase out ozone-depleting substances under the Montreal Protocol to protect the ozone layer (L. Schneider, Graichen, and Matz 2005). At the same time the Montreal Protocol has itself affected the Kyoto Protocol in various ways. On the positive side, the Montreal Protocol has informed the design of several aspects of the Kyoto Protocol and has contributed to climate protection by phasing out ozone-depleting substances (such as chlorofluorocarbons, CFCs) that are also powerful greenhouse gases. On the negative side it has led to a growing consumption of certain fluorinated greenhouse gases regulated under the Kyoto Protocol (Oberthür 2001). Interactions with further MEAs, such as the Convention to Combat Desertification and the Ramsar Convention on Wetlands, have been identified but not analyzed in detail (Oberthür 2006; van Asselt, Biermann, and Gupta 2004).

With respect to nonenvironmental institutions, most analyses have addressed interactions with economic institutions and, in particular, the WTO. In line with the traditional trade-environment debate, the WTO compatibility of multilateral or unilateral trade measures as a means for climate protection has been explored (e.g., Charnovitz 2003; Biermann and Brohm 2005). In addition, the market mechanisms of the Kyoto Protocol, most notably emissions trading, provide a particular angle for the trade-environment debate. In this context the question arises whether and to what extent international trading rules apply to trading in emission units created by the climate change regime. Furthermore, the relevance of international trade and investment rules and financial institutions has become an issue, particularly with respect to the implementation of climate protection projects under the CDM and Joint Implementation schemes of the Kyoto Protocol (Chambers 1998, 2001). Beyond the core economic and financial institutions, the analysis of the interaction of the climate change regime with the International Civil Aviation Organization and the International Maritime Organization (IMO) in regard to greenhouse gas emissions from international transport has highlighted the difficulties that can arise from regulatory competition and a lack of coordination among international institutions (Oberthür 2003, 2006). Further interactions of the climate regime with nonenvironmental institutions, such as the World Health Organization, have received less attention (van Asselt, Biermann, and Gupta 2004).

Ocean Governance

Ocean governance is a third area that has attracted considerable scientific attention. The prominence of relevant research is first of all obvious from the aforementioned discussion of both the WTO/MEA interplay and institutional interaction in climate governance, because ocean-related issues play an important role in both areas (e.g., WTO and fisheries agreements; IMO and climate protection). In addition, studies have focused on various subsets of the large number of institutions that interact in manifold ways in this area of governance. The large number of studies exploring fisheries governance is particularly striking (e.g., Stokke 2001a; DeSombre 2005; Stokke and Coffey 2006).

Research has in particular focused on a number of pertinent issues. A first focus has been on the exploration of the interplay of various institutions in particular geographical areas of ocean governance. Related studies have shed light on the interplay of various functionally differentiated institutions in the governance of particular regions such as the North Sea (e.g., Skjærseth 2000, 2006), the Arctic (e.g., Stokke 2007; Stokke and Hønneland 2007), and Antarctica (e.g., Stokke and Vidas

1996). The aforementioned studies on regional areas of ocean governance have frequently also addressed the effects of the nesting of regional
arrangements or functionally specialized institutions (e.g., fisheries agreements) into broader global institutions, most importantly the UN Convention on the Law of the Sea (Vidas 2000a, 2000b) and the UN Fish
Stocks Agreement (e.g., Boyle 1999; Stokke 2000, 2001a). Yet another
important research area has been the governance of particularly vulnerable marine species such as whales. In this regard it has turned out that
the existence of numerous functionally specialized institutions creates
opportunities for forum shopping that might be exploited by interested
actors. For example, the protection of whales, usually pursued within
the International Whaling Commission, might also be addressed under
CITES (Gillespie 2002).

Other Areas of Empirical Research
Noteworthy are two particular contributions by legal scholars. First, they
have begun to investigate the relationship and mutual influence of various courts and quasi-judicial procedures (e.g., Schiffman 1999; Shany
2003). A recent dispute between Ireland and the United Kingdom concerning the UK MOX plant in Sellafield has, for example, been addressed by procedures under the UN Convention on the Law of the Sea,
the OSPAR Convention, and the European Court of Justice (Lavranos
2006). Formal rules on jurisdictional delimitation and more informal
mechanisms (e.g., regarding information exchange) that minimize the
risk of contradictory judgments and jurisdictional competition exist to
some extent and could be further advanced to tackle these issues. Second, legal scholars have analyzed the consequences that norm conflicts
may have in general for the system of international law as well as the
means that are available in international law to resolve such conflicts
(Pauwelyn 2003; Wolfrum and Matz 2003). The resulting legal analyses
have highlighted that existing constitutional rules of international law,
such as the *lex posterior* and the *lex specialis* rules reflected in the Vienna
Convention on the Law of Treaties, are insufficient. The resolution of
norm conflicts frequently has to resort to a case-by-case approach of
clarifying the situation. As one result, many international treaties in international environmental governance explicitly address the relationship
with other treaties (M. Axelrod 2006). Jurisdictional norm interpretation
has also played an important role, for example, with respect to managing
the tensions between the WTO and MEAs. In other cases a resolution

has to rely on the political rather than the jurisdictional process of norm development and interpretation.

Other areas of environmental governance with possible interaction effects have received far less scientific attention. Only rarely studies have touched upon aspects such as the regional-global interactions concerning the North-South transfer of hazardous waste (Meinke 2002) and have addressed European air pollution as an empirical field (Selin and Van-Deveer 2003). Given the fact that virtually all areas of environmental governance are influenced by several institutions, there is furthermore room for many more empirical analyses of institutional interaction to shed light, for instance, on the governance of chemicals or the protection of species and biodiversity. Even with respect to the WTO-MEA relationship, global climate governance, and ocean governance, there is an enormous scope for further interplay analyses. In none of these areas have existing studies yet provided a comprehensive picture of the problems and promises of interaction. Also, studies of large numbers of cases that could provide a basis for comparative analyses have so far remained rare. To our knowledge our own research is the only example of such a large-*n* study to date (Oberthür and Gehring 2006c), although some scholars have begun to investigate particular aspects of interaction by employing quantitative means (e.g., M. Axelrod 2006).

Synergy and Conflict

One of the most noteworthy results of recent empirical research concerns the relationship of synergy and conflict in the realm of institutional interaction. Whereas Keohane, Haas, and Levy (1993, 15–16) identified more interinstitutional synergy than they expected, early analyses of individual cases such as the relationship among the WTO and MEAs focused on conflict and supported the notion that institutional interaction is problematic. Evaluating 163 cases of environmentally relevant interaction, we found in our own study that synergy is, counter to frequent assumption, at least as common among international and European Union (EU) environmental governance institutions as disruption (Gehring and Oberthür 2006, 316–25). The majority of our cases of institutional interaction led to synergy, and only about a quarter resulted in clear disruption. Furthermore, disruption and conflict in most cases occur as unintended side effects rather than deliberate results. Undoubtedly conflict is not negligible and poses severe problems, especially in interaction among environmental and nonenvironmental regimes; however, synergy dominates overall. Hence, the larger-n study points to a selection bias toward the conflictive, more politically salient cases.

Moreover, collective action is taken much more frequently in response to disruptive than to synergistic interaction. Positive effects of institutional interaction are commonly "consumed" without further action, irrespective of the potential for further improvement that may exist. This phenomenon appears to be widespread (identified in about 30 percent of our cases). A potential for improvement where positive effects occur has been neglected much more frequently than in the case of negative (disruptive) outcomes. The higher salience so far of problematic cases of interaction may be explained by the fact that people generally react more strongly to the risk of losses entailed in conflict than to the advantage of additional benefits (Tversky and Kahnemann 1981, 1984) and by the presence of aggrieved actors struggling for change. This suggests that it may be worth investing effort to identify potential for improvement irrespective of whether the original effect of an interaction was synergistic or disruptive.

These empirical findings have important implications for current debates about the reform of international environmental governance. These debates have been widely based on the assumption that conflict is the prevailing feature of institutional interaction. Concerns about disruptive interaction (between MEAs and the WTO as well as among environmental regimes themselves), incoherence, and duplication of work have been important drivers of both calls for a World Environment Organization (WEO) (e.g., Biermann and Bauer 2005) and more cautious bottom-up proposals for strengthening coherence and environmental policy integration in global environmental governance (e.g., Chambers and Green 2005; Najam, Papa, and Taiyab 2006). The aforementioned empirical results require a review of the basis for discussion of synergy and disruption and specifically suggest the need for more emphasis on preserving and enhancing synergistic institutional interaction as compared to minimizing interinstitutional conflict.

Conceptual Progress: From Classification to Causal Mechanisms

The IDGEC project has facilitated a number of attempts to develop general research concepts. Sound concepts are a prerequisite for more systematic research on institutional interaction. Starting in the mid 1990s, the search for a reliable conceptual foundation for institutional interaction has moved from classification efforts to more general propositions about the driving forces of institutional interaction and the deductive identification of causal mechanisms, elucidating both the pathways through which influence can travel from one institution to another and the consequences of interaction.

Categories for the Classification of Institutional Interaction

The search for analytical concepts started with a number of categories for classification. These classifications are useful for a first-cut exploration of the field of institutional interaction and establish valuable distinctions. They do not, however, capture the forces driving interaction.

Preceding the IDGEC Science Plan, Young (1996) put forward four types of institutional interaction and began to explore their inherent dynamics. He observed that issue-specific regimes are usually *embedded* in overarching principles and practices, such as sovereignty, and that they trigger long-term processes of change in these overarching structures. Institutional *nesting* addresses instances of interaction in which specific arrangements are folded into broader institutional frameworks that deal with the same general issue area but are less detailed. An example is the nesting of the Multi-Fiber Agreement within the General Agreement on Tariffs and Trade (GATT)/WTO (Aggarwal 1983). In cases of institutional *clustering*, actors combine different governance arrangements in institutional packages even when there is no compelling functional need to do so, as occurred in the UN Convention on the Law of the Sea. Finally, *overlap* addresses linkages in which individual regimes formed for different purposes and largely without reference to one another intersect on a de facto basis, producing substantial impacts on each other in the process. Young drew attention to the fact that nesting and clustering are typically the result of intentional attempts to redesign the institutional landscape, whereas embeddedness and overlap reflect unintentional consequences of human action. In the preparatory stages of the Science Plan, King (1997) developed a taxonomy of different types of institutional interaction, which focused also on possible political responses to institutional interaction. Rosendal (2001a) conjectured, somewhat surprisingly, that interaction will create synergy, if the specific rules of the institutions involved are compatible, and conflict, if they prove to be incompatible, whereas the institutions' broader norms are less relevant. However, the development of general causal mechanisms of institutional interaction demonstrated later on that the broader norms reflecting the policy direction of two or more institutions can have a tremendous impact on the quality of effects.

The IDGEC Science Plan proposed to distinguish between horizontal and vertical interaction (Young et al. 1999/2005; Young 2002b, 83–138). Horizontal interaction occurs among institutions at the same level of social organization or the same point on the administrative scale. At the international level this kind of interaction originates from the high degree of fragmentation of the international system in which actors frequently choose to pursue their common interests by establishing new institutions rather than expanding existing ones. By contrast, vertical interaction addresses the influence of institutions across different levels of social organization or administration. For example, the institutional design of domestic political systems shapes state interests and thus exerts influence on the design of international and European institutional arrangements (Héritier 1999). And global or regional environmental governance requires an appropriate institutional underpinning at the national and local levels (see Galaz et al., chapter 5 in this volume).

Most importantly, the Science Plan put forward the distinction between political and functional linkages among institutions (Young et al. 1999/2005, 50; see also Young 2002b, 23). Juxtaposing political and functional linkages provides an initial idea of some fundamental forces driving institutional interaction, namely, deliberate political action and underlying properties of the governance targets for international institutions that escape human control. A functional linkage was conceived of as a "fact of life," "in the sense that the operation of one institution directly influences the effectiveness of another through some substantive connection of the activities involved" (Young et al. 1999/2005, 50). It would exist "when substantive problems that two or more institutions address are linked in biogeophysical or socioeconomic terms" (Young 2002b, 23; also 83–109). For example, action taken within the ozone regime on CFCs is immediately relevant for the climate change regime, because CFCs have ozone-depleting properties and are at the same time potent greenhouse gases. Political linkages, on the other hand, involve the deliberate design of the relationship between or among different institutions. They were believed to "arise when actors decide to consider two or more arrangements as parts of a larger institutional complex" (Young et al. 1999/2005, 50). For example, member states of the climate change regime assigned the operation of the financial mechanism of this institution to the Global Environment Facility, thus establishing a permanent working relationship between the two institutions (Yamin and Depledge 2004, chapter 10). The distinction between functional and political linkages adapts the concepts of functional and political spillover from neofunctionalist integration theory (Rosamond 2000, 59–68).

This approach, however, is burdened with considerable analytical difficulties (see also Stokke 2001a). It underspecifies the realm of institutional interaction, because not all instances of institutional interaction fit either type: unavoidable fact of life or totally deliberate political design. Consider that the difficult relationship between trade-restricting MEAs and the WTO is neither deliberately designed by the member states of either of the institutions involved, nor is it an unavoidable fact of life because it originates from intended political action. The distinction also overspecifies the realm of institutional interaction because the two categories do not denote mutually exclusive types. Young et al. (1999/ 2005, 53) take the protocols on SO_2, NO_X, and volatile organic compounds of the international regime on transboundary air pollution as an example of a functional linkage, even though all these protocols belong to one convention managed under the UN Economic Commission for Europe and are thus undoubtedly parts of a larger institutional complex.

In addition to functional and political linkages, other types of interaction can be identified if a number of key factors believed to be crucial for the identification of causal pathways are systematically varied (Gehring and Oberthür 2004, 253–67). These factors shed light on different facets of an incident of institutional interaction relating to the causes and consequences of regime interaction, the nature of the influence at work, and the possible policy responses. Interaction can take place not only because institutions are functionally or politically linked, but also because they comprise different memberships, so that interaction occurs, for example, between a regional and a global institution operating in the same issue area. Interaction patterns can be expected to differ profoundly depending on whether or not a regime can unilaterally affect the development of another regime without the consent, or even awareness, of the actors operating within the target regime. Moreover, political action in response to observed or anticipated interaction can occur within either or all institutions involved.

Altogether the classifications of interaction illustrate the wide variety of possible paths of inquiry and serve as useful initial distinctions to structure the field. The distinction between horizontal and vertical interaction is, like the distinction between synergistic and conflictual qualities of effect among institutions, now well established. Young's four classes of institutional interaction provide an analytical framework for more specific inquiries; however, they have not been employed to analyze theoretically the causal factors behind institutional interaction.

Causal Mechanisms of Institutional Interaction

A number of authors set out to investigate the forces that drive institutional interaction and to identify general pathways clarifying how the institutions involved are related to each other. These attempts have yielded insights into how and under what conditions an international institution can influence another institution. Pointing to factors that might be important for causal analysis, these insights constitute a promising foundation for the search for theoretical models that elucidate the causes and effects of interplay between or among institutions.

In a series of studies on international resource management, Stokke (2001a; see also 2000, 2001b) proposed a set of four causal pathways through which institutional interaction may influence the effectiveness of the regimes involved. These pathways are derived from the major theoretical approaches of international relations. Hence, "ideational" interaction (originally referred to as "diffusive" interaction) relates to "processes of learning" (Stokke 2001a, 10) and implies that the substantive or operational rules of one institution serve as models for those negotiating another regime. This may, for example, help understand the rapid spread of general normative principles such as sustainability, precaution, and ecosystem management. "Normative" interaction refers to situations where the substantive or operational norms of one institution either contradict or validate those of another institution (e.g., in the case of the relationship of the WTO and MEAs). "Utilitarian" interaction relates to situations where decisions taken within one institution alter the costs and benefits of options available in another institution. Interaction "management," finally, relates to the political management of interinstitutional influence, including the deliberate coordination of activities under separate institutions in order to avoid normative conflict or wasteful duplication of programmatic efforts.

Against this backdrop a group of European collaborators developed a number of theoretically derived models of causal mechanisms and more specific ideal types of interaction that demonstrate how influence can travel from one institution to another (Oberthür and Gehring 2006a).

These models provide an account of how given causes create observed effects (Schelling 1998). They presuppose that one institution (the source institution) exerts influence through a particular pathway on the normative development or effectiveness of another institution (the target institution). Causal mechanisms open the black box of the cause-effect relationship between or among the institutions involved (Coleman 1990, 1–23; Hedström and Swedberg 1998, 21–23) and provide a microfoundation for the analysis of institutional interaction (George and Bennett 2005, 135–45).

The causal mechanisms approach suggests that institutional interaction is driven by one of four mutually exclusive general causal mechanisms covering three levels of effectiveness of governance institutions: namely, *output*—collective knowledge or norms prescribing, proscribing, or permitting behavior; *outcome*—behavioral change of relevant actors; and *impact*—the ultimate target of governance (Underdal 2004, 34, and chapter 2 in this volume). Two causal mechanisms are located at the output level and exert influence on the decision-making process of the target institution. A third causal mechanism is located at the outcome level, involving changes of behavior of relevant actors, while the fourth causal mechanism occurs at the impact level. The latter two mechanisms do not modify decision making of the target institution but rather its effectiveness within its issue area. The four causal mechanisms are believed to cover the full range of fundamental rationales that may drive institutional interaction. More specific ideal types are needed, however, to derive hypotheses about the conditions under which institutional interaction is expected to occur and its consequences for environmental governance.

Cognitive Interaction   Institutional interaction can be driven by the power of knowledge and ideas. The causal mechanism of cognitive interaction is based purely on persuasion and may be conceived of as a particular form of interinstitutional learning (similarly Stokke 2001a, 10). If the rationality of actors is "bounded" because information-processing capacity is limited (Simon 1972; Keohane 1984, 100–115), or if relevant information is not entirely available, the actors will be prepared to adapt their preferences to new information (Checkel 1998; Risse 2000). The decision-making process of an international institution will be influenced if information, knowledge, and/or ideas (P. Haas 1992b) produced within the source institution modify the perception of decision makers operating within the target institution. For cognitive interaction to occur, the source institution must generate some new information, such as a report, revealing, for example, new scientific or technological insights or an institutional arrangement solving a particular regulatory problem, which is subsequently fed into the decision-making process of the target institution by an actor. The information must change the order of preferences of actors relevant to the target institution and in this way affects the collective negotiation process and the output of the target institution. Depending on whether an interaction was triggered intentionally or not, we can distinguish two ideal types of cognitive interaction.

If cognitive interaction is unintentionally triggered by the source institution, members of the target institution voluntarily use some aspect of the source institution as a policy model. For example, the compliance system under the Montreal Protocol on Substances that Deplete the Ozone Layer influenced the negotiations on the compliance system under the Kyoto Protocol on climate change because it provided a model of how to supervise implementation and deal with cases of possible non-compliance (Oberthür and Ott 1999, 215–22). This type of cognitive interaction can occur between any two institutions, because international institutions share a number of functional challenges related to monitoring, verification, enforcement, and decision making. Also, numerous types of actors may pick up the information or idea and feed it into the decision-making process of another institution. Learning from a policy model can generally be expected to strengthen the effectiveness of the target institution, because it presupposes that the members and subjects of the target institution collectively consider the model to be useful. Policy models, however, are frequently modified or adapted to ensure their fit with the particular needs of the target ("complex learning"; see E. Haas 1990). The policy-model type of interaction highlights how members of an institution can improve the effectiveness of their governance efforts through the cognitive interaction involved in learning from other institutions.

If cognitive interaction is intentionally triggered by the source institution, it takes the form of a request by the source institution for assistance from the target institution. For example, the World Customs Organization adapted its customs codes in response to a request by CITES, thus supporting the implementation and enforcement of the latter's trade restrictions (Lanchbery 2006). A request for assistance requires that the issue areas involved overlap, because adaptation by the target institution would otherwise be meaningless for the source institution. Moreover, it will usually be successful only if the requested adaptation is either beneficial for, or at least indifferent to, the effectiveness of the target institution. Members of an institution cannot be expected to act upon external requests that harm their own institution. Whereas a successful request for assistance will generally produce synergistic or at least neutral effects for the target institution, it is intended to create a positive feedback effect on the source institution. Intentional cognitive interaction enables an institution to draw on other institutions in order to enhance its own effectiveness, even if it cannot exert pressure on the target institution to adapt its rules. The result is an instrument for furthering effective international governance.

Interaction through Commitment   Normative commitments may also provide the power behind interaction based on the premise that international obligations create at least some binding force on those they address. For this form of interaction to occur, an institution must adopt a prescription or proscription that formally or informally commits its member states. Subsequently this commitment must affect the preferences and negotiating behavior of these actors in another institution, a target institution, in ways that influence that institution's collective decision-making process and output. For example, the WTO commitment not to discriminate against imported goods renders it more difficult for WTO members to adopt trade sanctions within MEAs that would reinforce the effectiveness of these institutions (Brack 2002). Activation of this causal mechanism requires that both memberships and issue areas overlap at least partially. Without overlapping memberships, no member state of the target institution would be committed to obligations established under the source institution. And without overlapping issue areas, commitments established under one institution could not redefine preferences related to issues dealt with under the other institution.

If the membership of one institution forms part of the membership of another institution, a formally independent institution is "nested" in another institution with similar objectives and governance instruments. Interaction between nested institutions constitutes a mechanism for policy diffusion within the same policy field and creates synergies among the institutions involved. It is typically easier to reach agreement within a smaller (e.g., regional) than in a larger (e.g., global) institution (Snidal 1994). States committed within the smaller institution may develop a common interest in transferring their obligations to the larger institution governing the same issue area. For example, the ban of trade in hazardous wastes was more easily reached in a number of regional agreements than in the global Basel Convention on the Control of Transboundary Movement of Hazardous Wastes and Their Disposal, but the latter was subsequently heavily influenced by the regional agreements governing the same issue area (Clapp 1994). Interaction between nested institutions provides opportunities for "forum shopping" (exploration by actors of opportunities offered by different institutions to pursue their own interests). Its underlying rationale suggests that it will largely support the effectiveness of the target institution and occasionally also of the source institution. The identical objectives of the institutions generate compatible priorities and render disruptive effects highly improbable, if not impossible.

If a group of states addresses the same issues within two institutions pursuing different objectives, interaction through commitment creates mutual disruption of the institutions involved and, therefore, a demand for the delimitation of jurisdictions. Typically, institutions with different objectives will appraise a policy measure differently, so that disputes about the appropriate regulation arise. Environmentally motivated trade restrictions may be appraised as undesirable obstacles to free trade or as desirable instruments supporting environmental cooperation. In situations of this type, the members of the institutions involved possess a general interest in some sort of separation of jurisdictions in order to avoid fruitless regulatory competition; however, conflicting preferences regarding the appropriate solution make it notoriously difficult to solve such problems. Jurisdictional delimitation cases pose the governance challenge of identifying measures honoring the basic objectives of both institutions involved. This does not necessarily require an overarching institutional structure but may be achieved through mutual adjustment of institutional structures or even through careful implementation of obligations by the addressees.

If a group of actors pursues the same objectives within institutions controlling different governance instruments, interaction through commitment will produce synergistic effects because it activates an additional means. Such interaction occurs in two stages. First, actors committed under one institution transfer an obligation to another institution. Second, incorporation of the transferred obligation must mobilize an additional governance instrument, such as a particular form of law or a specific enforcement or assistance mechanism that provides an additional incentive to implement the obligation. For example, political agreement achieved at the high-level International North Sea Conferences paved the way for the acceptance of identical obligations enshrined in hard law within the regime for the protection of the North-East Atlantic (OSPAR) (Skjærseth 2006). Such interaction will regularly raise the effectiveness of both institutions involved, because the additional governance instrument benefits the implementation of both institutions simultaneously.

Behavioral Interaction   Institutional interaction may also be based on the interconnectedness of behavior across the domains of institutions. Behavioral interaction will occur if behavioral changes triggered by the source institution become relevant for the implementation of the target institution. This form of interaction is located at the outcome level and affects the performance of an international institution within its own domain. Relevant states and/or nonstate actors must adapt their behavior in response to the output produced by the source institution. The behavioral changes must affect implementation behavior under the target institution in ways that are relevant for the target institution's effectiveness. If the Kyoto Protocol, for example, creates incentives to plant fast-growing trees in ways that encroach upon biodiversity, this undermines the performance of the Convention on Biological Diversity (Jacquemont and Caparrós 2002). Behavioral interaction requires that the issue areas governed by the institutions involved as well as the direct and indirect addressees of institutional obligations are close enough to matter to each other. It does not depend on a collective decision within the target institution, because it occurs as the aggregate result of the behavior of actors operating within the two issue areas involved.

Implications of behavioral interaction for global governance depend, again, on whether the institutions involved differ predominantly in their memberships, objectives, or governance instruments. If different (usually overlapping) groups of actors address a given set of issues within institutions with similar objectives, behavioral interaction will always create synergy. Because of the matching objectives, behavioral changes will automatically benefit both institutions. If a group of actors addresses a set of issues within two institutions that pursue different objectives, interaction will tend to result in disruption of the target institution, because behavioral changes triggered by the source institution are easily at odds with the objectives of the target institution and may thus undermine the latter's performance.

Impact-Level Interaction Institutional interaction may also rest on the interdependence of the ultimate governance targets of the institutions involved. In impact-level interaction the ultimate governance target of one institution, such as economic growth or the ozone layer, is directly influenced by side effects originating from the ultimate governance target of another institution. Consider a stylized example: as cod eat herring, successful protection of cod by one institution, resulting in a growing population of this species, will unintentionally decrease the population of herring protected by another institution. In contrast to behavioral interaction, interinstitutional influence in this case does not depend on any action within the target institution or its domain but rests on the "functional linkage" (Young 2002b, 23, 83–109) of the ultimate governance targets of the institutions involved at the impact level. It is increased population of cod, not human behavior, that leads directly to a decreasing population of herring. While impact-level interaction may rely on stable interdependencies of the biophysical environment, as with cod and herring, functional linkages may themselves be subject to possible long-term change. For example, economic growth promoted by the WTO and the resulting growth in international transport currently lead to increased emissions of greenhouse gases, thus undermining the effectiveness of the global climate regime. This kind of functional interdependence, however, might one day be overcome by technical progress or changes in production methods.

The value added by the general causal mechanisms and their subtypes is twofold. First, the models provide a promising foundation for the development of an elaborated theory of institutional interaction. They allow for the formulation of meaningful hypotheses about the preconditions for institutional interaction and in regard to the effects of interaction for global environmental governance. Second, they provide analytical tools for use in structured analysis of empirical interaction cases, which can help explain how influence travels from one institution to another as well as which groups of actors might be involved in this process. Such models, however, do not replace the empirical exploration of existing interaction cases. They do not relieve the researcher from establishing the causal relationship between the (potentially) interacting institutions and exploring alternative causal pathways. Moreover, they do not provide precise descriptions of all properties of relevant interaction cases. Being deductively derived, they cannot be empirically right or wrong (Snidal 1985). Like game-theoretic models, they reflect the relevant components of the different causal pathways that a case of interaction may follow and thereby assist the empirical analysis of real-world situations.

Principal Research Strategies on Institutional Interaction

Research on institutional interaction adopts different perspectives. The new field of inquiry has not yet produced one or more standard approaches. Meaningful studies on institutional interaction, like research on any other subject of the social sciences, have to be founded on some basic assumptions about the dependent and independent variables and their relationship. Choices made in this respect influence the research questions that can be pursued in a particular study.

Explicitly or implicitly, research design on institutional interaction has to be based on decisions about the role of actors and institutions. Systemic approaches address the causal relationship among institutions so that both the dependent and the independent variables are located at the macro level of institutions, rather than the micro level of actors. Many studies of institutional interaction, including many legal analyses of overlapping and conflicting jurisdictions, focus entirely on the systemic level and bracket the activities of actors. In contrast, actor-centered research strategies address actors either as the independent variable or the dependent variable, locating the other variable at the macro level of institutions. Relevant research may start from a given interest of one or more relevant actors and explore the opportunities to exploit institutional interaction as an instrument to pursue these interests effectively (forum shopping). Alternatively, it may focus on the undesired side effects of institutional interaction that actors must take into account when establishing or redesigning a given institution. The exploration of the effects originating from institutional interaction and regime complexes (Raustiala and Victor 2004) on the behavior of relevant states and non-state actors also reflects an actor-centered strategy.

Research on institutional interaction can also focus on different units of analysis. It may focus on specific dyadic cases of interinstitutional influence in which one institution affects the normative development or performance of another institution (Oberthür and Gehring 2006a, 26–31). This perspective may require the decomposition of complex interaction situations. Even a comparatively narrow interaction situation like the interplay between the WTO and MEAs with trade restrictions may turn out to be composed of several component cases running in different directions and passing through different causal mechanisms (Palmer, Chaytor, and Werksman 2006). Research, however, may also take as its unit of analysis the overall patterns emerging from complex interaction situations, which might involve several institutions and possibly many individual cases of interaction. It will then seek to develop an integrated view on a complex phenomenon like the relationship between MEAs and the WTO or the institutional setting affecting the Antarctic environment (Young 1996). This approach has therefore been called integrationist (Young 2007).

Squaring these two dimensions, we get four different research strategies. Each of them is particularly well suited to address certain research questions and ignore others. Table 6.1 illustrates the four strategies and indicates their core research question.

Table 6.1
Key research questions of different perspectives on institutional interaction

		Unit of analysis	
		Case of interaction	Complex interaction setting
Level of analysis	Systemic	I. How, and with what effects, does an international institution influence another international institution?	II. How, and with what effects, does an institutional interaction affect the institutional structure of the international system?
	Actor-centered	III. How can and do actors exploit opportunities arising from institutional interaction or avoid undesired interaction effects? How does institutional interaction frame policy choices of actors?	IV. How, and with what effects, do actors change the institutional structure of the international system through institutional interaction?

Inquiries located at the system level and focusing on one or more specific cases of interaction (cell I) address the core question of how, and with what effects, an international institution can and does influence another international institution. The focus is on institutional interaction effects rather than on actors' behavioral changes. The combination of a systemic perspective with a case-oriented approach is particularly well suited for rigorous analysis of the causal mechanisms and effects of specific incidents of institutional interaction. Causal analysis requires identifying a clear direction of causal influence running from one institution to another, which is difficult in complex situations in which the origins and targets of influence are not readily discernible or in which feedback effects occur. This research strategy has so far proved particularly popular and has supported significant theoretical development reflected in the determination of causal mechanisms and more specific ideal types driving cases of interaction (see "Causal Mechanisms of Institutional Interaction" above). Empirical studies of institutional interaction (as explored above in "The Growth of Empirical Analyses") have also (implicitly) employed this strategy. Likewise, studies analyzing the specific legal implications of one sectoral legal system for the *interpretation* of another one usually follow this research strategy (Wolfrum and Matz 2003).

Systemic inquiries exploring complex interaction settings (cell II) tackle the core question of how, and with what effects, institutional interaction affects the institutional structure of the international system. Because of the complexity of the empirical subject of inquiry, this research strategy will frequently start from empirical observation and description of complex settings or with a classification of interaction patterns. In contrast to case-specific research, it stays closer to the actual appearances of real-world interaction patterns, but it may be limited in its analytical grip on the forces generating the observed effects. Both conceptual work and empirical work employing this research strategy are still rare. The taxonomy of four different types of interaction put forward by Oran Young (1996; and see "Categories for the Classification of Institutional Interaction" above) and the analysis of the emerging division of labor between the WTO and MEAs with trade restrictions (Gehring 2007) provide tentative examples for this approach.

The study of specific cases of institutional interaction using an actor-centered approach (cell III) examines how interested actors can and do seek to exploit opportunities arising from institutional interaction or to avoid undesired interaction effects. In contrast to research falling into cells I and II, this strategy allows the application of existing theoretical and methodological tools for the analysis of collective-action problems to the issue of institutional interaction. Interaction effects are treated like any other effects originating from an international institution. This research strategy is particularly well suited for exploration of the ways in which actors deal strategically with expected or anticipated institutional interaction in specific situations and how they exploit related opportunities for forum shopping. For example, Skjærseth, Stokke, and Wettestad (2006) examined how actors interested in enhancing the effectiveness of North Sea pollution control established the North Sea Conferences to exert influence on the existing Oslo-Paris Commission. M. Axelrod (2006) investigated actions of interested actors to protect the WTO agreements from undesired interaction effects originating from the newly negotiated Cartagena Protocol on Biosafety. Likewise, studies assessing the options for improving an interaction situation generally follow this research strategy (e.g., Biermann 2001b; Oberthür 2001; L. Schneider, Graichen, and Matz 2005).

Actor-centered studies focusing on more complex interaction patterns (cell IV) seek to investigate how the efforts of actors to employ institutional interaction change the institutional structure of the international system. They reflect that all institutional structures originate from interdependent human action and affect human behavior. Studies following this research strategy, however, must bridge a particularly wide gap between actors and institutions. The institutional structures of the international system emerging from institutional interaction are only an indirect consequence of human action that feeds into institutional interaction. Thus, cell IV research almost inevitably includes aspects of cell III and cell I research. Raustiala and Victor (2004) partly adopted this strategy in their study on the regime complex for plant genetic resources when examining the overall implications of postnegotiation implementation decisions adopted within international institutions dealing with legal inconsistencies of the normative systems involved. Their study demonstrates that this research strategy may imply going beyond a traditional understanding of institutions as resulting from the rational design of actors attempting to realize a common interest. Expanding traditional research on the effectiveness of institutions and studies exploring the combined effects of institutional complexes on the behavior of relevant states and nonstate actors also belong to this research strategy (Andersen 2008).

The choice among these research strategies depends primarily on the particular research interest. Although the combination of two or even more strategies in a single project is not excluded, it renders the construction of a reliable research concept more ambitious. Unless the different components are convincingly integrated, conceptual broadness may restrict analytical and theoretical depth. At the same time the different strategies are neither mutually exclusive nor antipodes. For example, research focusing on the exploration of individual cases of interaction (cells I and III) may well provide a sound basis for the exploration of complex interaction settings (cells II and IV). Likewise, cell III research will usually include insights from cell I inquiries. The research strategies therefore may well be employed in complementary ways.

Implications for the Understanding of International Institutions and Global Environmental Governance

What are the implications of the progress made in knowledge about institutional interaction for the understanding of governance institutions? What insights can be derived for policy making?

Understanding International Institutions

The study of international governance institutions has been dominated by the collective-action approach. This approach focuses almost exclusively on formal international institutions (Keohane 1993) and their rational design against the backdrop of well-defined preferences and constellations of interests of relevant actors (Koremenos, Lipson, and Snidal 2001). These institutions fulfill auxiliary functions depending on the characteristics of the underlying socially problematic situation (Oye 1985). In prisoner's dilemma situations, for example, institutions serve to define what is collectively considered as cooperation and as defection to produce transparency about the cooperators' behavior, and—possibly—to organize sanctions in order to preclude free riding and stabilize cooperation (Martin 1993). The collective-action approach implies a top-down perspective where actors implement valid regime rules (unless free riding occurs). The research on the effectiveness of international environmental governance adopts a stimulus–response perspective (Miles et al. 2002).

By comparison, in the social practices perspective, institutions are seen as reflecting social expectations of appropriate behavior and as shaping actors' preferences and identities (Young 2002b, 31–32). Institutions constitute social practices that are not collectively decided upon, nor formally established, but produced, reproduced, and changed in a permanent interaction process of relevant actors (Wendt 1987). If actors behave according to existing practices, they will reproduce them. If actors deviate from these practices, they will contribute to their modification or breakdown. Hence, social practices reflect "spontaneous" institutions that emerge from action (Young 1982a), whereas formal institutions and their "rational design" constitute but one among several ways to change an established social practice.

Important aspects of institutional interaction can better be grasped analytically by the social practices approach to institutions. If the normative structure of one institution is significantly influenced by other institutions, it cannot simply be traced back to existing preferences of relevant actors and the resulting constellation of interests. Two of the causal mechanisms uncovered (see "Causal Mechanisms of Institutional Interaction" above)—namely, cognitive interaction and interaction through commitment—demonstrate how actors' preferences regarding issues dealt with by one institution can be affected by another institution. Similarly, Raustiala and Victor (2004, 296) have pointed out that power, interests, and ideas do not map directly onto institutional decisions because they are also shaped by other institutions. At a minimum, institutional interaction, in addition to exogenous interests, thus significantly affects and shapes the preferences of actors. Accordingly, preference formation cannot easily be separated from institutional analysis.

Institutional interaction also creates new institutional structures that are difficult to design rationally, because they evolve gradually from, and are continuously shaped and reshaped by, numerous decentralized interaction occurrences. Interaction may lead to a particular division of labor of the institutions involved or to the mutual reinforcement of their effectiveness, as an emergent effect that is not reflected in either of these institutions. Such interlocking structures (Underdal and Young 2004, 374–75) do not arise from collective bargaining or institutionalized decision making at the aggregate level. Whereas virtually all institutions in international environmental governance comprise their own permanent decision-making centers, if only in the form of a conference of the parties, no such decision-making bodies exist with respect to interaction between international institutions. Although the EU and domestic political systems possess unitary institutional frameworks that can address related issues, the international system lacks a similar capacity. To the extent that overarching institutions like the Vienna Convention on the Law of Treaties or the International Court of Justice exist, they play a limited role at best. Under these circumstances interaction emerges from, and is influenced by, decentralized decisions made within any of the institutions involved and the behavior of individual actors. Far from being designed, interaction thus evolves and is produced and reproduced through the practices of relevant actors.

If institutional interaction affects the implementation of obligations established under international institutions, it will modify the meaning of these obligations. The causal mechanism of behavioral interaction demonstrates how an institution can affect the effectiveness of another institution at the outcome level (see "Causal Mechanisms of Institutional Interaction" above). Even if the formal rules of the target institution remain unchanged, their effects and their meaning as reflected in the social practices of relevant actors change significantly. Similarly, Raustiala and Victor (2004, 302) suggest that interacting institutions may address legal inconsistencies by means of mutual adaptation during implementation. Whereas the collective-action approach assumes from a top-down perspective that actors implement fixed regime rules (unless free riding occurs), institutional interaction highlights that the social practices emerging in the implementation of one institution may also be shaped by other institutions. The top-down implementation perspective may thus provide a valuable first cut, but it does not encompass the effects of institutional interaction at the outcome level.

Implications for Policy Making

The progress of research on institutional interaction achieved so far has several implications for policy making. First, institutional interaction requires that policy making take into account the broader policy implications of particular governance projects. Research of the past decade has demonstrated the importance of interinstitutional effects at all three levels of effectiveness: output, outcome, and impact. It is now established that environmental governance is frequently the result of several institutions and that an institution often has implications for other institutions. Skillful policy making will have to consider the existence of several institutions cogoverning an issue area. Accordingly, the institutional environment of the institution in which a policy initiative is launched will most likely have repercussions for its prospects of success regarding acceptance by other actors and effective implementation. And vice versa: the assessment of the impact of a policy initiative on an institution should take into account "side effects" on and from other institutions.

While to some extent constraining policy making, institutional interaction offers a wealth of new opportunities. Since the normative development of an institution can be influenced not only from within that institution but also by other institutions, actors may engage in forum shopping (Gillespie 2002; Raustiala and Victor 2004, 299–300). To the extent that issue areas overlap, actors can choose the most suitable institution for a policy initiative. They can develop integrated strategies for the pursuit of their preferences that take into consideration the potential of the varying institutions affecting an issue area for both norm making and implementation. Interested actors might even establish a new institution with the sole purpose of influencing an existing one, as the North Sea riparian states did with the establishment of the International North Sea Conferences directed at strengthening the existing OSPAR Commission (Skjærseth 2006). Moreover, they may create "strategic inconsistency" (Raustiala and Victor 2004, 301), causing disruption of an unwanted institution or regulation in order to increase the pressure for its revision or cancellation.

The research results have important implications for discussion about the reform of international environmental governance and the political management of institutional interaction. This discussion has so far focused mainly on the potential for institutional coordination and integration at the international level, most importantly by establishing a WEO (Biermann and Bauer 2005; Chambers and Green 2005; Najam, Papa, and Taiyab 2006). Findings of research on institutional interaction challenge this debate in several ways.

First, synergy among institutions has been found to be at least as common as disruption (see "Synergy and Conflict" above). This finding contradicts the presumption of most contributions to the debate on reforming international environmental governance that institutional interaction might primarily constitute a problem because it creates interinstitutional conflict and tension. If this presumption is revised, both the rationale for reform proposals and the yardstick for assessing their effectiveness need to be adapted. In particular, institutional reform proposals will have to demonstrate that they can, in addition to mitigating conflict, preserve and enhance synergy among institutions.

Second, institutional interaction research suggests that the institutional fragmentation of international environmental governance may constitute a strength rather than a weakness. Institutions with large regulatory overlaps appear to create substantial added benefit if they employ complementary governance instruments, represent different memberships, or provide for significantly different decision-making procedures. What may at first sight appear as a "duplication of work" or "redundancy" arising from institutional fragmentation, which is commonly deplored by policy makers and in the relevant literature, is in fact frequently a sign of effective governance. Slight differences in the instruments or procedures employed or the memberships of the institutions can make two (or more) institutions contribute in complementary ways to effective governance, as is best illustrated in the ideal type of interaction activating an "additional means" (see "Causal Mechanisms of Institutional Interaction" above). Regulatory competition among different forums can help prevent institutional sclerosis and provide an important driver of overall progress. Before pursuing a reduction of seeming "duplication of work," for instance, through a WEO or through the clustering of functionally related institutions or elements of institutions in global environmental governance (Oberthür 2002; von Moltke 2005), policy makers and analysts would be well advised to check carefully the "hidden" added value of the current fragmented arrangements.

Third, research indicates that disruption among international institutions is mainly rooted in competing institutional objectives, as is apparent in the jurisdictional delimitation type of interaction through commitment and the corollary type of behavioral interaction (see "Causal Mechanisms of Institutional Interaction" above). Accordingly, reform proposals would have to show how they promise to mitigate and minimize interinstitutional disruption and to reconcile diverging objectives of the institutions involved. For example, building a unitary institutional framework in the form of a WEO does not as such promise to resolve the trade-off between the competing environmental objectives of climate change and the protection of biodiversity regarding forest management. It would also require further clarification of how a WEO or other reform proposals would help mediate trade-offs with nonenvironmental objectives pursued by institutions such as the WTO.

Finally, recent research results challenge the conventional wisdom of the hegemony of the WTO vis-à-vis MEAs. The jurisdictional delimitation type of interaction demonstrates that power is involved when it comes to defining the division of labor among institutions with competing objectives. Environmental institutions have proved remarkably strong in comparison with the WTO. Several environmental institutions have successfully created "strategic inconsistency" by regulating particular areas of international trade as such or employing trade measures as an enforcement tool. As a result they have limited the implications of the existing free-trade rules and have carved out certain areas of the regulatory authority of the WTO (see "The World Trade Organization and Multilateral Environmental Agreements" above).

Future research on institutional interaction holds the promise of further valuable input to policy debates. In particular, knowledge about effective interaction management has remained sharply limited to date. As research on institutional interaction advances, it could provide a more solid basis for exploring options for such management.

Vertical Interaction

Frequently environmental governance involves institutions located at different levels of social or administrative organization, most importantly the international, the national, and the local levels. This creates a vertical dimension of institutional interaction as identified in the IDGEC Science Plan (Young et al. 1999/2005) as well as in related publications (Young 2002b). Vertical interaction has been studied almost entirely separately from horizontal interaction, although this separation may be predominantly the result of research interests and scholarly discourses rather than theoretical considerations. The causal mechanisms discussed above may turn out to provide an instrument for the theoretical integration of the two perspectives.

Studies on the vertical interaction between the national and the local levels draw upon and expand the discussion on the preservation of the local commons. The "tragedy of local commons" (Ostrom et al. 2002; also Ostrom 1990) and the social problems of local communities trying to establish reliable institutional solutions for the management and preservation of commons such as water resources or common fishing grounds through self-organization have been studied for a long time. Case studies treat national measures such as the introduction of property rights that were found to interfere with local solutions as undesired external factors. The vertical-interaction perspective addresses such interference as interaction between local and national institutions (Young 2002c, 266–76). As in the case of horizontal interaction, vertical interaction can be disruptive or synergistic, and authors have been primarily preoccupied with cases of disruption, mainly of well-operating local institutions by national institutions. In many cases national political institutions resulting in centralization of decision making, nationalization of resources, increased participation in markets, and priority for development policies have indeed been found to affect established local institutions adversely and to lead to the degradation of the local commons that had been effectively preserved in the past (Lebel 2005). In the face also of the "tragedy of the commons," with its implication of incentives for free riding, local communities nevertheless may also benefit from support of institutions located at a higher level of social organization (Berkes 2006b). Intervention by national institutions is reported to strengthen or rejuvenate local-level institutions, for example, by state recognition of local institutions, development of enabling legislation, cultural revitalization, capacity building, and local institution building (see Berkes 2002, 296–300).

Although the literature has so far predominantly focused on the top-down influence of national on local institutions, vertical interaction conceptually covers a broader realm. It broadens the research agenda to encompass interinstitutional influences of all sorts across all levels of social and administrative organization. For example, national political systems may both benefit from and be harmed by regional or global institutions.

Vertical-interaction research is particularly related to the issue of scale (Gupta, chapter 7 in this volume) but should not be confused with it. Determining the appropriate level of institutional action stays central to the discussion of the appropriate "scaling" of an environmental problem (Young 2002b; Cash et al. 2006). The issue of scale raises concerns of effectiveness (at which level is a problem to be addressed to be solved effectively?) as well as power and interest (at which level do particular actors want it to be dealt with?). Although the lower levels of social organization may be closer to the environmental targets and the related human activities, effective solutions of many problems require cooperation at higher levels of social organization. In any event, scaling must not be conflated with vertical interaction. Even if the scaling up of an issue to a higher level of social and administrative organization will almost inevitably cause vertical interaction between or among institutions located at different levels, vertical interaction addresses the distinct issue of interinstitutional influence.

Institutionalized comanagement has been the preferred solution to conflictual interaction between national and local institutions identified in the literature. The primary solution observed by researchers for the management or mitigation of such conflicts involves comanagement initiatives with formal power sharing. Many comanagement arrangements, sometimes including stakeholder bodies, exist in the areas of fisheries, wildlife, protected areas, forests, and other resources in various parts of the world. They range from joint forest management in India to the implementation of aboriginal resource rights in the United States, Canada, New Zealand, and Australia (Berkes 2002, 301–7). From a more conceptual perspective, possible solutions that do not rely on comanagement have received less attention. These include the gradual separation of the jurisdictions of the institutions involved, their merger, or the dominance of one of the interacting institutions (Young 2006). It is not clear, however, whether, or under which conditions, the effects of these solutions are malign or benign for environmental protection.

Interactions between or among local and national institutions dominate the discussion. Vertical interaction at higher levels of social organization occurs particularly between the national and the international levels (Young 2002c, 276–83). Independently from the relatively new framework of vertical interaction, the bottom-up influence exerted by domestic political systems on the shape and development of international institutions has been addressed under the "cooperation under anarchy" heading (Keohane 1984; Oye 1985). This perspective holds that opportunities for cooperation depend on the constellation of interests of the actors involved. Although states are here conceptualized as unitary actors whose interests may be shaped by national-level institutions, they constitute group actors that are, in fact, themselves institutions. Research on policy making within the EU revealed that national administrations frequently seek to establish their own domestically institutionalized solutions within the higher-level institution (Héritier, Knill, and Mingers 1996). The influence of international institutions on national political systems and institutions had also been intensely discussed long before issues of interaction appeared on the agenda (Chayes and Chayes 1993; Cowles 2001). The implementation of international rules has been found to depend not least on the compatibility of international commitments with domestic institutions (see Galaz et al., chapter 5 in this volume). It follows that the concept of multilevel governance becomes an applicable lens for examination of the increasingly dense interaction between the EU and the political systems of its member states (Hooghe and Marks 2003).

While research on vertical interaction is still at an early stage, components of a common analytical framework and research agenda are evolving. Existing studies have so far at best focused on limited numbers of cases of institutional interaction, and there is a lack of larger comparative studies. Efforts have been made, however, to reexamine existing case studies in a comparative manner in order to extract more abstract and conceptually founded insights. In particular, the demand for support of local institutions by institutions located at higher levels of social organization (Berkes 2006b) and existing institutional solutions for malign interaction problems have been assessed (Berkes 2002). Likewise, Young (2006) has made attempts to develop a comprehensive analytical framework addressing the relationships between or among the interacting institutions, their core differences, the causal mechanisms that drive vertical interaction, and the consequences of that interaction (see also Cash et al. 2006). Although this work will have to be expanded to develop a theory of vertical interaction, it provides a solid foundation for this endeavor.

Whereas very few links have been made between work on vertical and work on horizontal interaction, the two research areas overlap empirically. The two research communities have so far almost entirely ignored each other's activity. Neither our own approach toward horizontal interaction (Oberthür and Gehring 2006c) nor the most important conceptual contributions to vertical interaction (Berkes 2002; Cash et al. 2006; Young 2006) cite a single publication of the other domain. Likewise, Young (2002b) discusses horizontal and vertical interaction within his elaboration of the IDGEC Science Plan in two separate chapters. The *empirical* interest in vertical interaction overlaps, however, particularly where the focus centers on interplay between or among global and regional institutions. Our comparative study addressed the vertical relationship between the EU and international institutions (Coffey 2006) as well as between global and regional international institutions. Also, certain types of the causal mechanisms of interaction through commitment and behavioral interaction are particularly relevant for this relationship (see "Causal Mechanisms of Institutional Interaction" above). Interaction among nested institutions and interaction activating an additional means are particularly prominent types of vertical interaction between international institutions and EU legal instruments. Obligations agreed at the EU level provide a solid foundation for EU leadership in international institutions so as to internationalize the EU standard, and implementation of international obligations into EU law activates the particular supranational enforcement powers of the EU, which supports compliance by EU member states (Gehring and Oberthür 2006). Other studies have explored vertical interactions between regional and global institutions in several areas of environmental governance (Stokke 2001a; Meinke 2002). Investigating from another angle, Berkes (2006b) discusses regional institutions for the protection of certain fish stocks as intermediaries between the global institutions and the national and local ones.

Despite some differences, there is no theoretical reason to believe that vertical interaction operates fundamentally differently from horizontal interaction. Institutions located at different levels of social organization are hierarchically ordered, with a local institution operating in the shadow of a national one and a national one in the shadow of an international one. In contrast, international institutions, especially those that interact horizontally, are usually formally established independently of each other. Formal (jurisdictional) hierarchy must not be conflated with influence per se, as is seen in the well-known resistance of local or national institutions to the implementation of higher-order commitments. Equally, the frequent formal independence of institutions in horizontal-interaction settings does not imply the absence of influence. Although the particularities of influence may differ considerably, vertical interaction may be expected to resemble horizontal interaction in many respects.

Accordingly, lessons may be drawn from one strand of research for the other. It may turn out, for example, that vertical interaction frequently runs in both directions, rather than predominantly targeting lower-level institutions. As has been found in research on scale, vertical interaction may also open opportunities for the deliberate choice of an appropriate level as a particular form of "forum shopping" (see Gupta, chapter 7 in this volume) if regulation at different levels of social and administrative organization becomes, to some degree, functionally equivalent.

Future Research Directions

Recent advances in knowledge about institutional interaction provide fertile ground for future research. As outlined, research on institutional interaction has made important headway over the past decade or so. Rather than exhausting the field, this progress enables us to identify a wealth of new research opportunities.

The Development of a Theory of Institutional Interaction

Theory development in this area has just begun. More reliable theoretical knowledge on important aspects of institutional interaction is needed. To be able to detect hidden instances of interaction and formulate reliable advice for policy makers requires a theory of the conditions under which institutions tend to influence each other's normative development or effectiveness. The existing theoretically derived causal mechanisms and their subtypes may provide a promising foundation for the development of an expanded theory of institutional interaction. For this purpose the concept needs to be enlarged and elaborated in at least two directions. First, the models do not yet contain reliable information about the sufficient conditions under which the respective causal mechanisms are triggered. Second, knowledge about the development of institutional interaction situations is waiting to be systematically developed. Do the actors involved tend toward full exploitation of the synergies inherent in a situation, or do such opportunities remain unexploited? Do actors succeed over time in minimizing or avoiding disruption among institutions with different objectives, or does conflict tend to prevail? The patterns of the many cases of institutional interaction that have as yet received little attention could also be more intensively studied.

Empirical Knowledge

Such knowledge is still largely lacking in a number of important areas of institutional interaction. First, as observed above, the majority of existing case studies on instances of institutional interaction has focused on a limited number of interaction settings, including the WTO-MEA interface, interactions involving the climate change regime, as well as issues related to the governance of the oceans and the broadly discussed foundation of a WEO. Effects of institutional interaction in other areas, such as governance of chemicals or the preservation of biodiversity, have received far less attention. While analysis of interaction in these fields can use existing analytical tools, it might reveal yet unknown patterns of interaction and thus contribute to the progress of generalized institutional knowledge. Second, still very little is known about the significance of institutional interaction both generally and in specific cases. The efficient management of interaction situations depends on a more precise assessment of the significance of interaction effects. Finally, we need more comparative and large-*n* studies that allow systematic comparison of a smaller or larger number of interaction cases or situations. Such comparative studies promise to generate inductively generalized knowledge. Theoretical insights on such issues as the development of patterns of interaction situations can hardly be derived deductively. They must be founded on the systematic and comparative assessment, or even on quantitative studies, of an appropriate number of cases in a structured and focused manner.

Interaction Management
So far, interaction management remains underresearched. Besides a number of contributions looking into the general legal instruments available (e.g., Wolfrum and Matz 2003), the exploration of the kinds of policy responses that are, or could be, applied by actors in order to enhance synergy and mitigate or prevent conflict is still at its very beginning (van Asselt 2006). Perhaps surprisingly, relatively little is known about what policy responses various actors have applied at the various levels so far and how they have performed. More empirical research into existing policy responses and their performance over time may provide the most suitable starting point for thinking about further policy options for enhancing synergy and mitigating conflict as well as conditions for their successful implementation. The systematic assessment of interaction management (Stokke 2001a) will have to focus on different sorts of policy responses. Actors may respond unilaterally to institutional interaction issues in the implementation of institutional commitments. Members of one institution involved in an interaction situation may also collectively attempt to manage related challenges, as is evident in the ideal type of an interinstitutional request for assistance (see "Causal Mechanisms of Institutional Interaction" above). Actors may even strive for the coordination of interaction management in an overarching framework spanning several or all of the institutions involved in a certain situation (see also Gehring and Oberthür 2006, 314–16).

The management of impact-level interaction constitutes a particularly challenging task. This type of interaction addresses the functional interlinkage of the ultimate targets of the institutions involved. Whereas this interlinkage relies in some cases on barely modifiable biophysical facts, in other cases it may be subject to long-term change that might be influenced by skillful management. For example, environmental protection will in the long run depend not least on the successful decoupling of economic growth (the ultimate target of the WTO) from the global climate (the ultimate target of the climate change regime). Such management will have to occur at least partially outside these institutions and within one or more other institutions, fostering, for example, energy efficiency or the development of new technologies, or governing traffic.

Institutional Complexes and Broader Governance Structures
These wider topics have so far largely escaped theoretically guided research. Exploring systematically the nature, evolution, and consequences of sets of institutions that cogovern particular issue areas promises more integrated understanding (Raustiala and Victor 2004). Most important will be knowledge of the particular division of labor that develops over time among a number of institutions cogoverning an issue area or of institutions with overlapping issue areas. It is one thing to examine how the WTO affects relevant MEAs, or vice versa, and quite another to explore how the overlapping area of environmentally motivated trade restrictions is jointly governed by these institutions. Unlike the sector-specific institutions involved, such interlocking structures (Underdal and Young 2004) are not the product of more or less rational design, since they emerge tacitly from interaction among several international, regional, and even domestic institutions. Cases of interaction may form sequential chains so that an individual case gives rise to a subsequent case that feeds back on the original source institution or influences a third institution. Cases of interaction may also cluster around certain issues and institutions. In this way a number of institutions jointly address a particular problem and contribute to the effectiveness of governance of a certain area. Complex interaction situations raise the problem of "emergent" properties because they may be affected by so many cases of interaction in ways so unexpected that new properties emerge that are not inherent in the single cases. The analysis, then, of complex interactions could start with an assessment of the coexistence of the single-interaction cases involved (Gehring and Oberthür 2006, 358–67).

Future Research on Vertical Interaction

This dimension of interaction has to date received much less systematic attention than horizontal interaction. The theoretical exploration of the origins, types, and consequences of cases and instances of such interaction has only just begun. It is still largely based upon the secondary assessment of existing case studies. It would thus benefit from the systematic comparison of well-selected cases of vertical interaction across levels, including a comparison of cases linking the local to the national level with cases linking the local to the global level and cases linking the national to the global level. Eventually the aim would be to develop theoretical models of the causal mechanisms and types of interaction that reveal information not only on how causal influence is transferred, but also on the conditions of its occurrence and its consequences for environmental governance. Also needed are theoretically sound and empirically reliable conceptions of the different types of division of labor between or among institutions located at different levels of social organization as well as the implications for environmental governance. In some respects the study of vertical interaction might be advanced by employing, or adapting, the analytical tools developed in the area of horizontal interaction. The result could be a more encompassing theory of institutional interaction that accounts for both horizontal and vertical interaction.

We do not claim that this list of research topics is exhaustive. It is meant to identify a number of core avenues that future research efforts may travel in building on past research. Research in the indicated areas promises to advance not only our knowledge about institutional interaction as such but also our understanding of environmental governance more broadly.

Note

1. A diversity of terms can be found in the literature to denote the phenomena subsumed here under institutional interaction, including *interplay*, *linkage*, *interlinkage*, *overlap*, and *interconnection* (see, e.g., Herr and Chia 1995; Stokke 2000; Young et al. 1999/2005; Young 2002b; Raustiala and Victor 2004; Young et al. 2008). We use the term *interaction* in this chapter.

7

Global Change: Analyzing Scale and Scaling in Environmental Governance

Joyeeta Gupta

Introduction

Of fit, interplay, and scale, the three analytical themes identified by the project on the Institutional Dimensions of Global Environmental Change (IDGEC) (Young et al. 1999/2005), scale has only recently begun to receive attention within the research community. The literature from different disciplines on concepts of scale and scaling and their role in environmental governance focuses on a range of research questions and uses a wide variety of definitions and methods. Therefore, any effort to synthesize the material is unlikely to do justice to the depth and breadth of the research. Two research areas in particular can be singled out, however: the politics of scale and the challenges and opportunities in the transferability of problem definitions and solutions from one level to another. This chapter examines these two strands; analyzes whether the concept of scale has exploratory, explanatory, and predictive value; and identifies research questions for the future, as well as policy implications.

Definitions and Terms

As stated in IDGEC (Young et al. 1999/2005), Gibson et al. (2000), and Young (2002b), scale is treated as a ruler along which relative magnitude can be measured. There are many different scales. Two of the most important in environmental governance are the administrative scale and the timescale. Other scales assign levels to space, rules, groups, financial resources, and technologies (Cash et al. 2006; Lebel and Imamura, forthcoming; Bali Conference discussions). Levels are points along a scale. Scaling is the act of moving up or down from one level to another on a particular scale. On the administrative scale, for example, this involves moving up or down among local, regional, national, and international governance systems (see figure 7.1).

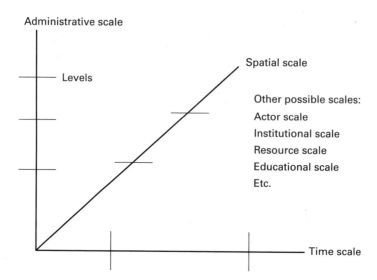

Figure 7.1
Different scales and levels.

Scaling problems up and down is intricately linked to the scaling up and down of proposed solutions. However, although individual actors may regard the process of formulating the problem as closely related to the process of defining solutions, in reality, as a result of the bargaining process among actors, problems and solutions may not correspond. In other words, although a problem may be scaled up or down, the eventual solution at the new level may not be a scaled version of the original successful solution—a result that occurs because of the politics of the process.

Institutional solutions have traditionally focused on problems involving common-pool resources (Young et al. 1999/2005). This chapter looks beyond these resources to the politics behind the scenes of environmental problems and solutions. It looks at how norms, principles, concepts, instruments, and tasks become important factors in defining problems and creating institutional solutions. It considers the actors and networks engaged at different levels and in different ways, from individuals, communities, (transnational) bureaucracies, and epistemic communities to industries, multinational corporations, and nongovernmental environmental agencies. Depending on the environmental issue involved, each actor may play a different role, one that becomes more complex as

Table 7.1
Scalar vocabulary

Scale	A ruler against which relative magnitude is measured
Level	Point along a scale
Grain; resolution	Degree of detail within information
Cross-level	Linking between different levels
Cross-scale	Linking between different scales
Economies of scale	Reduced costs resulting from increasing the level of magnitude of the activity
Level-related characteristics	Particular features of specific levels on a scale
Scalar jumping; Scalar mobility	The tendency of actors to bypass specific levels in order to influence policy processes, or to operate at different levels simultaneously
Scalar lens/analysis	Examining problems and policy processes by focusing on issues of scale and level
Scalar shopping	The deliberate choice by actors of a level or levels from a range of levels as a way to gain influence over a problem
Scalar strategy	Developing a strategy using scale and/or level to influence a problem
Scalar/territorial trap	The tendency to think in terms related only to specific scales and levels
Scale limited	Unable to be shifted to other levels (as in some problems and solutions)
Scale mismatch	Unsuitable level of the organizational framework for the problem and vice versa
Transferability	The capacity of a problem definition or solution to be scaled up or down successfully

one or more of the actors, intentionally or unintentionally, seeks to scale a problem definition or an institutional solution up and down to suit its own purposes.

The literature reveals a number of terms—*scalar jumping, scalar shopping*, and so on. These are presented briefly in table 7.1.

The Problem of Scale

The concept of scale is problematic because it is used differently in different disciplines, generating insights that are not always mutually consistent or complementary. Whereas political geography, economics, and ecology focus explicitly on different elements of scale, environmental

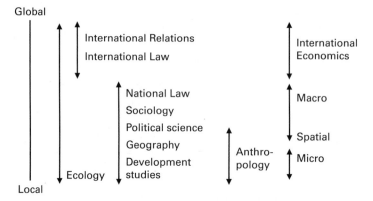

Figure 7.2
The spatial/administrative scale covered by different (sub)disciplines.

governance looks more implicitly at scale-related issues. The concept also has a strong unifying effect, however, as concentration on different levels of scale and ways of scaling (see figure 7.2) produces cross-disciplinary fertilization and richer analysis (compare Cumming, Cumming, and Redman 2006).

Different disciplines also tackle scale using different perspectives. While for legal scholars a state-centric approach is logical, political geographers often question this level of scale as a starting point. While natural scientists and ecologists quite naturally focus on a range of scales, social scientists tend to stay with administrative processes. While economists often use the *ceteris paribus*—all other things being equal— approach in scaling up propositions, anthropologists seek resolution and detail in all other things. While international relations scholars tend to simplify by focusing on aggregative forces that shape policy making, sociologists look at patterns that manifest themselves at smaller scales. The concept of scale is problematic also because each discipline has its own particular research questions and issues. Economics, for example, focuses on economies of scale and scope; some subbranches of economics question the theoretical validity of linear scaling up and down through modeling exercises (Van den Bergh, forthcoming; compare Daly 2003); political geographers look at the politics of scale (Bulkeley, forthcoming); legal scholars, at the allocation of responsibilities to different administrative levels and to new pluralistic forms of governance such as bureaucratic networks and self-governance by industry, as well as how or whether they meet these responsibilities (see Winter, forthcoming); institutionalists, at the circumstances under which solutions can be transferred from one level to another. Anthropologists and development scientists further question transferability on the basis of their hands-on experiences in specific countries (Spierenburg and Wels, forthcoming).

The IDGEC project (Young et al. 1999/2005) focused on understanding the conditions under which institutional solutions can be scaled up or down. While acknowledging that ecosystems and social systems differ, the project aimed to determine if similarities within these systems nonetheless allow for effective scaling up or down (Young 2002b, 26). Few articles in the literature, however, actually focus directly on these questions. Some that do, end by asking whether the question was useful (Berkes 2006a).

The Research Questions and Approach

Against this background, this chapter addresses three research questions: What does the research tell us about the motives of actors and networks in considering the option of scaling? What are the findings on transferability of institutional solutions from one level to another? What is the exploratory, explanatory, and predictive value of the concept of scaling?

A potential area for exploration under the first question is the common confusion of scaling up or down with the concept of "fit" (see chapters 1 and 5 in this volume). Objectively, from the research point of view, and subjectively, from the practitioner perspective, this can be a difficult but enlightening topic to untangle. The key objective of the "fit" theme is to study "the congruence or compatibility between ecosystems and institutional arrangements" (Young 2002b, 20). Our question leaves aside the degree to which the rules of the game are suited to the field of play. Instead, it looks at how actors and networks at different levels of scale frame problems and what motivates them to do so. As a consequence, this chapter does not address cross-scale mismatches, for these indeed belong essentially to the theme of fit (e.g., Cumming, Cumming, and Redman 2006).

This chapter analyzes the literature inductively—finding and developing common patterns and themes. It builds on the papers that are being prepared for the manuscript entitled *The Politics of Scale in Environmental Governance* edited by J. Gupta and Huitema. The emergent themes have evolved over time and have been presented at three international workshops, two of which were financed by the Netherlands Royal Academy of Sciences. They have been modified and nuanced on the basis of the rich response produced during the IDGEC Synthesis Conference in December 2006 (Berkes 2006a; Dhakal 2006; Lebel 2006).

The Focus and Limits

We look at developing country issues and North-South relations, since concepts and theories that apply in various development contexts in different parts of the world are of more value than those limited to situations in the developed world. We review the issues using only the administrative scale as the scale at the center of most of the social science literature on this theme (compare Cash et al. 2006). The administrative scale accommodates a view of governance at all levels, including the self-organizing structure of local communities and the self-regulatory systems established by nonstate actors, and/or within systems established by state actors. The scaling of institutional solutions at the state level is of particular interest, but at the same time we recognize that some researchers question the relevance of actors at this level. This chapter does not focus on cross-level linkages (e.g., Cash et al. 2006), as this falls within the separate topic of vertical interplay—interactions among institutions at different levels of the administrative scale.

The scope of this chapter has been limited to the work of researchers interested specifically in scale. It does not aim to be a systematic, rigorous, comparative analysis of disciplinary approaches or comparative case studies. It attempts rather to bring together themes emerging from different disciplines and case studies and to detect any convergence that could convey key messages about the relevance of scaling for institutional analysis as an approach to solving global environmental change problems.

The Politics of Scale

How do actors and networks frame environmental problems in terms of scale? The social science perspective shows that there are a number of reasons behind the way actors frame issues.

We consider a problem as global when the direct and/or indirect causes and/or impacts of the problem occur worldwide, or when the problem arises all over the world, or when it affects the common good (as in Catholic social theory [Benson and Jordan, forthcoming]). Thus, when a problem involves the hydrological cycle or the global climatic system, it is viewed as a global problem. A problem can also be thought of as global when it arises locally but results from a global cause, say, an ideology or economic and political dynamics that prevail worldwide (Agarwal et al. 1992). When a problem is seen as global, however, it needs also to be defined in terms of the way it manifests at national, state, and local levels. This helps to ensure the legality, legitimacy, and effectiveness of institutional solutions. Global problem solving, then, necessarily entails the development of complementary instruments designed for governance at different levels. For example, although climate change is defined as a global problem, it takes different forms at different levels on the administrative scale from local to international (J. Gupta, forthcoming). Wilbanks and Kates (1999) submit that climate change can be unpacked into distinct sets of problems at each level.

Some argue that the emissions problem is a global problem while the impacts aspect and adaptation are local challenges (Bodansky 1993). This perspective is useful as justification for a global regime to reduce the emissions of greenhouse gases while limiting liability for impacts. Others believe that redefining climate change as a local problem has certain advantages because ultimately it must be addressed at the local level (Bulkeley, forthcoming). The potential dilemmas involved in designation of levels were analyzed in 1985 by Clark (21), who asserted that "this need not be a problem, so long as participants in debates about the interactions of climates, ecosystems and societies concede that causal explanations, variables and generalisation relevant to one regime [level] are unlikely to be appropriate at others. The challenge is not to establish the pre-eminence of any particular [level] but rather to match scales of explanations, processes, and patterns in a realistic and effective way."

Analysis of the development of water governance provides an example where the challenge may not yet have been met. Studies over time reveal an incremental tendency to give preeminence to higher and higher administrative levels in relation to water problems, and to larger levels on the ecological scale, from hydrological basins to ecosystems. Water problems have also become measured over longer segments of the timescale. Freshwater governance sees various levels of problem definition. Development workers and scholars typically define it and take action at local levels. National governments consistently refer to the national dimensions and create national policies (see Lebel, Garden, and Imamura

2005). The rise of the concept of international river basin management has focused attention at this level and raised the argument that national boundaries do not take into account the hydrological integrity within the river (see Hooper 2005). Within the Global Water Systems Project some researchers argue that water problems can be seen holistically from a global level because water anywhere in the world is part of the hydrological system and thus has an impact at the global level; the driving factors underlying a number of water problems often cannot be regulated at local, national, or basin level because local phenomena may contribute to significant global trends that require a global-level response (Pahl-Wostl, Gupta, and Petry 2006). Lebel, Garden, and Imamura (2005) submit that water resource problems are being rescaled at will, an example of the continuous process of upscaling issues that reflects the political interests of the actors and networks involved. They also point out that this may have significant impacts on communities and countries vulnerable to water problems.

Tienhaara (forthcoming) examines the deforestation debate and finds that actors and networks play a critical role in articulating the level at which they wish to define the deforestation problem. For example, states with high forest management standards may support addressing the problem at a global level (Humphreys 2005). Such positions may also change over time. Thus while the Canadian and Brazilian governments continue to call for deforestation to be defined at a global or local level respectively, the United States, the environmental nongovernmental organizations (NGOs), and Malaysia have changed their positions over time (see figure 7.3).

Environmental problems frequently involve direct causes and effects that are essentially local in nature but that from a cumulative perspective may hold global implications. Desertification, deforestation, biodiversity loss, and water pollution are the principal examples. Although desertification was initially not identified as a global problem (see early drafts of the Global Environment Facility documents, e.g., World Bank, UNDP and UNEP 1991), many in the South argue that its absence reflected a political choice (Shiva 1992; J. Gupta 1995). Noise pollution, by contrast, appears to have few global-level effects and may be considered large-scale only to the extent that it is caused by technologies produced by a few large multinationals.

The shifts seen in designation of and preferences for different levels of decision making for environmental problems suggest the involvement

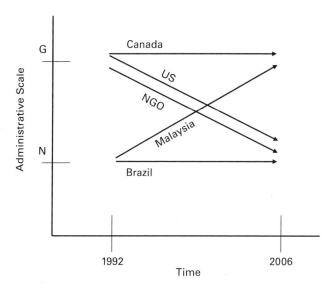

Figure 7.3
Changing positions over time of actors regarding their preferred level (from national—N—to global—G) for addressing forest-related issues.

of a large degree of choice. The choice may be dictated by values and ideas of the actors and networks as influenced by the social context in which they operate. Most environmental problems are closely inter-locked. Small and short-term systems are nested in larger and longer-term systems and altogether make up the environment as a whole. Defining the "appropriate" administrative level for problem definition and solution is often dictated by practical and political considerations that lead actors to engage in scalar shopping. Most scholars tend to agree that the search for an optimal level at which policies should be crafted to deal with a particular problem is a futile effort, as most prob-lems tend to operate at multiple levels and solutions need to be devel-oped accordingly (e.g., Hahn, forthcoming). As with climate change and water issues, the challenge is to unbundle and down- or upscale the problem, defining and addressing it at all appropriate levels.[1]

The Politics of Scaling Up Problems
If defining a problem at a specific level of governance is a conscious choice made by actors and networks engaged in policy making, we must ask: what are the motivations that underlie these choices? Anthropologists

argue that scaling a problem is rooted in cultural perspectives (Spieren-burg, forthcoming). Ecologists demonstrate that the scalar lens often changes the way we interact with landscapes and ecosystems as, for instance, the application of the acre as a unit in farming has helped to change their shape (Vermaat and Gilbert, forthcoming). Some political scientists and geographers meanwhile argue that scaling is often a function of interest and power (Bulkeley, forthcoming; Gareau, forthcoming).

There are arguably four clusters of arguments and reasons (not necessarily matching or voiced, depending on the actor and the circumstances) for scaling up problems. The first cluster focuses on enhancing understanding of a problem. One of the most important arguments identified for upscaling problems is that it takes externalities into account. All problems have a number of externalities—indirect causes and indirect impacts—that under normal circumstances are not included in problem solving. Scaling up helps ensure that a greater number of influential factors are covered in the definition and analysis of a problem, thus helping to produce better solutions (e.g., Van den Bergh, forthcoming). Another key reason for scaling up is to determine if there are global impacts or thresholds of a problem. Do local trends reflect or cause major cumulative problems at the global level? For example, to understand when climate change could become dangerous to the human race, rather than specific populations, the global impacts of climate change problems must be understood (e.g., J. Gupta and van Asselt 2006; Schellnhuber 2006). In addition, the driving forces behind a problem may be rooted in ideological beliefs whose roots become visible when the problem is scaled up, making it easier to design appropriate solutions (Agarwal et al. 1992; J. Gupta 1997). This occurs, for instance, when "ideological beliefs have only too easily permeated development thought," precluding disciplined analysis of the relation between state and market (Meier 2001, 6).

A desire to improve the effectiveness of governance provides the basis of a second cluster of reasons. Motivation here embodies a normative approach, where the motivation for scaling up is to conform to or is based upon a principle of right action binding on actors involved and serving to guide, control, or regulate acceptable behavior. Scaling up issues may be consistent with a perceived need to be inclusive and create greater legitimacy in the policy-making processes. This can ensure that all potential actors are engaged from the start in understanding if not in addressing the problem. Scaling up may also be motivated by notions such as the common good (Benson and Jordan, forthcoming) or the need to promote sustainable development. The latter reasons inevitably link environmental problems to many contextual issues and entail a requirement for a critical mass of actors in favor of addressing cumulative and systemic problems (Parto 2004).

The third set of reasons behind scaling up a problem concentrates on domestic or local interests. Actors and networks often engage in scaling up (or down) to change the level of decision making as a way to postpone possible decision making and avoid taking responsibility at a domestic level. In such cases shifting the level at which a problem is to be dealt with may not bring any advantage from enhancing (or narrowing) the scope of analysis and address of a problem (e.g., Tienhaara, forthcoming, on forestry; Compagnon, forthcoming, on climate change; and Gareau, forthcoming, on ozone depletion). Another reason for this sort of scaling up is a desire to make domestic policy implementation more cost-effective and prevent the loss of competitiveness of domestic industry (e.g., Tienhaara, forthcoming, on forestry; Gareau, forthcoming, on the ozone regime; and J. Gupta, forthcoming, on climate change). As a corollary, domestic actors may also wish to increase pressure on entities in other countries or regions to tackle commons problems, create a level playing field, and minimize free riding. Motivation for scaling up can also come with the aim of increasing the adoption of cleaner and/or more efficient technologies. With globalization and liberalization, the need to be competitive is ever more present. This can drive actors and networks to adopt the cheapest technology, which can in turn have serious environmental impacts. Those who wish to counter such a "race to the bottom" often argue in favor of scaling up related environmental problems in order to promote specific standards and policies to increase market penetration of quality technologies through sales or access.

A fourth group of reasons revolves around a perceived need to promote extraterritorial interests. Scaling up problems often helps actors gain power over resources that strictly speaking exist within the territorial limits of a different actor or actors. With globalization and increasing consciousness of the need to preserve the capacity of the environment to contribute to human welfare, many actors and networks scale up issues in order to protect environmental resources. In some circumstances this may incur a loss of power over resources by actors at a lower-scale level. Spierenburg (forthcoming) argues that when concerned outsiders see a local inability to address local problems, they are often moved to internationalize the problems and use science-based solutions to develop international policies, which are then (often unsuccessfully) imposed on local actors. Spierenburg and Wels (forthcoming) show in their analysis of transfrontier conservation areas how such power politics play out in southern Africa. Another extraterritorial motive is seen in scaling up with the aim of bypassing governments (a form of scalar jumping) that are distrusted (Compagnon, forthcoming). This often occurs because local actors feel their governments are either unwilling or incapable of addressing a problem or that problems are more likely to be effectively addressed at a higher level (Benson and Jordan, forthcoming). In these cases actors with vested interests may have greater powers at the local level (Lebel and Imamura forthcoming); such actors may work to prevent decision making at a higher domestic level that could potentially threaten their interests. It is also argued that scaling up is a tactic used by developing countries in order to create a level playing field. Here, scaling up provides an opportunity for developing countries to generate joint political power to address a problem, for example, through negotiating in multilateral rather than bilateral discussions.[2] Smaller members of the European Union (EU) often also use this tactic. Finally, scaling up may be motivated by a desire to broaden the decision-making space as a means to improving the opportunities for win-win solutions. Techniques employed here include integrative bargaining and issue linking as methods that enlarge the scope for trade-offs (Cowell 2003; Bulkeley forthcoming; see table 7.2).

Many of these arguments embody a state-centered bias. For some of us the state will remain central to economic and fiscal policy and to democracy. In the international legal context, only states may negotiate and be held accountable for international treaties; however, there are also signs of increasing pluralism and decreasing state centricity on the global stage. Although clear roles can be assigned to specific new actors, such as multinationals and nongovernmental actors (Bulkeley, forthcoming; Hahn, forthcoming; Winter, forthcoming), the open question here is how will scaling up be undertaken when the world is no longer largely a collection of states but a network of transnational capital and information flows?[3]

The Politics of Scaling Down Problems

The literature also reviews a number of reasons and arguments that favor scaling down global environmental problems. These reasons and arguments tend to be hybrids of those for scaling down the causes and

Table 7.2
Reasons for scaling up problems

Type of motivation	Motivation
To enhance understanding of a problem	To take account of externalities (indirect causes) as influential factors To determine global impacts and thresholds of a problem To understand the ideologies driving decision making
To improve effectiveness of governance	To include countries and other social actors, creating greater political legitimacy To protect the common good; to attain sustainable development
To promote domestic interests	To postpone decisions or avoid taking measures at the domestic level To make domestic policy implementation more cost-effective and to prevent the loss of competitiveness of industry; to pressure actors in other countries or regions to create a level playing field and minimize free riding To avoid a race to the bottom and promote the use of cleaner and/or more efficient technology
To promote extraterritorial interests	To gain influence over resources in another location despite potential loss of control over resources by actors at a lower-scale level To bypass an agency because of its lack of either motivation or capacity To create a level playing field through facilitating joint action that enhances the power of cooperating actors To increase the decision-making space, thus enlarging the scope for trade-offs

impacts of and responses to a problem. Four groups of arguments and reasons to scale down result.

The first group focuses on downscaling as a means to improve resolution and enhance the understanding of the problem. When information is scaled up, often a situation arises in which unrealistic assumptions are made "regarding substitutability, homogeneity of inputs and outputs and linearity" in order to provide aggregate economic information (Van den Bergh, forthcoming). In the process, specific critical local patterns and processes may become lost (Vermaat and Gilbert, forthcoming), often leading to unrealistic analysis. Hence, many authors argue in favor of scaling down problems to increase the detail of information about critical influential factors (actors and networks; contextual elements such as ideologies, customs, institutions, mandates), thus producing a better problem definition.

A second cluster involves downscaling as a means of increasing the effectiveness of governance to address a problem. One reason for this is the desire to make use of existing problem-solving institutions, whether administrative or otherwise, as these often include processes designed to ensure legitimacy, legality, transparency, and accountability. Another argument for scaling down a problem is to mobilize the knowledge and capabilities of local people. When problems are defined at higher administrative levels and on the basis of aggregate facts, local people often lose interest and motivation to help design and implement solutions. Defining problems at lower levels brings the process closer to local actors, making the problem more tangible or "real" and a more likely focus for mobilization (J. Gupta and Hisschemöller 1997; Bulkeley, forthcoming; Hahn, forthcoming). Some see such processes as inherent in achieving sustainable development since context, and therefore local action, is highly relevant to the concept of sustainable development.

A third cluster of reasons for downscaling hinges on the need to promote domestic interests. Scaling down a problem can be considered desirable by some as it externalizes specific impacts and allows for decision making based on lower-level interests only. This helps some actors avoid being held liable for externalized impacts (see Bodansky 1993; J. Gupta 1995, 2005). Efforts to scale down a problem to national and local levels using sovereignty and subsidiarity arguments may be motivated by the desire to manage and protect national and local interests (compare Lebel, Garden, and Imamura 2005).

A fourth cluster aims to promote extraterritorial interests. Some actors find that scaling down can increase opportunities for divide-and-control

Table 7.3
Reasons for scaling down problems

Type of motivation	Motivation
To enhance understanding	To enhance problem understanding through greater resolution and grain regarding critical local and contextual elements
To improve effectiveness of governance	To use existing problem-solving institutions and thereby to take advantage of built-in processes designed to ensure legitimacy, legality, transparency, and accountability To mobilize local people in designing and implementing solutions, using their knowledge and capabilities
To promote domestic interests	To avoid liability for externalized effects To manage and protect national and local interests
To promote extraterritorial interests	To divide and control or include and exclude To bypass an agency that is perceived as a hindrance

tactics. Room is created also, when a problem is scaled down, to divide and control actors in other parts of the world. Scaling down may also help to include or exclude specific actors and/or administrative levels (Bulkeley, forthcoming). Downscaling may also bypass a nation-state perceived as corrupt, inefficient, or simply incapable. Such approaches may be well-meaning but may serve to destabilize authority rather than help support institution building in the affected country (Compagnon, forthcoming; see table 7.3).

Inferences
The social science literature suggests that levels considered appropriate for specific problems cannot be taken for granted because, based as they are on reasons as outlined above, designations of level are socially constructed and politically produced and contested (Brenner 1998; Swyngedouw 2004; Lebel 2005; Bulkeley, forthcoming). The literature also reveals political dimensions in the way actors and networks use scalar strategies to meet their own interests when they frame problems. Scaling is a critical instrument for framing environmental problems as it places them in a specific context. The way scaling limits or expands the scope of the problem shapes potential solutions. Scaling as a strategy has the

power to legitimize or delegitimize certain approaches. New political aspects arise at each scalar level, creating the necessity for great care when transferring problem definitions and solutions from one level to another.

Actors and networks try to assume or shy away from responsibility by shifting the level at which problems are defined. Scalar shopping in regard to forestry is the phenomenon underlying figure 7.3. Nonstate actors try to jump scale to influence policies, seeking administrative levels where they can be of greatest influence. Nonstate actors in richer countries have more scalar mobility and are better able to jump scale than those in poorer countries (Gareau, forthcoming).

The research finds a stronger tendency toward and a greater number of reasons for scaling up problems than for scaling them down. This may be driven partly by broader sustainable development goals that may call for upscaling. Another factor behind this trend could be the fact that globalization means few locations are now untouched at least by cumulative effects. This leads to the argument that "it seems that for a locale to exert influence on events to which it is subjected or to affect the direction of macro events at the local scale, a certain socio-economic and political critical mass is required" (Parto 2004, 95). Actors may try to upscale when lower-level agencies show no sense of ownership of a problem or when they have no control over, say, the sources of pollution, as studies show is the case with city governments (Berkes 2006a; Dhakal 2006).

The reasons for downscaling do not always run parallel to those for upscaling. They also involve fewer parameters. Often problems are not truly downscaled as much as they are strategically positioned out of the influence of agencies at a higher level. Cases of true downscaling frequently involve the unpackaging of problems into different issues. In other cases downscaling may share many similarities with implementation, as may occur when policy makers decide that specific policy responses can be taken at the local level and allocate such responsibilities accordingly to local authorities.

Scaling Solutions Up and Down

Three main subquestions arise regarding scalar transferability: first, what sort of problem definitions and solutions are scaled up and down? Second, what does the literature say about the politics and practical aspects of scalar transferability? And third, are there level-specific characteristics that might make particular levels appropriate for specific problem-solving tasks or that might serve as indicators to suggest how problem definitions and solutions can be transferred?

The Politics of Scaling Solutions Up and Down
This subsection looks specifically at problem definitions and solutions for governance, including good governance and the rule of law, certain environmental principles, concepts such as intellectual property rights and traditional knowledge, emissions trading in climate change, private sector participation in the water regime, the transfer of investment provisions, and stakeholder and public participation.

Good Governance, Including the Rule of Law   In recent years there has been a tendency for research and policy documents to argue that problems in developing countries and in Central and Eastern European countries arise from the lack of good governance and the rule of law in those countries. This has led to a situation in which governments (the EU and the United States) and development banks (e.g., the World Bank) are actively promoting the adoption of governance solutions in other countries or suggesting that assistance should be limited to those countries that exhibit good governance (Carothers 1998; Santiso 2001). Pressure, then, to upscale problem definitions and solutions of governance is being brought to bear horizontally from one state to another, this despite criticism that the technocratic manner in which solutions are imposed, and the way this forces scaling up in the target countries, may have counterproductive results (Anders 2005; J. Gupta 2005; Santiso 2005). Research finds that good governance practices can be successfully transferred to such countries when scientists and policy makers stop supplying standardized instruments and instead promote practices tailored to the political economy of these countries on a case-by-case basis (Compagnon, forthcoming).

The literature is ambiguous, however, about the need for promoting governance at the international level. Whereas one strand in the literature argues in favor of promoting governance solutions at the international level, another suggests that the best that can be achieved at the international level is some minimal norm of procedural rule of law. The minimal participation of the United States in several multilateral agreements is an example (see Gareau, forthcoming).[4] Some researchers see the rule of law at the international level as an impossible goal, given that states with power do not wish to be subject to the rule of the majority of states (Baum 1986; Koskenniemi 1990; Watts 1993). Although the rule of law calls for procedural and substantive fairness, the former protects the status quo while the latter calls for substantive change (Franck 1995). Substantive change in the global order is not always in the interests of powerful actors. Some American scholars promote procedural fairness at the international level (Esty 2006), although the concept of procedural fairness needs further interpretation.

The United States is an example of a powerful actor willing to move problem definitions and solutions horizontally on the same level to other countries but reluctant to scale them up to the international level. If the rule of law and good governance are made necessary preconditions for policy making at the domestic level, however, how is it possible to deny the need to upscale the need for good governance to the international level (J. Gupta 2005)? To paraphrase the words of Fitzpatrick (2003), is the rule of law not needed to ensure the legitimacy of global politics? Scholars in developed countries often argue that developing countries are so corrupt and vested interests so strong that the chance of effective scaling of good governance to these countries is low (Carothers 1998, 96). Can a similar case be made regarding good governance in the international arena—that the vested interests of major national actors will prevent successful upscaling of good governance solutions at the international level (see Albright cited in Whittell 2002; Alston 1997; Annan 2004)?

Environmental Principles   The Brundtland Commission proposed a number of environmental principles (World Commission on Environment and Development 1987); several were adopted in the soft-law document of the Rio Declaration on Environment and Development (Rio Declaration on Environment and Development 1992). An examination of the related environmental treaties, however, shows that some principles are more easily scaled up than others.

Recent treaties have not only incorporated the cost-effectiveness principle; they have translated it into market-based instruments, thereby giving this principle more teeth than others such as the precautionary principle, the common but differentiated responsibility principle, and the development-oriented principles. While some environmental and economic principles are subject to processes scaling them up to the international level, development principles and social issues are subject to much weaker scaling-up processes (Schrijver 2001; French 2002). Some of the literature questions the usefulness of principles such as common but differentiated responsibilities at the international level (see Weisslitz 2002); while the "polluter pays" principle is seen as important at the national level, it is not really being upscaled to the global level. For example, the Rio Declaration recommends the polluter-pays principle for national governments but not for the international level. No global agreements have incorporated this principle thus far (J. Gupta 2007).

Intellectual Property Rights versus Traditional Knowledge   As knowledge increasingly becomes a source of revenue, those in possession of knowledge seek to protect it through intellectual property rights. A complicated institutional framework has been developed in most Western countries to protect intellectual property. Many indigenous societies have developed cumulative knowledge over the centuries that was and is freely available to members of these societies. Institutions to protect the use of indigenous knowledge have not existed and still do not really exist, whereas the opposite is true of institutions to share such knowledge. With the elaboration of the biodiversity and trade regimes, a key issue that has come to the fore is the difference in the way intellectual property rights and traditional knowledge are dealt with at the international level. Abu Amara (forthcoming) argues that the concept of intellectual property rights has been transferred to the international level, but the concept of traditional knowledge, although recognized at the international level, does not get the same treatment. A lower degree of protection results for indigenous knowledge, providing space for "biopiracy." The upscaling of intellectual property rights, however, had not been free of political maneuvering. J. Gupta (1997) argues that developing-country actors were very upset at the way intellectual property rights discussions were moved from forum to forum, shifting influence during the process of scaling the concept up to the international level.

Differentials in the transferability of concepts from one level to another are a factor in scaling, along with not only power politics but also the degree of maturity of the concept, problem, or solution under consideration. The lack of power of developing-country governments involved in upscaling intellectual property rights for traditional knowledge has been seen in their inability to negotiate effectively, even though there is a strong, long-established proposal for recognizing such rights. Unfortunately, systems designed to protect indigenous knowledge are far less mature. Community knowledge is more complex than private knowledge. Büsscher and Critchley (forthcoming) demonstrate the practical problems of identifying the owner of community knowledge, of shifting problem definitions and solutions relating to such knowledge to different administrative and social levels, and of rewarding actors for this kind of knowledge.

Emissions Trading in the Climate Change Regime   The emissions trading system in the climate change regime provides an example of a controversial case of scaling up. The system is based on limited past domestic experiences in the United States. When designed and implemented carefully, the instrument has the capacity to reduce pollution in a manner that achieves financial cost-effectiveness and very low transaction costs. When the concept is transferred to the international level, however, two problems arise. The first is the political problem of the allocation of emissions permits among rich and poor countries. The question here is whether a system based on allocating rights to those who are polluters should be promoted, or whether such rights need to be allocated on the basis of specific principles such as the per-capita principle. Can the politically easier method at the national level of allocating rights to polluters be justified at the global level, since it will privilege those who pollute more over those who have thus far polluted less? Given that the current system is designed using grandfathered instruments and that follow-up systems within the EU are being developed along this principle, the issue becomes important. The second problem is one of implementation. Emissions trading is a highly sophisticated instrument that functions within some of the richest countries in the world. Can it be implemented as an effective solution in countries with much weaker governance systems and monitoring mechanisms? In fact, the scaling up of emissions trading may well reflect the interests of rich countries as opposed to those of poorer countries (Young 2002b; J. Gupta 2005, 2007).

Private-Sector Participation in the Water Regime   Private-sector participation in the water regime represents another example of an environmental solution favored by many actors as a candidate for upscaling. In the past decade the lack of investment in the water sector has been ascribed mainly to the inability of governments, especially in the developing world, to channel financial resources into this sector. The widely held assumption that the public sector is generally inefficient, and the private sector generally efficient and wealthy, has led to worldwide promotion of the idea that private-sector participation in the water sector is essential to achieving sustainable use of the resource.[5] Widespread revisions to national legal systems and lending schemes established by international banks have facilitated private-sector involvement. A number of cases of such involvement now exist (Gleick et al. 2003). Although some of these cases of private participation address key issues of water supply, there are also controversial cases such as the participation of Bechtel in the water services of Cochabamba in Bolivia, which eventually led to international arbitration. The initial lesson from such investments is that the returns on investment required by the private sector are not easily generated by supplying water (or energy) to the poorest of the poor (Schouten and Schwartz 2006; see also Petrella 2001; Barlow and Clarke 2002; Shiva 2002). As a consequence, the private sector is gradually withdrawing from the water field. In this way private-sector involvement in the water sector shows how scaling up is often strongly influenced by ideology and less by contextual information.

Investment Provisions in the Investment Regime and Their Impact on the Environment   The cases of emissions trading and private-sector involvement in the water sector raise implications for commercial law concerning international investment. As the private sector is increasingly invited to participate in the management of natural resources and environmental issues—whether water or climate change—more and more private and public-private contracts are being drawn up. Such contracts are typically subject to private international or commercial law. Given the complexity of such law, there has been a recent trend to promote upscaling of investment principles to the international level. Although multilateral efforts to do so within the Multilateral Agreement on Investment failed to reach completion (Werksman and Santoro 1998), in the past fifteen years more than 2,300 bilateral investment agreements have been negotiated. These agreements aim to protect the foreign investor from the changing domestic policies of new (and often corrupt) governments (Wälde 2006). The effect of this, inter alia, is that national governments may not be able to adopt environmental regulations that apply to such foreign investments or risk being taken to court on such issues (see, for more details, Chalker 2006; Sornarajah 2006; Tienhaara 2006). Here, too, neoliberal concepts that protect specific interests of powerful actors have been more easily scaled up than instruments aimed at protecting the commons.

Stakeholder and Public Participation   The notion of stakeholder participation is currently popular and is expected to address the problems of democratic deficit and technocratic decision making in the developed world. This concept, a governance solution, has been subject to scaling-up processes in a number of international agreements. It is recommended as a key element of good governance and as an ingredient of integrated water resource management. It is included in a number of policy documents of lending organizations. The question remains, however, whether upscaling is appropriate for this concept in developing countries. Low local literacy levels depress stakeholder involvement (Ankersmit 1998; Turton, Schreiner, and Leestemaker 2000) in these countries, where it is also often difficult to engage women and youth (United Nations Environment Programme 2000). Differences in such contextual features between developed and developing countries require sophisticated analysis and serious consideration in conjunction with power politics when scaling stakeholder solutions in governance (Upadhyay 2003).

Reasons for Upscaling or Downscaling Solutions   The research reveals four reasons that motivate actors to scale problem definitions and solutions up or down, including norms and instruments. A key reason is the desire to promote good practices and processes, best available technologies, efficient instruments, and better institutional design. These are often identified as successful solutions in peer-reviewed scientific publications, which actors then select as solutions to be upscaled beyond their original context.

A second reason is the desire of political actors and networks to promote their own or national interests. In some cases scaling up can help demonstrate consistency of political direction. In others the aim is to show national success of the norm or instrument scaled up with the further goal that the same solutions become accepted internationally. If this occurs, the original actor saves costs and time since it will have no need to change its existing institutions and instruments to conform to new international practices. Promotion of upscaling motivated in this way often ignores local realities in other countries (Dhakal 2006).

Ideological commitment represents a third motive behind upscaling. Proponents of certain ideologies, such as neoliberalism, become convinced that certain solutions should be applied all over the world. Research shows that this type of reason lies behind some instances of promotion of upscaling for market mechanisms like emissions trading, intellectual property right instruments, and investment agreements.

A fourth reason behind scaling up or down is a desire for strategic influence. Actors and networks may seek to scale up norms and instruments to the international level to protect their own interests by influencing domestic actors in other regions (e.g., the case of investment agreements). Thus, whereas some actors assume that the international level is chaotic and anarchic, and hence do not propose upscaling solutions to this level, others argue that scaling up to the international level will, for example, bypass high levels of corruption, elite pacts, and inefficiency in developing countries. Scaling to the global level can create open-ended legal implications, but within the EU scaling up is often a means of "developing 'policies by stealth'" by promoting technical standards (Benson and Jordan, forthcoming). There are more examples of scaling up norms and instruments than scaling down as a result of globalization and the need to promote good practices and policies at other levels. Differences in the bargaining power of countries will affect whether strategic influence is achieved. Certain norms and instruments become scaled up or down, and articulated and adopted or not, at the international level because of such differentials brought to bear in the negotiating process (see table 7.4). Gareau (forthcoming) argues that the United States supported its strawberry producers by requesting a delay in the phaseout of methyl bromide in the negotiations on the depletion of the ozone layer; the local industry stood to lose its competitiveness vis-à-vis developing countries that were allowed to phase out methyl bromide some years later. This case supports his argument that both state-centric actions and movements of global capital and knowledge are used to promote scalar jumping or scalar shopping as a means of promoting the interests of particular actors and networks.

Hypotheses on Transferability versus Transformability of Problem Definitions and Solutions

The potential for a problem definition or solution to be scaled up or down successfully is called its scalar transferability. Problem definitions

Table 7.4
Reasons for scaling solutions up or down

Types of motivation	Explanation
To adopt the latest and best scientific solutions	The focus of social science on, inter alia, good practices and processes, best technologies, and good institutional design leads to the recommendation of such practices for scaling to levels beyond those focused on in the original science.
To demonstrate leadership and effectiveness	Scaling up or down demonstrates consistency of political direction; or scaling up following the national success of a norm or instrument saves costs and time since existing institutions and instruments will conform to new international practices.
To express ideological commitment	Ideas and experiences rooted in specific ideologies are promoted to different parts of the world or lower administrative levels in the belief that they are generally applicable.
To exert strategic influence	Actors and networks seek to influence domestic actors in other regions, or lower-level actors; bargaining differentials are a major factor here.

and proposed solutions may be more transferable along a scale when at least some of the patterns and processes involved are similar across levels. Sufficient similarity across levels of the following factors is needed: the nature of the environmental resource, the nature of the problem in terms of distribution of costs and benefits, the nature of the science and knowledge in terms of certainty and controversy, the formal legal framework, the contractual environment, and the values and beliefs (cultural contexts) (J. Gupta et al. 2003).

Caution is needed in determining similarity because aggregation may blur individual patterns and processes (Van den Bergh, forthcoming; Vermaat and Gilbert, forthcoming). In sum, it appears that a sufficient degree of homogeneity in terms of ideological and cultural inclinations, polity, policy, and politics among levels is required before problem definitions and solutions become transferable.

Expansion of the EU has brought large-scale experimentation with up- or downscaling propositions to deal with common problems through harmonization. One can observe a large-scale horizontal transfer of ideology, norms, principles, concepts, and instruments to the new members of the EU. This effort at homogenizing the European continent, creating an EU culture, may lead to effective transfer of concepts. The jury is still out as to whether the large-scale transfer of a whole package of propositions makes it easier to implement these effectively or not. Earlier experiments with horizontally scaling up market mechanisms to post-Communist regimes have not always been successful because of the very different polities involved. It is possible, however, that the whole process of Europeanization in these countries will lead to the accelerated development of factors and institutional dynamism that will make effective implementation of EU policies easier.

Some research has reviewed the internal struggles over scaling within the EU. Although the principle of subsidiarity is expected to guarantee that only international issues are scaled up to EU level, in effect, because it is difficult to define objectively what an international issue is, all problem definitions and solutions can be scaled up to the EU level (Benson and Jordan, forthcoming). Furthermore, an analysis of the implementation of EU water policy (which is legalistic in nature) in the Netherlands (which is oriented toward the generation of social consensus) shows that this case of downscaling leads to less relevant and more expensive problem-solving approaches (Huitema and Bressers, forthcoming).

Given the poor consequences of scaling without transferability, the need arises to *transform* problem definitions and solutions. The aim here is to ensure that attention is paid to local and national aspirations, cultures, ideologies, power and bargaining inequalities, adaptability, flexibility, transparency, accessibility, and ownership of implementation (Büsscher and Critchley, forthcoming; Compagnon, forthcoming; Spierenburg, forthcoming; Spierenburg and Wels, forthcoming) as well as compatibility with the enabling environment in which it is to be implemented (Hahn, forthcoming). For example, the contextual determinants for the successful transference of market mechanisms can be postulated to include a liberal, free-market ideology and tradition; a regulatory framework that respects contractual agreements, creates the rules of the game, and protects the interests of producers, traders, and consumers; widely available information; and a social structure that compensates those who lose out in the process and cushions against other negative impacts of competitive behavior. It follows that where free-market ideology and tradition is of more recent origin, where the regulatory framework is much weaker and information widely contested, and where social cushioning is lacking, the prognosis for transferability of market mechanisms will be poor.

At any level where the ideological and cultural context is different, any efforts to scale a problem definition, or a norm, principle, concept, or instrument, may require that such items are transformed for them to be effective in the new context. Such transformations, however, need to take into account the different and new factors at the new level. Anders (2005) demonstrates this very effectively in his analysis of how the implementation of good governance solutions in Malawi led to a multiplication of the opportunities for corruption.

Investigation is always needed of ways in which problem definitions and solutions can be made appropriate for the context found at the level to which transfer is desired. The significance of scaling to the diversity of cultures and ideologies found at different administrative levels finds a parallel in the importance of spatial heterogeneity in ecosystems, a factor that adds to the complexity of ecosystem management (Compagnon, forthcoming; Spierenburg and Wels, forthcoming; Vermaat and Gilbert, forthcoming). An additional finding has been that scaling up and down often involves compartmentalized knowledge and may be less relevant in traditional cultures (Büsscher and Critchley, forthcoming).

Level-Related Properties and Implications

Besides contextual differences, levels also exhibit particular properties (analogous to emergent properties of ecosystem levels) that can be used to indicate when tasks are appropriate for other levels or to determine how problem definitions and solutions should be transformed. If we distinguish four major levels—global, regional, national, and local—can we also find characteristics that are specific to each? The limited literature on the subject indicates that the global level shows a higher degree of aggregation of factors, a greater number of actors, higher stakes, greater cultural diversity, larger uncertainties, and a generally weaker institutional framework (see Young 2002b, 140–59). At regional levels problems may be somewhat easier to identify; cultural harmony may be greater, fewer actors involved, and a range of strengths of institutional frameworks in play, as where, for example, the EU has a much stronger institutional framework than the African Union. A key characteristic at the national level is the concept of sovereignty, which shapes and in turn is shaped by the behavior of states at the international level. At lower levels problems often become simpler with fewer actors and less uncertainty. At these levels social norms and stakeholder participation in decision making may become more critical (Cash et al. 2006; Hahn, forthcoming). Some argue that leadership may be more critical at lower levels than at the international level (Cash et al. 2006; Hahn, forthcoming).

Some characteristics of nonstate actors have also been ascertained. Transnational scientific communities work toward globally relevant science and typically cite peer-reviewed journals. Industry has access to its own knowledge and trade secrets, and it typically accounts to shareholders. Transnational nonstate actors are not accountable to any party at the international level.

Where sovereignty is seen as a problem, actors and networks often promote cooperation between transnational actors to make progress. For example, Winter (forthcoming) examines the problem of dangerous chemicals and concludes that although the issue is of global importance, a global agreement is unlikely to be forthcoming. He then proposes the allocation of different tasks to different levels, arguing that data generation and risk assessment are the tasks of industry, and that rules for collecting such data, on assessment methodology, testing, good laboratory practices, classification, and labeling, should be made by transnational bureaucracy networks. He then submits that risk management is a task for national governments using the risk assessments also conducted at the national level (see figure 7.4). These tasks are assigned based on implicit assumptions about the level-related characteristics of each actor.

At the same time, Compagnon (forthcoming) submits that actors sometimes wish to avoid the national level in developing countries because of factors arising out of sovereignty and elements such as corruption, lack of capacity, and bureaucracy. He submits, however, that strategies to jump the national scale may not be effective. Instead, policy support, based on full cognizance of how the state functions, its limits, and its mandate, may not only help to transform instruments into appropriate tools for the context of each developing country but also help to strengthen the state. He argues that a well-functioning state is essential to guarantee the space in which local actors can effectively operate.

The problems of societies are inherently complex. Solutions do not lie in general recommendations such as decentralization or centralization, or scaling up or down. Research points rather to the allocation of responsibilities at different levels according to the specific characteristics of

Administrative scale

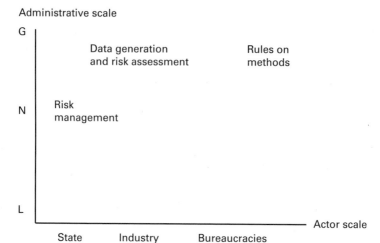

Figure 7.4
Assignment of different tasks at different levels to different actors in the chemical sector.

those levels; and to a focus on multilevel governance, comanagement, or adaptive management (Cash et al. 2006; Benson and Jordan, forthcoming; Compagnon, forthcoming; J. Gupta, forthcoming; Huitema and Bressers, forthcoming; Spierenburg and Wels, forthcoming).

Practical Guidance on Scaling Up Innovations   Very little has been written on the practical elements involved in upscaling innovative problem definitions and solutions. A proposal for agricultural communities suggests a ten-step method for scaling up. The steps involve identification of innovations made by farmers using participatory rural assessments, verification of the innovation, recruitment of farmer innovators, analysis of the innovation, establishment of networks of farmer innovators, monitoring and evaluating the process, site visits to other farmer innovators, study tours for these innovators to develop the innovation further, testing of the adapted innovation in the new location, and dissemination of the knowledge.

Although much of the proposal is based on empirical experience in Africa, particular concerns arise in scaling up problem definitions and solutions generated by a community. Since ownership of community knowledge is unclear, it remains unclear who should be responsible for communication of such knowledge and who should reap the rewards of such creation and ownership of knowledge. Scaling up also requires sensitivity to local customs and behavioral patterns. The process is neither as mechanical nor as automatic as a reduction, say, to ten steps might suggest (Büsscher and Critchley, forthcoming).

The Exploratory, Explanatory, and Predictive Power of Scale

Exploration of the concept of scale provided a preliminary platform for attempts to unite different disciplinary approaches. Scalar analysis could help to bridge "level-limited" disciplinary approaches in economics, politics, law, geography, international relations, development studies, and anthropology and provide a common platform for governance analysis. In this way theories developed in individual fields as diverse as technology transfer (Büsscher and Critchley, forthcoming) and world system theory and capitalist theory (Gareau, forthcoming) might find applications beyond their original scope.

Exploratory work reveals that scale offers a powerful metaphor that can shape, expand, and limit the contours of discussion. It shows how local, regional, and global levels along the administrative scale are often referred to loosely in scientific efforts and governance processes, and how analysts often fall into territorial traps either because of their disciplinary focus or because of preset ideas (Bulkeley, forthcoming; Tienhaara, forthcoming). Territorial traps limit thinking about a problem and the possible array of solutions. Incorporating multiscalar analysis helps to ensure recognition of complexity in a problem definition and of the appropriate levels for solutions or specific components of solutions (Lebel and Imamura, forthcoming). Exploration of scale also shows how actors and networks adopt scalar strategies, such as scalar jumping, in order to promote their own agendas.

In terms of explanation, research into scale reveals a number of possible explanations behind attempts by actors and networks to scale individual problem definitions and solutions. The larger number of reasons to scale up reflects several aspects of modern society: the impacts of globalization (e.g., trade, multinationals, international communication, the global media, and so forth), the realization that small-scale systems with short time frames are nested in larger systems with longer time frames, the capacity of local trends cumulatively and systemically to contribute to global horror stories, the need to understand the "bigger picture" to achieve sustainable development, the need for a critical mass to achieve effective action. Such aspects reflect practical issues as well as power politics and the perceived need to access and control beyond spatial borders. The scalar discourse even raises questions about the ideological foundations of modern society. While much of the research highlights the limits to scaling norms, principles, concepts, and instruments on North-South and sometimes West-East dimensions (revealing an implicit bias, given that the terms used here are not *South-North* and *East-West*), the West alone yields a number of examples of effective scaling that occurs on a day-to-day basis in the United States and Europe. Yet even here a struggle arises between efforts to scale up problem definitions and solutions in order to promote harmonization of policy, and attempts to scale down to try to ensure that responsibility for problems is assigned at the lowest level with the capacity to resolve them (as seen with the subsidiarity principle in the EU) (Benson and Jordan, forthcoming). In addition, difficulties occur in both directions as scaling encounters characteristics specific to different levels. This is seen even within the relatively homogeneous context of the EU, for example, as scaling down policies sometimes clashes with the need for policy to be coproduced by consensus among social actors—a key characteristic of Dutch society, for example (Huitema and Bressers, forthcoming). Tensions surface among actors working to scale in opposite directions, particularly as the motivation behind scalar strategy is often political, aimed at controlling an agenda and the nature of responsibilities assigned to individual actors. Furthermore, it is important to be aware of the fact that while scaling down increases the grain, texture, or degree of detailed relevant information available, scaling up enhances the bird's-eye perspective of large-scale cumulative and systemic trends. There is a trade-off between these two when focus centers on either one or the other.

The research shows that actors and networks are motivated in different ways to transfer problem definitions and solutions from one level to another. It explains some critical, level-related characteristics that affect what is being scaled and the allocation of tasks. Scaling may be an effective tool within homogeneous societies, landscapes, and ecosystems where explanations, processes, and patterns repeat at different scalar levels. But when scaling is undertaken in heterogeneous societies, landscapes, and ecosystems, malfunction arises. Effective scaling within heterogeneous systems must take into account level-related characteristics, including the ideological and cultural context and the types of polity, policy and politics, landscape, and ecosystems involved.

Future Research

This chapter has presented a number of insights for future research to test further. Additional studies could also take a less state-centric approach to enhance analysis of the motivations, methods, and consequences associated with scaling in a world increasingly conscious of its pluralistic nature.

A focus on scalar transferability to the international level may be premature. Should the world become "flat," as Friedman (2006) hypothesizes, in terms of increasing uniformity of culture, science, and communication, or should the world become ideologically one, as Fukuyama (1992) predicts in his end-of-history analysis, scaling will need to take into account far fewer level-related properties. Such uniformity is a long way off, however, as Fukuyama increasingly acknowledges and as Friedman fears. It remains useful, therefore, for researchers to continue to understand how level-related features and context-relevant institutional capacity are likely to necessitate the transformation of problem definitions and solutions as they are transferred from one level to another.

This chapter has looked primarily at the administrative scale, but future research needs to look at other scales, particularly that of time (Berkes 2006a). The literature on slow and fast variables (e.g., Gunderson and Holling 2002) and the role of different institutional designs in dealing with slowly evolving problems like HIV/AIDS and ones evolving faster like those related to the tsunami of 2004 may provide interesting research questions (Adger et al. 2006). Finally, more research is needed on cross-scale linkages (Lebel and Imamura, forthcoming).

Policy Implications of Scale

First, policy makers, like scientists, are often caught in territorial traps. To those operating at the local level, the local level can often seem all-important, and it is similarly so at other administrative levels. Awareness that there is no optimal level for a problem along a scale is critical. A key message arises: levels are identified through a political framing process, a process that itself changes the nature of the problem, the way it is measured, and the menu of possible solutions.

Second, scalar analysis can help to explain the importance of scales and levels. For a policy maker it is not enough to be confronted with facts about mean global temperature rise or sea level rise. It is far more relevant to understand how this will manifest itself in terms of increased variability and reduced predictability at specific levels and within specific contexts relevant to the policy maker's jurisdiction (Krupnik and Jolly 2002; Berkes 2006a).

Third, scalar analysis produces a unifying effect in showing that for global and globally recurring problems, solutions need to be both multi-level and multiscale; they must also be mutually consistent and context relevant, taking into account specific features, as applied at different levels along each scale. This has been emphasized in a number of policy documents. For example, the implementation plan of the World Summit on Sustainable Development points to the significance of the multilevel nature of such problems by "encourag[ing] relevant authorities at all levels to take sustainable development considerations into account in decision-making" (United Nations 2002, section III.19). The phrase *at all levels* is mentioned eighty-one times in the sixty-two-page document.[6] At the same time, tasks relating to only one solution may need to be assigned at different administrative levels and/or times. This requires assignment of different responsibilities to different administrative levels based on the specific level characteristics and the actors involved (Berkes 2006a). An emphasis on comanagement, complexity management, and adaptive management comes as a natural corollary.

Fourth, although the principle of subsidiarity has become a long-accepted governing principle in many arenas, research about how subsidiarity translates into practice shows it is difficult to guarantee its aim of local control over local issues. With globalization even the smallest of issues can be affected by decisions taken elsewhere and at larger scales. Subsidiarity has become unworkable and more an illusory panacea offered to local and national governments in return for loss of sovereignty.

Fifth, where solutions are downscaled, unless they are transformed endogenously to match the context for their lower-level application, they may be irrelevant and/or destroy the self-organizing capacity of groups, with the result of disempowerment of people at the lower level and a loss of their existing problem-solving approaches, which are often not "legible" to the outsider (Spierenburg, forthcoming). Thus, context-relevant solutions need to be crafted in close cooperation with local actors with context sensitivity by knowledge brokers (Büsscher and Critchley, forthcoming). Multilevel alliances are needed among actors and networks to promote the successful transformation from one level to another (Compagnon, forthcoming).

Sixth, where the goal is to protect ecosystems, a focus on larger ecosystems may often entail protection for smaller ecosystems. (This is generally true except where there are linear elements in a landscape, such as shorelines, that require more specific protections [Vermaat and Gilbert, forthcoming].)

Seventh, protecting single ecological functions may lead to a greater loss of biodiversity (Vermaat and Gilbert, forthcoming). Similarly, attempting to address individual environmental problems may lead to new environmental problems. Addressing the ozone depletion problem by introducing hydrochlorofluorocarbons has inadvertently exacerbated the climate change problem.

Finally, while addressing problems at a global level allows for dealing with all externalities, an institutional framework to address a problem does not always result in effective policy articulation and implementation. Those charged with implementation may be less motivated to follow through given the perception that global problems tend to be abstract, distant, and not immediately relevant to the local context; and that global science tends to provide an overall perspective and averaged statistics. However, while addressing problems at the local community level may lead to better policy articulation and implementation, since they involve a basis of more detailed information, there may be a tendency not to factor in indirect causes and impacts, resulting in a reduction of the effectiveness of problem solving.

The politics of scaling reveals a low level of joint concentration on problems by actors across the world, a low level of solidarity. This is especially remarkable given that solidarity is a key ingredient of achieving sustainable development. The environmental justice literature shows that the weak and powerless are more vulnerable than the strong and powerful to the impacts of environmental problems (Adger et al. 2006; Berkes 2006a). Scalar analysis shows how powerful actors and networks use scalar strategies to protect their own interests at the cost of the interests of others. Policy makers need to be aware of the equity implications and learn to use scaling to help bring about coordinated, consistent, and effective effort at all appropriate levels to solve globally prioritized problems.

Acknowledgments

I thank the Netherlands Science Foundation's VIDI for their grant (452-02-031), the Netherlands Royal Academy of Science for its support for

two workshops, all the authors cited, and my co-editor on the manuscript on scale, Dave Huitema.

Notes

1. I am indebted to Shobhakar Dhakal for this insight.

2. I am indebted to Agus Sari for bringing up this point.

3. This point was raised as a critical issue by Louis Lebel in Bali.

4. See, for example, Newman and Thakur (2006) for an examination of how multilateralism is under challenge and how the recent behavior of the United States in international negotiations shows its unwillingness to be subject to an international system with rules, a position that may have consequences for its domestic and international policies.

5. This has been done implicitly and explicitly in the Dublin Declaration of 1992 and the World Water Conferences of 1997, 2000, and 2003.

6. I am indebted to Fikret Berkes for this point.

IV

Policy Relevance and Future Directions

8

Contributing to the Science-Policy Interface: Policy Relevance of Findings on the Institutional Dimensions of Global Environmental Change

Heike Schroeder, Leslie A. King, and Simon Tay

Introduction

Arthur Schopenhauer is widely reported to have said, "All truth passes through three stages. First, it is ridiculed. Second, it is violently opposed. Third, it is accepted as being self-evident." Although this phenomenon occurs often, the forces behind the progression are not fully understood. The science-policy interface in the field of global environmental change offers a clear example. Critical moments arise when opportunities emerge to reframe issues and adopt new policies. But these opportunities are fleeting, so it is essential to be vigilant and prepared to take the initiative when they appear.

In this chapter we discuss the policy relevance of the findings of the Institutional Dimensions of Global Environmental Change (IDGEC) project in the hope that the work will enhance the capacity of policy makers to make good choices as opportunities arise to address large-scale environmental issues. We begin by providing a brief account of the nature of the science-policy interface. We then describe the strategy suggested by the project findings to reduce the gap between science and policy with regard to matters of governance, and we then offer general advice regarding policy making in the form of observations or insights derived from research conducted under the auspices of IDGEC.

The Science-Policy Interface

The gap between science and policy has stimulated concern in both communities and given rise to the following questions: How can science better inform and improve policy making? How can policy makers influence research agendas to meet their knowledge needs? What are the obstacles to improved science-policy interaction, and how can we overcome those obstacles? One way to bridge the gap is for scientists to sharpen their understanding of the policy process and to identify points of entry where input can be effective in meeting the needs of policy makers. In other words, scientists should bring their tools of analysis and inquiry to the policy-making process and use their improved understanding to open the way for better interaction at appropriate stages in the cycle of policy making.

Many observers have identified distinct stages of the policy process, including issue identification and framing, agenda formation, championing the issues, adopting a policy, implementing the policy selected, and evaluating the results (Stone 2002). Whether an issue rises to the top of the political agenda and stays there long enough to become the focus of policy making often has more to do with its perceived salience, legitimacy, and credibility than with the science that elucidates the nature and extent of the problem (R. Mitchell et al. 2006). It is easier for scientists to influence the policy process at some stages (e.g., framing issues and establishing the relevant discourse) than at others (e.g., policy selection). It is also clear that more opportunities for interaction between scientists and policy makers would contribute to closing the science-policy gap.

Analysts have identified a number of significant obstacles to improved science-policy interaction that may be difficult to alleviate, including biases and stereotypical thinking on the part of members of the two communities; differences in language and culture; gaps in priorities, such as political concerns versus academic rewards and incentive structures; and—perhaps most important—different time frames associated with ecological and geological processes, intergenerational concerns, and electoral cycles. Unequal power relationships, divergent standards of evidence, and decoupling of decision making from knowledge at different levels in the political system also prove to be obstacles to bridging the science-policy gap. The science and policy communities are social enterprises with their own values, interests, norms, rules, habits, and organizations. These institutional and organizational differences act as powerful constraints on efforts to improve science-policy interactions.

Mind-sets are another key issue. Experts and scientists are comfortable exploring complex issues in a variety of ways, starting with observation first and perhaps foremost. In contrast, policy makers often begin with the end in mind, rushing toward a proposed solution and actions needed to implement it. Whereas experts and scientists often anticipate issues and enjoy the luxury of having time to conduct research to enhance their understanding of them, policy makers today are so busy and inundated that it often takes a crisis or an emergency before they can find time to focus on a particular problem, even if it has been on the agenda for some time. Policy makers most often live within a paradigm of reaction, whereas experts and scientists try to work within time frames and parameters of anticipation. The difference between the two communities is much like the difference between a stopwatch and a grandfather clock.

Rather than declaring defeat, however, many observers have focused on ways to improve the relationship or to provide intermediaries who can lower the barriers between the two communities. In the field of global environmental change, bridging the science-policy gap is a matter of applying knowledge to action for sustainability. A growing body of literature points to a number of roles that are important in this context, including those of knowledge brokers (Litfin 1994), boundary organizations (Cash 2001; Guston 2001), champions for specific issues (often nongovernmental organizations [NGOs]), science advisers, scientific committees and advisory bodies, funding agencies, science coordinators, and even individual scientists and policy makers who reach out beyond the confines of their own communities. Processes that can help in this setting include scientific assessments such as the periodic assessment reports of the Intergovernmental Panel on Climate Change (IPCC), the Arctic Climate Impact Assessment, and the Millennium Ecosystem Assessment. Capacity building for improved science-policy interaction can take the form of increased participation of scientists located in government agencies, scientists engaging in political processes, informal interaction among scientists and policy makers, and the engagement of policy makers in setting research agendas.

Although members of the IDGEC community developed a lively interest in a number of these procedures for improving science-policy interactions, the project developed a particular interest in the role of knowledge brokers in improving the science-policy dialogue. Those who play this role effectively are able both to communicate the findings of scientific research in a manner that is comprehensible to policy makers and to help scientists understand the nature of the policy process and how it differs from scientific activities. Successful knowledge brokers are highly skilled at communicating these matters in a manner that is free of technical jargon understandable only to specialists. IDGEC made a deliberate effort to identify individuals able to play the role of knowledge broker relating to the institutional dimensions of global environmental change and to draw these individuals into the activities of the project. Individuals identified as knowledge brokers played a pivotal role at the IDGEC Synthesis Conference in Bali in December 2006. They participated in discussions of the scientific findings and implications for improved policy making in the global change arena. This experience indicates that policy makers and scientists should devote time and energy to identifying and engaging such knowledge brokers at different stages of the policy process and assisting them to link scientists and policy makers. The ideas we present in the next two sections owe a great deal to the constructive contributions that these individuals made at key stages in the work of the project.[1]

Six Insights for Policy Making

Faced with the challenge of (re)designing institutional arrangements to address global environmental problems, scholars and practitioners have attempted to formulate and implement general design principles that prescribe institutional and organizational solutions for policy problems (Ostrom 1990). Although such design principles may appear useful at first sight, research conducted under the auspices of IDGEC demonstrates that there are no simple recipes when it comes to designing institutional arrangements to solve specific environmental problems; the devil is always in the details. Because of the dynamic and complex nature of socioecological systems, design principles have proved to be disappointing in terms of providing useful advice to policy makers. In the domain of institutional design, one size definitely does not fit all.

Rather than declaring that no design guidance is possible, however, IDGEC has developed an approach to institutional design that centers on the idea of institutional diagnostics. The diagnostic method involves a two-step process in which a problem is systematically diagnosed prior to devising a specific governance system to address that problem (Young 2002b, chapter 4 in this volume). The diagnosis encompasses an assessment of the nature of the problem at hand, the players involved, the politics of the relevant issue area, and the established practices in that area. The solution takes the form of a set of recommendations emphasizing the selection of institutional arrangements that are well matched to the characteristics of the issue at hand (Young, chapter 4 in this volume). This method does not constitute a cookie-cutter approach. Rather it provides guidance to those seeking to (re)design institutional arrangements in a way that is well matched to the problem to be solved, so that institutional mismatches (see Galaz et al., chapter 5 in this volume) can be avoided. The basic idea is to examine an issue area systematically and to take care in identifying its specific attributes before devising institutional arrangements and policies.

These IDGEC findings provide the basis for some general insights that should be helpful to policy makers and institutional designers endeavoring to solve large-scale environmental problems. In this section we build on the diagnostic findings identified in chapter 4 and apply those observations to policy issues. We identify six general insights and illustrate them with a few simple examples relating to environmental concerns.

Insight 1   The diagnostic method can be used to identify key features of the relevant socioecological system and to craft institutions that are well matched to that system.

As chapter 5 demonstrates, institutional misfits arise frequently as policy makers create institutions that are poorly matched to important biophysical and socioeconomic aspects of a problem. These misfits often occur as a result of mismatches along the scales of space, time, and social organization. Once created, these mismatches are extremely hard to undo. Policy makers may attempt to solve local problems through the imposition of national policies or treat global problems by taking initiatives at the local level that are not sufficiently comprehensive and fail to come to terms with the systemic or global character of the problem. The resultant mismatches between properties of socioecological systems and attributes of institutional arrangements or regimes created to solve problems are pervasive and often intractable. How can we alleviate these mismatches and improve the fitness of the institutional arrangements created to manage human activities affecting large-scale environmental systems (Cleveland et al. 1996; Berkes and Folke 1998)?

Research conducted under the auspices of IDGEC makes it clear that fitness—from the point of view of scale, and from both the project's original perspective of match with the biophysical aspects of the problem and its more recent, expanded view that also takes into account the socioeconomic elements of the level at which the problem exists—are major determinants of institutional performance. Generalized design principles cannot be employed in order to alleviate mismatches that have occurred as a result of applying a favorite prescription to problems with distinct socioecological characteristics. Many writers have advocated privatization, for example, as the solution to a wide range of environmental problems. But such simple prescriptions fail to recognize the complexity and diversity of environmental problems (von Weizsäcker, Young, and Finger 2005). In some countries privatization often compounds problems, given imperfections in market conditions and even collusion between governments and the private sector as well as failure to recognize the importance of the environment as a public good.

Rather than espousing any single approach to solve all problems, the diagnostic method provides a tool kit and some general findings regarding matters of fit, interplay, and scale that should assist policy makers in developing more targeted responses, avoiding pervasive institutional mismatches, and therefore improving the performance of regimes dealing with specific problems. Although this may seem commonsensical, policy makers, for reasons stated above, all too often end up adopting policies borrowed from other countries and contexts without adapting them to the situation at hand, much less coming up with innovative solutions.

The case of biodiversity illustrates the problems that occur when policy makers attempt to create uniform arrangements as applicable to all areas. Rather than finding a single, global approach, the challenge of biodiversity is to devise institutional arrangements that can deal effectively with a limited number—a few dozen at most—of "biodiversity hot spots" that are widely distributed in places like Madagascar and the central Amazon basin. Because these hot spots are located within the jurisdiction of a limited number of countries, there is no need for a global regime that seeks to impose a uniform set of regulations across the board. What is needed is an incentive system that appeals to policy makers in a position to make decisions relevant to those areas where hot spots are located. The Montreal Protocol Multilateral Fund, created to help developing countries phase out the production of ozone-depleting substances, may offer a useful precedent in this realm.

Similarly, in Southeast Asia there is the problem of recurring land and forest fires that cause regional haze pollution that has affected public health and caused billions of dollars in economic damages. These fires are started predominately in a handful of Indonesian provinces and are most harmful when located in degraded peatlands. Efforts to solve this problem need to focus on these specific areas and to appeal to local officials and communities. But most of the efforts to date have focused on the interstate level and have floundered, in part, because the Indonesian capital of Jakarta is not directly affected by the haze.

The problem of climate change requires a different approach. Whereas the previous two environmental problems are cumulative in nature, climate change is a systemic problem. Mitigating it will, in the end, require collaboration among all significant emitters. Also, it calls for alertness to interactions with many other environmental problems, including ozone depletion, loss of biodiversity, desertification, and water scarcity as well as socioeconomic changes, including various manifestations of globalization. International negotiations on climate have been hampered severely by the obstructive attitude of a small number of influential players. The best way to show leadership regarding this problem is to make reductions in emissions regardless of the actions of others, even though this seems counterintuitive given that the problem can be solved only systemically and by (near) universal cooperation. Given the synergies with other pressing environmental problems and the positive side effects from climate mitigation socioeconomically, countries can only benefit from early action.

Insight 2   The role of discourses in framing problems and determining the character of solutions can be significant.

IDGEC has played a prominent role in shifting the discourse regarding large-scale environmental issues away from a focus on the role of governments and toward a broader perspective that highlights *governance* (Young 1999b). In part because of this contribution, it is now common to consider options involving public-private partnerships and various forms of private governance in efforts to solve problems of governance. This is a particularly important contribution in the context of many developing countries in which the idea of strict state sovereignty has often been invoked as an excuse for inaction and the role of nonstate actors has been questioned and even denied.

The creation of the U.S. Climate Action Partnership—including an agreement among ten private corporations brokered by Environmental Defense—with the goal of meeting the targets of the Kyoto Protocol through private initiatives is illustrative of this growing trend. Shifting the discourse in this way brings into focus opportunities for creating institutional arrangements operating outside conventional multilateral environmental agreements created and implemented by governments.

The shift to a governance approach encourages consideration of ways of solving governance problems or fulfilling the functions of governance that do not require the creation of yet another public agency, cabinet ministry, or formal governmental committee. This leads to a focus on the nature of the problem and the sorts of arrangements that may best solve that problem. Embedded and sometimes competing mental models or discourses often limit the ability of policy makers to consider alternative solutions that may yield better results.

In the case of global deforestation, conflicts between discourses, such as between the managerial/rational and the pluralist discourses, are at play in efforts to understand the causes of the problem (Adger et al. 2001). Some observers target shifting cultivators in developing countries as the villains and argue that population growth, increasing demand for fuel wood, and a lack of concern with long-term consequences are the causes of deforestation. Others point to globalization and exploitation by multinational corporations that act as "roving bandits," clear-cutting and moving on with no regard for the future. Careful studies on the ground make it clear that neither of these mental models can account for the scale and speed of deforestation in actual situations on the ground. But it is easy to slip into the habit of relying on mental models of this sort. When it comes to the creation and implementation of institutions, the result is apt to be regimes that do not fit the problem well and consequently do not solve it.

The case of biosafety and trade presents a good example of how a shift in discourse—in this case involving the precautionary principle—can have a significant effect on designing the governance system, in this case on international trade in genetically modified (GM) crops. Pushed by the Europeans, who had already established the precautionary principle at the European Union level, the shift took place in the period 1997–2000 and strikingly in the context of the negotiations leading to the Cartagena Protocol on Biosafety. All members of the Convention on Biological Diversity took part in the negotiations, and the protocol was adopted by consensus. The precautionary principle thus became accepted by the international community as the proper vantage point for evaluating on a case-by-case basis new products or technologies whose adverse impacts on health have yet to be established (Andrée 2005). The Cartagena Protocol on Biosafety requires that countries exporting genetically modified organisms (GMOs) solicit an Advance Informed Agreement based on the results of a risk analysis from the importing country prior to the export of GMOs intended for introduction into the environment.[2] The protocol thus allows countries to restrict imports of GMOs on precautionary grounds when they believe they may be harmful to ecosystems or to human health (Andrée 2005). In the absence of a shift in discourse, the rules would unquestionably have been different. The establishment of the Cartagena Protocol on Biosafety may even have seemed redundant, as there would not have been an internationally accepted mechanism to protect against possible harmful impacts resulting from trade in GMOs.

Insight 3 Linear thinking in dealing with complex and dynamic systems in which changes are apt to be abrupt, discontinuous, surprising, and irreversible can be misleading.

The mental models of environmental managers and policy makers typically regard environmental problems as developing slowly and evolving in ways that are linear and relatively insensitive to initial conditions. This mentality underlies the common process of budget allocation for government agencies in which expenditure for one year is usually expected to follow the experience of the previous year, unless there are clearly exceptional circumstances. From this point of view, change is predictable in broad terms and unlikely to produce abrupt and irreversible consequences. But in reality problems often involve highly dynamic systems that undergo abrupt changes with irreversible consequences. Thus, real conditions often differ—sometimes dramatically—from those assumed implicitly. In reality managers have to confront situations featuring thresholds leading to regime shifts, nonlinear consequences, and emergent properties that are difficult to identify in advance. Policy makers are generally unprepared for such shifts. Even worse, so are implementing agencies and bureaucracies as well as systems for resource allocation.

Fisheries management offers clear examples in this realm. Why do those responsible for managing fisheries so consistently fail to understand, act on, or plan for population collapses, even when they are obvious? Again and again, collapses of fish populations take managers by surprise, though it is clear in retrospect that such collapses are consequences not simply of overfishing but also of pushing ecosystems to thresholds that lead to rapid change. A particularly excruciating example involves the collapse of northwest Atlantic cod stocks in the early 1990s. Both the United States and Canada failed to deal with the problem, because they were managing cod stocks instead of thinking about the

Atlantic as a complex and dynamic system. Both scientists and managers were unprepared to deal with driving forces such as the cod's life cycle, fishing methods, water pollution, climate change, the larger food chain, and changes in fishing activities in response to the implementation of Exclusive Economic Zones. As a result, they failed to recognize the impending collapse until it was too late (Harris 1998). One consequence of focusing on complex and dynamic systems is a constant need to cope with uncertainty as the interplay among numerous drivers is not well understood. This recognition leads to models and practices that differ from those created to understand deterministic systems. To be successful, institutions dealing with such problems need a lot of flexibility, early warning systems, and prospecting mechanisms.

In many developing countries the situation is exacerbated by the lack of capacity, technology, and resources for planning and response. Our example of the fires and haze in Southeast Asia further demonstrates that early warning, even when it is present, may not lead to effective response. The fires have been recurring since 1997; Singapore has provided Indonesia with weather forecasts and satellite images covering the epicenter of the fires, which can and do serve as early warning and timely detection mechanisms. Yet the institutions in Indonesia and ASEAN (the Association of Southeast Asian Nations), the relevant regional organization, have struggled to respond to the fires in a timely and effective manner (Tay 1999).

The rules of both the Cartagena Protocol on Biosafety and the World Trade Organization (WTO) reflect the current state of knowledge regarding effects arising from transboundary movements of GMOs. But knowledge is dynamic; it may change or develop at any time. The potential impacts of GMOs may be recognized as less harmful than currently feared by many consumers and governments, or they may turn out to be extremely harmful to ecosystems and human health and even irreversible in their effects. A major GM food scare could suddenly shift debates on GMOs. In short, the growth of trade in GMOs has the potential to spark nonlinear changes regarding ecosystem health, human well-being, and even ethical concerns.

Insight 4   Unintended consequences arising from institutional interplay can be severe.

The concept of institutional interplay was one of the major themes of the IDGEC project (Gehring and Oberthür, chapter 6 in this volume). Issue-specific institutions increasingly interact with one another, either at the same level of social organization or across levels. This, in part, is because of interconnections among environmental problems as well as the increasing density of institutions operating to address complex problems. The implication of the growth of institutional interplay is that managers can no longer afford to view problems and institutions in isolation. Because institutional interactions are widespread and increasing, unintended consequences of interactions among them are becoming more apparent and often conflictual. Institutional designers must therefore pay focused and conscious attention to by-products and unintended consequences of the institutional arrangements they create to deal with specific problems.

A striking example of the perils of ignoring the consequences of institutional interplay has arisen regarding the climate impacts of trifluoromethane (HFC-23), a potent greenhouse gas (GHG) and a natural by-product of chlorodifluoromethane (HCFC-22), which is also a GHG (*New York Times* 2006). Although it is banned under the Montreal Protocol for industrialized countries, it is still produced on a massive scale in countries including China, India, and Korea. Because it is cheap to capture and burn HFC-23, producers of HCFC-22 are in a lucrative business. Getting the cleanup accredited and paid for through the Kyoto Protocol's Clean Development Mechanism naturally provides incentives to continue with the practice of producing HCFC-22. The moral of this story is that it is important not to think about problems in isolation, but to pay careful attention to institutional interplay likely to arise in efforts to mitigate problems like climate change. In such cases coordination across institutions is needed.

The fires in Indonesia provide another good example of interplay. The Indonesian Ministry of Environment—together with local officials and NGOs—intends to address the fires and haze. But the Ministries of Forestry and of Agriculture have been largely quiescent or even resistant to this effort. In large part this is because the underlying cause of the fires relates to regimes governing land conversion that benefit forestry and agricultural interests. Thus many of the fire-prone peatlands are being opened for operations on the part of pulp and paper companies, and other areas are being developed for palm oil plantations. Although the Indonesian president has called for action, solutions will be hard to come by unless greater attention is paid to the unintended consequences of these forestry and agricultural policies.

Given the substantive differences between the WTO and the Cartagena Protocol on Biosafety with regard to objectives, rules, and norms and, at the same time, the degree of jurisdictional overlap between the two regimes, there is a high likelihood of unintended consequences arising from institutional interplay in this realm. The biosafety protocol is faced with the challenge of developing in a manner acceptable to both exporters of GMOs, who are eager to reap financial benefits from modern biotechnology, and importers, who seek to avoid or minimize the inherent risks. While the biosafety protocol reflects the concerns of importers, the perspectives of exporters are embedded in the WTO Agreement on the Application of Sanitary and Phytosanitary Measures, an arrangement dedicated to avoiding protectionist misuse of health and environmental regulations (Safrin 2002; Brack, Falkner, and Goll 2003; Isaac and Kerr 2003; Oberthür and Gehring 2006c; Schroeder 2008; Young et al. 2008).

Insight 5   Using analogies, especially in cases where they cut across levels of social organization, can backfire.

Reasoning by analogy, and adopting institutional arrangements that have proved successful in one setting or at one level of social organization, is all too tempting but can lead to mistakes in transporting solutions from one setting to another. There are a number of reasons for this. The players may be different, aspects of the social system may differ, and compliance measures that work at one level may be ineffective at others.

This is an acute problem for developing countries. In part it is driven by the assumption that developed countries provide examples of environmental policy that can be copied. This notion is made worse by any number of so-called policy experts from international organizations, developed countries, or even within the developing world who, consciously or otherwise, assume that what is good for the West is also good for the rest. As such, many developing countries swing between outright rejection of Western examples and experiences in the environmental field and a nearly blind adoption and devotion to imported policies and institutions. A more balanced approach of adaptation emphasizing the role of adaptive management would be preferable.

The cap-and-trade system introduced at the national level in the United States to deal with emissions of sulfur dioxide (SO_2) is also an example of bad analogies. This arrangement is widely regarded as successful. But various factors may cause failure of such a system in other settings and especially at the global level. The cap may be set too high; compliance may be hard to monitor; provisions relating to banking and offsets may generate unforeseen problems. With regard to SO_2 at the national level, emissions are monitored at the smokestack, and emissions data are fed directly into an electronic repository. Noncompliance is not an option. These measures would be difficult to replicate at the global level. Setting the cap and reducing it over time are politically difficult enough at the U.S. federal level; they would be even more difficult at the global level. Yet regional emissions trading schemes have the attraction of allowing participants to gain experience with carbon markets. They allow individual firms to gain experience in this realm while polishing their reputations for social responsibility (Kruger 2005). From the perspective of institutional design, moreover, they may open up opportunities for adaptive management in which lessons may be learned from actual experience with a variety of systems that differ from one another in significant ways. A particularly interesting feature of this situation is the opportunity it affords for various nonstate actors to assume key roles.

Insight 6   Form should follow function.

It is a common mistake to create a new organizational structure and then to decide what it should do and how it should operate, rather than focusing first on the problem and designing the institution to fit the problem in terms of scale, the nature of the problem, actors, and so forth. When faced with a new problem, policy makers and others tend to think in terms of (re)forming organizations. This makes sense given that institutions are intangible and that it is easier to imagine a new ministry, agency, or committee than to envision appropriate institutional arrangements. In applying the maxim that form follows function, policy makers should start by thinking about the nature of the problem to be solved or the goal to be fulfilled and then create administrative arrangements most appropriate to dealing with the problem or goal at hand. In short, we need to shift the focus from organizations to institutions.

Recent proposals calling for a world environment organization illustrate this point. Numerous policy makers as well as scholars have proposed the creation of a new or upgraded global environment organization, variously called the World Environment Organization or the UN Environment Organization. In either case, the result would be an intergovernmental organization to highlight the role of member states as opposed to nonstate actors and civil society. Would this be the right form to meet the demand for global environmental governance in the twenty-first century? Or would it close off or divert attention from many other approaches to meeting the challenge of governance?

IDGEC research highlights serious issues involved in such questions. The growing role of nonstate actors and, as a result, the emergence of private governance is undeniable. The need for institutions and the organizations created to manage them to be agile in adapting to nonlinear and sometimes abrupt changes is clear. And the importance of incentive mechanisms in contrast to conventional regulatory schemes is widely accepted. For all these reasons, in today's world it would be shortsighted to assume that conventional intergovernmental organizations offer the right mechanism for implementing governance systems designed to deal with large-scale environmental issues. What is needed now is innovation regarding suitable organizations rather than relying on arrangements that amount to business as usual.

The goal in negotiating the terms of the post-Kyoto climate regime is to reduce net GHG emissions substantially, regardless of the organizational arrangements needed to fulfill this goal. Remaining flexible regarding matters of form while pursuing this substantive objective may lead to important innovations regarding organizations. A more effective successor to the current regime might well accord an important role for municipalities, not only in implementing arrangements devised at higher levels but also in coming up with innovative mechanisms for reducing our dependence on systems that are large emitters of GHGs. Similar remarks are in order regarding opportunities for nonstate actors (e.g., associations operating at the level of civil society) to influence the behavior of those whose actions give rise to high levels of emissions.

Concluding Remarks

The IDGEC research and policy communities confirmed a serious communications gap between science and policy in the global environmental change arena. The project findings indicate that new institutions and the redesign of existing institutions are needed to confront emerging problems and sets of interdependent problems. Yet policy makers have not taken up findings about the role of institutions in causing and mitigating global environmental change, the effectiveness of current institutions, and the institutional mechanisms required to improve institutional effectiveness. Obstacles to improved science-policy interaction include vastly different time horizons, lack of opportunities for scientists and policy makers to interact, and lack of knowledge about the policy-making process and opportunities for scientific input and influence. Ways to overcome these obstacles include identifying and supporting knowledge brokers to link the two communities, increased opportunities in the research process and in funding decisions, and, for policy makers, making their knowledge needs known to the scientists and funding bodies.

Awareness of environmental problems has grown rapidly in recent years. Currently, concern about climate change is rising sharply. This concern is not limited to members of economic and social elites, nor is it confined to Western, developed societies. Concern about such matters is now on the policy agenda in many developing countries as well. Widespread awareness is producing a search for new institutions and institutional mechanisms that can lead to innovative ways to address these problems. IDGEC has highlighted the role of institutions not only in causing but also in mitigating large-scale environmental problems. The project does not offer simple solutions. On the contrary, a major finding is that one institutional size or design does not fit all global environmental problems. Governance systems need to be well matched in a range of key ways to specific problem contexts. IDGEC has provided a tool kit for institutional diagnostics and design that can help policy makers seeking to address global environmental problems to arrive at workable solutions on a case-by-case basis.

Notes

1. See the project's Web site at http://www2.bren.ucsb.edu/~idgec/abstracts.php.

2. In the Convention on Biodiversity and its Cartagena Protocol on Biosafety, GMOs are subsumed under the broader concept of Living Modified Organisms (LMOs).

9

Earth System Governance: A Research Agenda

Frank Biermann

Introduction

With this volume the long-term international research project Institutional Dimensions of Global Environmental Change (IDGEC) reaches closure. Using an extensive and ambitious research agenda on the role of institutions in global change, the project has analyzed institutions at all levels of human activity, from local communities to international relations. At the same time IDGEC has maintained its intellectual link to the larger social science community and in particular to the new institutionalism in the social sciences. All this has made the IDGEC project a success story, as this volume documents.

The question arises at this juncture whether there is a need for a new research effort that would build on IDGEC's legacy and if so, how this research effort should be structured. In 2004 members of the IDGEC community initiated a consultation process on this issue. The response from the community overwhelmingly indicated the need for a second phase of IDGEC and showed much enthusiasm to take up the venture. This chapter draws on these deliberations and provides a first outline of a new long-term research program as both a continuation and an extension of IDGEC.[1]

The second section discusses the overall perspective of a new research activity that would build on, but also further develop, the IDGEC legacy. In particular it sets out how research on institutions could respond to two new developments that have gained relevance over the duration of the IDGEC project: first, the new focus on governance systems, as opposed to single institutions; and second, the emergence of a worldwide Earth system science community that poses particular challenges for, and demands upon, scholars working on institutions and governance. Given these two new trends, I propose to frame the next research program under a new overarching research paradigm: "Earth system governance" (Biermann 2002b, 2005, 2007).

The third section lays out five analytical problems that I see at the core of a new research effort on Earth system governance. These analytical problems build on the former Science Plan of IDGEC and the decade of research that has followed. Yet they also further develop several aspects from IDGEC's legacy and add new foci that have gained political and scholarly relevance since IDGEC was first conceived.

Overall Perspective of a New Research Effort

From Institutions to Systems of Governance
The focus of IDGEC has been, as indicated in the project's name, on institutions defined, as seen in chapter 1, as clusters of rights, rules, and decision-making procedures that give rise to social practices, assign roles to participants in these practices, and govern interactions among players of these roles. There is wide support in the community to extend this focus in a new research effort to a broader perspective that looks at entire governance systems.

Governance has been defined in a variety of ways, and there is no consensus among scholars on the core elements of this concept (overviews in Alcántara 1998; Van Kersbergen and Van Waarden 2004). In most bodies of literature the notion of governance adds to the concept of institutions a dynamic perspective that looks at processes of governing; a stronger focus on governance systems that integrates research on interlinkages of single institutions; and a stronger emphasis on actors and especially on nonstate actors. At the national level governance usually denotes new forms of regulation that differ from traditional hierarchical state activity and implies some form of self-regulation by societal actors, private-public cooperation in the solving of societal problems, and new forms of multilevel policy. At the international level the term *global governance* is often used to describe processes of modern world politics, although here, too, no consensus on an appropriate definition has been reached.[2]

In addition to its analytical usage, the term *governance* is also used prescriptively as a political program to cope with problems of modernity, for example, in calls for global governance as a counterweight to globalization and for new and more effective international institutions, organizations, or financial mechanisms. Importantly, at both national and international levels, the concept of "governance" is not confined to states and governments as sole actors but is marked by participation of myriad public and private nonstate actors at all levels of decision making, ranging from networks of experts, environmentalists, and multinational corporations to new agencies set up by governments, such as intergovernmental bureaucracies. Governance systems thus also include widely shared belief systems or actor networks such as public policy partnerships.

The concept of governance is therefore broader than the concept of institutions. It covers a wide area of phenomena that are crucial for understanding steering systems in the field of human dimensions of global environmental change but that are not completely addressed through the notion of institutions. On the other hand, governance systems generally include one or several institutions, and therefore much of the IDGEC legacy on institutions will be an integral part of a future governance research agenda. In sum, a wider focus on governance systems—which would include institutions—would further enrich, rather than limit, the research agenda for a new program.

From Environmental Research to Earth System Science
In addition to the emergence of the discourse on governance, a new research effort in this field has to recognize the developments within global change research, in particular the evolution of integrating concepts such as Earth system analysis, Earth system science, and sustainability science.

The notion of "Earth system analysis" has emerged from the complexities of global environmental change that require the involvement of most academic disciplines at multiple spatial and temporal scales. Especially in the natural sciences that build on quantification and computer-based modeling, efforts have long been under way to combine and integrate models of different strands of research to gain understanding not of isolated elements of global change, but of the totality of processes in nature and human civilization. Integrated Earth system analysis as a scientific enterprise is the consequence of these efforts. Hans-Joachim Schellnhuber (1998, 1999), a key proponent of the concept, ascribes to Earth system analysis the status of a science *in statu nascendi*, because, as he writes (with Volker Wenzel), it has "1. a genuine subject, namely the total Earth in the sense of a fragile and 'gullible' dynamic system, 2. a genuine methodology, namely transdisciplinary systems analysis based on, inter alia, planetary monitoring, global modeling and simulation, 3. a genuine purpose, namely the satisfactory (or at least tolerable) coevolution of the ecosphere and the anthroposphere (vulgo: Sustainable Development) in the times of Global Change and beyond" (Schellnhuber and Wenzel 1998, vii).

Earth system analysis relates to "sustainability science," a closely connected concept, to integrate different disciplines and communities in the larger quest for a transition to sustainability.[3] As Robert Kates, William Clark, and colleagues argue (2001), the challenge of sustainable development is so complex that it requires a "sustainability science" as a new integrative field of study. A sustainability science would improve collaboration of natural and social scientists as well as deliver research designs that better integrate all scales from local to global.

These integrated notions are reflected in the Earth System Science Partnership, an initiative of four global change research programs: the biodiversity sciences program diversitas, the International Geosphere-Biosphere Programme, the World Climate Research Programme, and the International Human Dimensions Programme on Global Environmental Change (IHDP).[4] The partnership builds on a holistic concept of the Earth as a complex and sensitive system regulated by physical, chemical, and biological processes and influenced by humans. It focuses on anthropogenic change, including through integrated approaches and advanced modeling technologies. To this end the partnership supports joint projects that cut across the various global change research programs, such as the Global Carbon Project, the Global Environmental Change and Food Systems project, the Global Water System Project, and the Global Environmental Change and Human Health project. Another recent type of crosscutting activity is represented by the regional integrated programs, such as the Monsoon Asia Integrated Regional Study.

A better understanding of governance mechanisms and institutions is crucial for the success of the joint projects within the Earth System Science Partnership. There is thus a growing interest in organizing research on institutions and governance as a crosscutting theme that would run through most programs and projects under the partnership. Furthermore, many researchers in the field of integrated Earth system analysis and sustainability science have become interested in incorporating governance and institutions into their models and research programs.

These developments may therefore advise better linking of institutional and governance research to the overarching concerns of the Earth System Science Partnership and recognizing this link through developing a research theme that focuses explicitly on Earth system analysis and governance.

The 2001 Amsterdam Declaration on Global Change, in which the four global change programs called "urgently" for "an ethical framework for global stewardship and strategies for Earth System management,"[5] could be considered a starting point. The notion of "Earth system management" appeals to parts of the policy world and is used in a number of natural science–oriented programs.[6] For most social scientists, however, the term is problematic. In social sciences "management" often relates to notions of hierarchical steering, planning, and controlling of social relations. For most social scientists, Earth system management as an analytical or normative concept would be both infeasible and—given its connotation of hierarchical planning—undesirable. Global stewardship for the planet is different from centralized management. Instead, it must be based on nonhierarchical processes of cooperation, coordination, and consensus building among actors at all levels. It must include state and nonstate actors. It must include complex architectures of interlinked institutions and decision-making procedures, but also different forms of collaboration, such as partnerships and networks.

In a world of diversity and disparity, Earth system "management" is no option. Instead, I argue, we observe the emergence of a different new paradigm: *Earth system governance*.[7]

Earth System Governance as an Overarching Research Paradigm
Earth system governance can be thought of first of all in phenomenological terms: it describes an emerging social phenomenon that is expressed in hundreds of international regimes, international bureaucracies, national agencies, local and transnational activist groups, local community initiatives, and expert networks. At the same time I understand Earth system governance as a political project that engages more and more actors who seek to strengthen the current architecture of institutions and networks at local and global levels. In both meanings, Earth system governance stands as a demanding and vital subject of research for the social sciences. It can be defined as the interrelated and increasingly integrated system of formal and informal rules and actor networks at all

levels of human society (from local to global) that are established in order to influence the coevolution of human and natural systems in a way that secures the sustainable development of human society.

Defined in this way, Earth system governance is as much about environmental parameters as about social practices and processes. Its normative goal is not purely environmental protection on a planetary scale, as this would make Earth system governance devoid of its societal context. Environmental targets within Earth system governance—such as control of greenhouse gases—can be reached through practice of global and local governance through different means with different costs for actors in different countries and regions. Earth system governance is thus about environmental protection as well as social welfare; it is about effectiveness as well as global and local equity. The normative aspiration of Earth system governance must hence be sustainable development framed within the triangle of ecological, economic, and social sustainability.[8]

Analytically, Earth system governance bridges traditional levels of analysis in governance and policy studies. On the one hand it goes beyond traditional environmental policy analysis as it emerged in the 1970s, with its focus on managing environmental problems of industrialized countries. The anthropogenic transformation of the Earth system encompasses more puzzles and problems than have been traditionally examined within environmental policy studies, problems that now range from changes in geophysical systems to the global loss of biological diversity. Key questions—such as how Bangladesh could adapt to rising sea levels, how deterioration of African soils could be halted, or how land-use changes in Brazil could be analyzed—have barely been covered by traditional environmental policy research. Yet they are inevitably part of the study of Earth system governance.

On the other hand, Earth system governance covers more than problems of the "global commons," but also local problems from local air pollution to the preservation of local waters, waste treatment, or desertification and soil degradation. Earth system governance thus exceeds the academic disciplines of international relations and international law. Contributions here remain crucial, however: hundreds of international regimes now regulate the environmental behavior of governments, and understanding the processes of these regimes and their interlinkages is ever more important. The international relations and law communities have produced a vast literature in this field. Yet these are still largely related to theory development within their own disciplines, less to research on domestic politics and to the larger global change research community. Earth system governance requires the integration of all these strands of research and must bridge scales from global to local. This need of integrated multilevel analysis is widely agreed upon in principle. It needs further efforts in practice.

As a research program as much as a political problem, Earth system governance poses daunting challenges.[9] Researchers and practitioners have to cope with persistent uncertainties regarding the causes of global environmental change, its impacts, the interlinkage of various causes and response options, and the effects of policies. Researchers have to deal with intergenerational dependencies that pose exceptional governance challenges, since causation and effect of Earth system transformations can be separated by decades, often by generations. Researchers must also respond to the functional interdependence of Earth system transformation and of potential response options, which links problems and solutions in one area with myriad other policy domains. Moreover, researchers and practitioners have to cope with new forms and degrees of global spatial interdependence that interlink policies in different locations. Eventually researchers have to deal with, and practitioners prepare for, extraordinary degrees of possible harm for which current governance systems might not be fully prepared. All this makes Earth system governance one of the most challenging, probably one of the most difficult, but at the same time one of the most fascinating areas in the social sciences.

Earth System Governance and Earth System Science

How can Earth system governance, as a social science research program, relate to the broader notion of Earth system science? From the perspective of integrated Earth system analysis, research on institutions and governance mechanisms is often viewed as part of the integrated effort and is formally included in most theoretical conceptualizations in this field. The physicist Schellnhuber, for example, has formalized the notion of a "global subject" S, which he conceptualizes as part of the human civilization H together with the anthroposphere A (the totality of human life, actions, and products that affect other components of the Earth system). Translated into social science language, this "global subject" S could be seen as the political system at the global level, including its national and subnational subparts, all of which share the collective ability to bring the "human impact" in line with the needs of the ecosphere (Schellnhuber 1999, C20–C22). Likewise, the Earth System Science Partnership asserts that "the core" of its activities is the "in-depth analysis and advanced modeling of the Earth System as a whole, incorporating data and information from the diverse fields represented by the four global change programs."[10]

In practice, however, it remains unclear to what extent institutional and governance research can contribute to, and integrate with, the more model-driven research programs, apart from problem-oriented, issue-specific collaboration. Quantifiable hypotheses and computer-based modeling are problematic for most students of institutions and governance—and are likely to remain so (Young, Lambin, et al. 2006). Social science research groups that attempt to use computer modeling and quantification as a tool for integrating governance research into larger models have still to provide convincing results. Qualitative modeling projects to analyze international governance processes and institutions are in their infancy (Eisenack 2003; Eisenack, Kropp, and Welsch 2006). Major problems in modeling governance processes remain. These include, to name a few, the complexity of relevant variables at multiple levels, human reflexivity, and difficulties in quantifying key social concepts such as "power," "interest," or "legitimacy."

Given this mismatch between formalized methods and fuzzy social realities, proponents of an integrated Earth system analysis often relegate governance research to an auxiliary, advisory, and essentially nonscientific status. Quite typical is the conceptualization of social science in the twenty-three overarching questions put forward by the Global Analysis, Integration and Modelling task force of the International Geosphere-Biosphere Programme for the Earth system analysis community (Schellnhuber and Sahagian 2002).

Some of these questions relate to the social sciences. These questions, however, are not viewed as "analytical" (questions exclusively related to natural science) but as "strategic" (for example, question 23, "What is the structure of an effective and efficient system of global environment and development institutions?") or "normative" (for example, question 18, "What kind of nature do modern societies want?"). The value of institutional research as an *analytical* program of inquiry is relegated to its policy-oriented, advisory dimensions. It appears that this is a logical outcome of an Earth system analysis program that is motivated by computer modeling and quantification.

Consequently, students of governance should resist subjecting their governance and institutional analysis of human-nature interactions to epistemological uniformity and to methods that are infeasible to implement and impossible to trust in the social sciences. Instead, social scientists will need to continue to develop *independent research programs* that are interdisciplinary across the different social sciences—for example, linking international relations and law—but that follow the internal logic and particular theoretical, epistemological, and methodological approaches of the social sciences and the humanities, which are essentially qualitative, case based, context dependent, and reflexive.

One overarching theme for such a research program is Earth system governance. The study of Earth system governance is thereby part of the larger project of global change research, yet it must also remain autonomous in its distinct methodological and theoretical development.

Global change research therefore rests on two theoretical and methodological pillars. One is Earth system analysis driven by an integrated, computer-based approach that brings together models and modules of natural sciences, as well as of some social sciences that are able to contribute models and quantified data, such as economics and some strands of geography. The other pillar is the development of an Earth system governance theory. This effort unites those social sciences that analyze organized human responses to Earth system transformation, in particular the institutions and agents, at all levels, that are created to steer human development in a way that secures a "safe" coevolution with natural processes. Both pillars are crowned by a common, collaborative roof that organizes issue-specific cooperation between the pillars.

The major examples for such issue-specific cooperation are the various joint projects of the Earth System Science Partnership, such as the Global Environmental Change and Food Systems project, the Global Water System Project, the Global Carbon Project, or the Global Environmental Change and Human Health project. These joint projects serve as flagship activities in which natural science and social science theories, methods, and approaches can be combined to analyze real-world problems. At the same time, these areas of focused cooperation can serve as a breeding, experimenting, and testing ground for methodological progress in the issue-specific combination of natural and social science approaches.

This methodological elaboration and experimentation through joint projects could be fruitful especially in two areas. First, the joint projects within the Earth System Science Partnership allow scholars of governance and institutions to experiment with linking their findings to computer-based modeling projects, to explore possibilities and problems of such types of integration, or to generate hypotheses on governance and institutions through formalized models. Several programs in this direction are under way, including some in the fields of qualitative modeling, agent-based modeling, game theory, and scenario development; it seems crucial to explore further the analytical value of these approaches.

Second—and in a sense leading in the opposite direction—the close interdisciplinary collaboration in joint projects allows social scientists to reemphasize the "social" aspects of global change research, that is, the social construction of knowledge, the cultural and temporal embedding of the researcher, and the reflexivity of social knowledge. This is especially important regarding the normative uncertainty prevailing in Earth system governance. What governance systems and governance outcomes future generations might want is unknown. This calls for particular forms of participatory research and assessment that integrate lay experts in academic research programs (Hisschemöller et al. 2001; Siebenhüner 2004; Van de Kerkhof 2006). Important advances have been achieved in the field of the participatory appraisal of research and of policies, which have not yet, however, been systematically integrated in the Earth System Science Partnership.

Research Problems

What would be the key research questions for a renewed long-term program on governance and institutions? Within the IDGEC community there are currently three broad clusters of research questions. First is the IDGEC Science Plan, approved by IHDP in 1998 and implemented since then. As explored in this volume, the Science Plan laid out three research foci of institutional research—namely, causality, performance, and design—and put forward three analytical themes: the problems of fit, interplay, and scale. The plan suggested a focus on certain regions, notably Southeast Asia and the polar regions. Second, in the course of implementation of the IDGEC Science Plan, a number of new issues emerged that were subsequently integrated into the IDGEC project as its so-called Extended Framework. This included in particular research on vulnerability, adaptation, and resilience; on trade and environment; and on water resources. Third, in 2004 IDGEC invited ideas for new research under the IDGEC New Directions initiative. Themes that emerged as new and forthcoming include the role of nonstate actors and private governance, the questions of legitimacy and accountability of governance and institutions and eventually of "democratic governance beyond the state," questions of institutional learning and change, and the question of distributive effects of institutions and the influence of different allocation mechanisms for the effectiveness of institutions.

A long-term research project on Earth system governance within IHDP could combine the existing and forthcoming topics under one line of five core analytical themes outlined in the following sections. These are the problem of the overall *architecture(s)* of Earth system governance, the problem of *agency* beyond the state, the problem of *adaptiveness* of governance mechanisms, the problem of *accountability* and legitimacy of Earth system governance, and the problem of *allocation* in Earth system governance.

The Problem of Architecture

The first major research and policy concern of Earth system governance is its overall "architecture." Most research in this field has focused on single institutions. A better understanding now exists of the creation, maintenance, and effectiveness of international environmental regimes, as well as better methodological tools to study these questions.[11] It has been shown, for example, that different international norms and verification procedures, compliance management systems, and modes of regime allocation as well as external factors, such as the structure of the problem, all influence regime effectiveness (see Mitchell, chapter 3 in this volume; Underdal, chapter 2 in this volume; Young, chapter 4 in this volume). Most studies have focused on the effectiveness of single institutions, often within larger comparative projects. More recently the increasing number and scope of international environmental institutions has led to new research on their interaction, for example, in studies on regime interlinkages, regime "clusters," or regime "complexes."[12] Institutional interplay has also been one of the three analytical themes of IDGEC (Schroeder 2008; on the results, see Oberthür and Gehring 2006c and Gehring and Oberthür, chapter 6 in this volume).

These approaches to understanding the effectiveness and the interaction of different institutions had to be methodologically reductionist to be successful. Distinct institutions, sometimes distinct institutional elements of larger institutions, and their effectiveness and relationship to other institutions or institutional elements have been analyzed. The macro level—that is, the system of institutions that address aspects of Earth system governance—has remained largely outside the focus of the major research programs. Given the advances in regime theory and institutional analysis, it appears that further progress now requires a complementary research program that analyses this macro level and the overarching research puzzles. This can be called the "architecture" of Earth system governance, a term increasingly used in policy circles; the term *global governance architecture*, for example, is now (February 2007) mentioned on 317 Web sites largely related to policy institutions and advocacy groups.[13] The architecture of Earth system governance refers, then, to the entire interlocking web of widely shared principles, institutions, and practices that shape decisions by stakeholders at all levels in this field. The concept of architecture can also be analytically useful to conceptualize subfields of Earth system governance, for example, with regard to a climate governance architecture. A renewed research effort on the problem of architecture will continue the current expansion of the IDGEC research program in four ways.

First, the problem of architecture entails looking *beyond single environmental institutions*. This includes continuation of research under conceptual headings such as institutional interplay, institutional interaction, institutional complexes, and institutional constellations (see Gehring and Oberthür, chapter 6 in this volume). A number of chapters in this volume have come to parallel conclusions: that more work is needed to understand the performance of single environmental institutions within larger settings and to understand the performance of entire clusters of institutions, which can be described as governance architectures.

Second, the problem of architecture requires looking *beyond environmental institutions*. This includes, for one, a focus on the environmental consequences of institutions that do not primarily address environmental policy. This work will entail a continuation in more detail of research on the environmental consequences of global trade or of World Bank programs, but also on new areas of conflict that are emerging, such as the environmental consequences of bilateral investment treaties (Tienhaara 2006).[14]

Third, the problem of architecture entails an exploration of *vertical institutional interaction* and the role of institutions within multilayered institutional systems. In international relations and political science research, this problem is generally understood as the problem of multilevel, or multilayered, governance. Global standards are implemented and put into practice at the local level, and global norm setting requires local decision making to set the frames for global decisions. This results in the coexistence of policy making at the subnational, national, regional, and global levels in more and more issue areas, with the potential of both conflicts and synergies among different levels of regulatory activity. The international regulation of trade in genetically modified organisms is a prime example for such multilevel governance, where the "global is local" (A. Gupta 2004). The more elements of governance above the state emerge, the more relevant multilevel governance as a research problem will become.

Within the IDGEC community, vertical governance interaction has been addressed through two different lenses. One, the conceptualization of interplay, as advanced in the IDGEC Science Plan, included the notion of vertical interplay. The extent to which the concept of vertical interplay has become a fruitful part of the research agenda or remains conflicted with the concept of multilevel governance and the related literature is still contested (see Gehring and Oberthür, chapter 6 in this volume). In addition, the concept of "scale,"[15] in its eventual formulation through a team of IDGEC researchers (J. Gupta, chapter 7 in this volume), has added the insight that the scale of any given institution is contingent on political processes and that actors deliberately try to choose scales of governance that best fit their interests and strategies. Both the concepts of vertical interplay and of scale indicate the need for further research in this area, which can be seen as part of the problem of architecture.

Fourth, the problem of architecture goes beyond the study of institutions and of their interaction. It also covers, for one, the inquiry into *noninstitutions*, that is, conflict areas where no institutions have been agreed. This inquiry addresses the recurrent problem in social science of case selection based on the dependent variable, in this case on the explanation of the emergence and performance of institutions through the analysis only of issue areas where institutions have been agreed (see Underdal, chapter 2 in this volume, for more detail on this problem).

Fifth, and finally, the problem of architecture entails more research on overarching metaprinciples and norms. Given the density of governance mechanisms and the emerging overarching system of institutions as the "architecture" of Earth system governance, there is an increasing need to understand better the principles and norms that run through all, or through a large number of, institutions and governance mechanisms. In a more general sense, this is the problem of deciding on universally accepted constitutional principles and basic norms in Earth system governance. The political behavior of states is guided not merely by calculations of material interest and power, but also by international norms that prescribe and prohibit types of behavior and create an international society that "socializes" states—including new governments that have not participated in the original creation of norms.[16] For such norms to be effective, they must be relatively simple, cross-culturally appealing, and sufficiently clear and unambiguous. For example, the success of the world trade regime in liberalizing trade and phasing out most customs duties within half a century is partially attributed to the simplicity and general acceptability of its basic principles, notably reciprocity and the most-favored-nation clause. Another example is the development of human rights norms in the course of the twentieth century (Risse, Ropp, and Sikkink 1999). Similar basic norms for Earth system governance are emerging, such as the principle of common but differentiated responsibilities of nations. Others are still disputed, such as the notion of interstate liability in the area of global climate change. Analyzing such universally accepted constitutional principles is hence a key research challenge for scholars of both international relations and international law regarding the problem of architecture.

The Problem of Agency beyond the State
Many vital institutions of Earth system governance are today inclusive of, or even driven by, nonstate actors. Activist groups, business associations, and research institutes provide research and advice, monitor the commitments of states, inform the public about the actions of diplomats at international meetings, and give these diplomats direct feedback. Carefully orchestrated campaigns of environmentalists have changed foreign policy of powerful states or initiated new global rules.[17] International networks of scientists and experts have emerged, in a mix of self-organization and state sponsorship, to provide complex technical information that is indispensable for policy making on issues marked by analytic and normative uncertainty.[18] Business has taken a more prominent role in international decision making,[19] for example in the Global Compact that major corporations have concluded with the United Nations (Cutler, Haufler, and Porter 1999; Higgott, Underhill, and Bieler 1999; Hall and Biersteker 2002; Pattberg 2005, 2006, 2007). Increasing also are the role and relevance of international bureaucracies, ranging from the specialized agencies of the United Nations to the hundreds of international bureaucracies set up for issue-specific management func-

tions as secretariats to international treaties (Biermann and Bauer 2004; Biermann and Siebenhüner 2007).

The activities of nonstate actors in Earth system governance are no longer confined to lobbying or advising governments in the creation and implementation of international rules. Increasingly, nonstate actors participate in rule setting with states or set their own rules. Private actors have joined governments to put international norms into practice, for example, as quasi-implementing agencies for development assistance programs administered by the World Bank or bilateral agencies. Private actors also participate in global institutions and at times negotiate their own standards, such as in the Forest Stewardship Council or the Marine Stewardship Council, two standard-setting bodies created by major corporations and environmental advocacy groups without any direct involvement of governments (Pattberg 2005, 2006, 2007). Public-private cooperation has received even more impetus with the 2002 Johannesburg World Summit on Sustainable Development and its focus on partnerships of governments, nongovernmental organizations, and the private sector—the so-called Partnerships for Sustainable Development. More than three hundred such partnerships have been registered with the United Nations around or after the Johannesburg summit (Andonova and Levy 2003; Glasbergen, Biermann, and Mol 2007).

The effectiveness of such initiatives, however, is as yet insufficiently understood. Most literature still builds on single-discipline case-study research with case selection often influenced by practical considerations or flawed through case selection on the dependent variable, in particular where only "success stories" are chosen. The major effort of the 1990s on analyzing intergovernmental environmental regimes needs to be complemented by a similar research program on "global participatory governance" that explores the public-private and private institutions in Earth system governance. This research program could address, first, the key conditions that explain the *emergence* of public-private and private-private governance mechanisms at global and regional levels. Second, research could focus on the political *effectiveness* of private governance. Many explanatory variables are conceivable, some of which might be similar to variables identified in the literature on intergovernmental regimes. For example, the effectiveness of private institutions could be influenced by their organizational structure; funding mechanisms; coordination, decision-making, and management mechanisms; or compliance mechanisms. Problem structure is likely to influence the effectiveness of private institutions, too. Private institutions could also be more effective the more their policies are tailored to the needs and capacities of targeted actors and to the national administrative and regulatory structures of the country in which agreements are to be implemented. Yet no comprehensive research findings on these hypotheses are yet available. In sum, this field still awaits research programs that systematically analyze the emergence and effectiveness of private institutions in Earth system governance.

Another important area for further research regarding nonstate actors is the role of science and of scientists. The way Earth system governance has unfolded has been strongly influenced by scientists, in particular through major long-term and worldwide expert assessments such as the Intergovernmental Panel on Climate Change. Such assessments are not only about science; they are boundary instruments between science and policy and are thus, in essence, to be understood as political activities. Understanding how knowledge and institutions interact and which institutions might best produce knowledge is therefore central to effective Earth system governance, and the role of expertise and the politics of "big science" will thus remain—despite a number of larger project reports in this field (Jasanoff and Long Martello 2004; R. Mitchell et al. 2006)—important elements of the problem of agency beyond the state.

The Problem of Adaptiveness
In addition, a new research effort on Earth system governance needs to address the problem of adaptiveness. This term can be combined with two separate yet interlinked discourses.

First, new research needs to go beyond the current predominant research on institutions that mitigate global environmental change but must focus more strongly on the governance of adaptation. Most studies on international as well as national environmental policy have focused so far on institutions designed to mitigate environmentally harmful activities, such as emissions of pollutants, trade in harmful substances or endangered species, or destruction of habitats. Scholars have only recently begun also to study governance mechanisms for adaptation to the impacts of global environmental change and to investigate the extent to which local institutions and governance systems allow for adaptation. This emergent stream of work will eventually need to evolve from local adaptation research into a research program on the core functions of public policy and of the state itself. Whereas past work has theorized extensively about the environmental state or the green state, new thinking is needed on defining the core functions of the "adaptive state": a state able to adapt internally and externally to large-scale transformations of its natural environment.[20]

Eventually this research will need to extend to the level of the adaptiveness of the *global* governance architecture. While natural scientists predict widespread harm if current trends of Earth system transformation continue, governance scholars need to understand better what kind of mechanisms at the global level can assist in adaptation efforts at national and local levels. In addition, knowledge is still lacking on how the system of global governance as such can be made more resilient against the impacts of global environmental change. For example, how can the UN system address streams of refugees that may result from land degradation and sea-level rise, two prominent effects of climate change? What are the implications of possible climate-related extreme events for global and regional economic systems, and how can these systems be made more adaptive? Finally, what are the security implications of Earth system transformation?

Research in this area will require particular attention to methodology. Adaptive Earth system governance requires analysis and design of governance systems that respond to emergencies that are merely predicted for the future but may exceed in scope and quality most of what is known today. Adaptive governance systems that take account of changes in monsoon patterns, large-scale breakdowns of ecosystems, or modifications in the thermohaline circulation will need to deal with magnitudes that are unprecedented. Whereas traditional social science builds on the development and testing of theories and hypotheses through historical experience, Earth system governance, which is inherently future oriented, has to rely on new forms of evidence and new forms of validity and reliability of empirical knowledge.

Second, the problem of adaptiveness extends not only to adaptation governance but also toward an improved understanding of the adaptiveness of institutions and governance systems as such. Thus, the problem of adaptiveness includes numerous social phenomena and scholarly discourses, and it covers all forms of change within systems of governance that have at times been described under an array of terminology, such as *institutional dynamics*, *institutional change*, and *social learning* (e.g., Parson and Clark 1995; Webler, Kastenholz, and Renn 1995; Social

Learning Group 2001; Siebenhüner 2004, 2005). It also could cover a variety of recent analytical frameworks that have been advanced to understand interconnected social and ecological systems, including social-ecological resilience, adaptive comanagement, and adaptive governance (Janssen and Ostrom 2006; Galaz et al., chapter 5 in this volume). All these concepts share a common core research question that could be at the center of the problem of adaptiveness in a research program on Earth system governance: how can governance systems change in reaction to changes in their environment?

The Problem of Accountability

Most research within the IDGEC community has focused on the assessment and explanation of institutional performance. Equally important, however, is the question of the accountability and legitimacy of institutions and systems of governance, both in their own right and with a view of accountability and legitimacy as intervening variables that affect overall institutional effectiveness. In the twentieth century, legitimacy and accountability were problems of national governments. In the twenty-first century, with its new needs of Earth system governance, accountability and legitimacy appear in a different context. Eventually this problem comes down to the quest for *democratic* Earth system governance. There are two broad types of research needs.

The first is a theoretical one: in purely intergovernmental norm-setting processes, legitimacy derives indirectly through the accountability of governments to their voters. Likewise, international bureaucracies can derive legitimacy through their principals, the governments. Such long lines of accountability, however, have been questioned in recent years.[21] Many authors see a solution in the participation of private actors in global governance. David Held (1999), for example, recognizes " 'new' voices of an emergent 'transnational civil society' ... in the early stages of development ... [that] point in the direction of establishing new modes of holding transnational power systems to account, that is, they help open up the possibility of a cosmopolitan democracy" (108).

The accountability and legitimacy of private actors themselves, however, are problematic. Private organizations may derive legitimacy through their members or donors or from the environmental good they seek to protect. Yet few citizens have the means to donate time and money to philanthropic organizations. Given the financial requirements of participation, more rights and responsibilities for nonstate actors in Earth system governance may privilege representatives of industry and business at the cost of other groups. In the international context, with its high disparities in wealth and power, accountability and legitimacy of private actors are even more complex. Most philanthropic organizations are headquartered in industrialized countries, and most funds donated to their cause, both public and private, stem from the North.

All this leads to the second, practical challenge: because of these disparities, researchers need to design, and practitioners to develop, institutions that guarantee participation of civil society in Earth system governance through mechanisms that vouchsafe a balance of opinions and perspectives. For example, networks of transnational private actors can seek to balance views and interests through self-regulation, including financial support for representatives from developing countries. This is done, for instance, through North-South quotas in meetings and alliances of nonstate activists within the UN Commission on Sustainable Development or in the Intergovernmental Panel on Climate Change.

Another option to increase legitimacy and accountability of Earth system governance by strengthening private participation in a balanced way could be a "quasi-corporatist" institutionalization. For example, the representation of labor unions and employers' associations in the International Labor Organization (ILO) has been discussed as a model for achieving a balance in participation of private actors from North and South in order to make Earth system governance more representative and legitimate. In the ILO each state is represented with four votes, two of which are assigned to governments and one each to business associations and labor unions. Concerning more far-reaching proposals, the Commission on Global Governance has proposed an international Forum of Civil Society within the United Nations, which would comprise three to six hundred "organs of global civil society" to be self-selected from civil society, or even a global parliamentarian assembly (Commission on Global Governance 1995, 257–58).

In sum, the problem of accountability comes down to three specific research questions: First, what are the different sources of accountability and legitimacy of institutions and governance systems in the field of Earth system governance, at national and international levels? Second, what is the effect of different forms and degrees of accountability and legitimacy for the performance of governance systems at national and international levels? Third, what concrete institutional designs can produce the accountability and legitimacy of Earth system governance in a way that guarantees balances of interests and perspectives?

The Problem of Allocation

Politics is about the distribution of resources and values, and Earth system governance is no different. Long-term credible, stable, and inclusive Earth system governance requires the agreement of all stakeholders that the allocation of costs and benefits is fair. With the increasing relevance of Earth system governance in the twenty-first century, allocation mechanisms and criteria will become central questions for both social scientists and decision makers. At stake are not only the costs of mitigation. Given the potentially disastrous consequences of Earth system transformations, questions of fairness in adaptation will arise (Adger et al. 2006). Compensation and support through the global community of the most affected and most vulnerable regions, such as small island states, will not only be moral responsibilities but also politically and economically prudent. Climate change, for example, has raised questions of litigation and legal liability. In sum, allocation modes are needed that all stakeholders in North and South perceive as fair and will support over the course of the twenty-first century.[22]

The problem of allocation can be framed in two ways. First, allocation can be conceptualized as an independent variable; that is, different modes of allocation in systems of governance can be studied in comparative research programs with regard to their influence on eventual performance indicators. Second, and related to the first point, allocation can be seen as one of the fundamental output variables for all sorts of governance systems and for Earth system governance in particular. Eventually the problem of allocation can then be translated into the quest for fair and equitable Earth system governance.

The problem of allocation relates to all levels of governance. Allocation is key when governance systems for forest management or local water management are designed. Allocation is central also to the global level of Earth system governance, probably the area to which scholars have paid the least attention so far. While domestic allocation challenges remain important, especially unprecedented and contested in Earth system governance are questions of allocation among nations (see, e.g., Tóth 1999). At present, different modes of allocation in global Earth system governance coexist, and little research in the social sciences has yet been addressed to the comparative appraisal of these different mechanisms.

For example, costs can be globally allocated through intergovernmental agreement and implemented through public funds under the authority of the community of states. The 1990 London amendment of the Montreal ozone protocol, for example, saw the creation of a multilateral fund to reimburse the full agreed incremental costs of developing countries in implementing the treaty and in phasing out ozone-depleting substances (Biermann 1997). A similar mechanism is the Global Environment Facility. Alternatively, costs of mitigation and adaptation can be allocated through market-based mechanisms that are under public control and based on international agreement. There are few examples of public markets that allocate environmental mitigation costs. Apart from an early form of joint implementation of commitments among industrialized countries in the ozone regime, there exists currently only one intergovernmental system that trades mitigation obligations: the flexible mechanisms under the Kyoto Protocol to the climate convention. But similar mechanisms along these lines are conceivable, for example, with a view to global markets on biodiversity protection certificates (Whalley and Zissimos 2001). A third mode of allocation in Earth system governance works through environmentally motivated restrictions on international trade that force producers and investors in some countries to change their process and production methods according to the standards of their trading partners. In other words, environmentally motivated trade restrictions either reduce the market share of exporting countries or force them to adjust their product designs and production processes. These restrictions can be compulsory—that is, enshrined in governmental regulations that ban certain products from a market—or voluntary—for example, through labeling schemes that leave the choice to the consumers in the importing market.

These three allocation modes in Earth system governance represent different principles of allocation. The ozone fund and the Global Environment Facility build on state-based, universal decision making. They come closest to domestic modes of allocation; on the revenue side, the contribution to the funds is linked to the relative wealth of countries. Regarding the expenses, the funds are governed by decisions of state representatives in a way that grants both the developing and the developed countries a de facto veto right. The disbursement of the funds is largely based on need; the funds reimburse the incremental costs of poorer countries that take action to mitigate global environmental change. Public markets for mitigation obligations—to put it differently, for emissions entitlements—also build on state-based decision making inasmuch as governments decide on the allocation of mitigation obligations. Overall the market structure will guarantee an efficient allocation of mitigation costs and may induce technological innovation. The eventual distributive effect of this system depends on the initial allocation of mitigation obligations and hence can differ from the basic principle of international funds. Allocation through environmental restrictions of the global trade in goods is based on the principle of consumer authority: consumer markets are empowered through trade restrictions to globalize their own preferences and production standards and to define and shape the production standards in producer countries. There would be no reimbursement of incremental costs for environmental policies in producer countries and no right of codecision by their representatives. From the perspective of poorer, smaller nations, private allocation through environmental restrictions of global trade may therefore be less preferable.

In sum, research in this field has been scarce in the past, particularly in regard to empirical research programs that could lend substance to the more policy-oriented, philosophical treatises on equity. The causes and consequences of different allocation mechanisms in Earth system governance are still not sufficiently understood. Few research efforts have yet been directed at understanding the causal pathways that lead to specific allocation mechanisms. Little systematic analysis has been devoted to studying allocation as an independent variable and to analyzing allocation mechanisms in relation to variant effectiveness of the core institutions of Earth system governance. Given the growing relevance of Earth system governance in the course of the twenty-first century in terms of both mitigation and adaptation costs, allocation is certain to become a major concern for researchers and practitioners alike.

Conclusion

This chapter has sketched the outline of a possible new long-term research effort on the institutional dimensions of global environmental change. It argues that this research effort should build on IDGEC's legacy but should also go further by broadening the research focus from institutions to larger systems of governance, and by conceptualizing this field as a research area of Earth system governance. Five analytical themes have been outlined: the problems of architecture, of agency beyond the state, of adaptiveness, of accountability and legitimacy, and of allocation. These analytical themes draw on existing work within the IDGEC program, but they also refocus the debate based on the findings of IDGEC and add new problems and perspectives, notably the role of nonstate actors, of accountability and legitimacy, and of the allocative effects of institutions and systems of governance.

Taken together, the problems of Earth system transformation make Earth system governance one of the most challenging, but thus also one of the most exciting, research objects in the social sciences. As a political program Earth system governance is no less daunting. Politics appears often to be determined more by economic stagnation, short-term interests, and reemerging nationalism than by global governance and collective stewardship of the Earth. The bolder visions of the earlier philosophers, such as Seneca's idea of a *res publica* whose boundaries would be "the sun alone" (*De Otio* IV, §1) or Kant's proposal of a global federation of states for "the eternal peace," seem hardly more realistic now than they were in their days. Yet Earth system governance is emerging. More than nine hundred international environmental agreements are in force. Many harmful substances, such as the ozone-depleting chlorofluorocarbons, have been phased out through international cooperation and local action. Mitigation and adaptation projects against global warming are mushrooming in many places, from India to the Netherlands, often inspired, guided or coordinated by global collaborative programs.

Yet how to create an effective global architecture for Earth system governance that is adaptive to changing circumstances, participatory through involving civil society at all levels, accountable and legitimate as part of a new democratic governance beyond the nation-state, and at the same time fair for all participants—this research and governance challenge still lies ahead.

Acknowledgments

This text is based on the author's inaugural lecture at the Vrije Universiteit Amsterdam in October 2005. Many thanks for useful comments on earlier versions of this chapter to Michele Betsill, Ries Bode, Joop de Boer, Jan Boersema, Klaus Dingwerth, Klaus Eisenack, Nicolien van der Grijp, Aarti Gupta, Dave Huitema, Louis Lebel, Robert Marschinski,

Hans Opschoor, Philipp Pattberg, Heike Schroeder, Uta Schuchmann, Bernd Siebenhüner, Frans van der Woerd, and Oran R. Young, as well as to many participants of the IDGEC Synthesis Conference held in Bali, Indonesia, December 6–9, 2006. Parts of this chapter draw on an earlier shorter article in *Global Environmental Change* (2007).

Notes

1. In March 2007 the Scientific Committee of the overarching International Human Dimensions Programme on Global Environmental Change (IHDP) appointed a scientific planning committee to develop a science plan for a new research program to be finalized in early 2008. See http://www.earthsystemgovernance.org on this initiative.

2. See, for example, Young 1994a, 1999a; Commission on Global Governance 1995; Finkelstein 1995; Rosenau 1995; Gordenker and Weiss 1996; Smouts 1998; Kanie and Haas 2004; Dingwerth and Pattberg 2006.

3. Key texts are available at http://sustsci.aaas.org/. See also Schellnhuber et al. (2004); Clark, Crutzen, and Schellnhuber (2005); as well as Friibergh Workshop on Sustainability Science 2000.

4. See the partnership's Web site at http://www.essp.org.

5. See http://www.sciconf.igbp.kva.se/fr.html.

6. One finds the term *Earth system management* mostly in relation to natural science programs, for example, when it comes to providing data on Earth system parameters that are influenced by human action. For instance, Earth system management is one of the three research foci of the natural science–oriented Centre for Marine and Climate Research in Hamburg, Germany, there defined as provision of models and methods as instruments for information, planning, and legislation on global, regional, and local scales. Tellingly, the first time the term has been used—to my knowledge—was at the 7th International Remote Sensing Systems Conference in Melbourne in 1994 by a representative of the UN Environment Programme, Noel J. Brown, in his presentation "Agenda 21: Blueprint for Global Sustainability: New Opportunities for Earth System Management" (Heiner Benking, personal communication, August 2005).

7. The concept was first developed in Biermann 2002b, 2005, 2007.

8. For a recent overview of definitions and conceptualizations of sustainable development, see Kates, Parris, and Leiserowitz 2005.

9. This has been elaborated in Biermann 2007, 329–31.

10. See the partnership's mission statement at http://www.essp.org.

11. For recent overviews and discussions see R. Mitchell 2002a and chapter 3 in this volume; Underdal, chapter 2 in this volume; Young, chapter 4 in this volume. Important contributions are, for instance, Haas, Keohane, and Levy 1993; R. Mitchell 1994a; Young 1994a, 1997, 1999a, 2001b; Bernauer 1995; Keohane

and Levy 1996; Brown Weiss and Jacobson 1998; R. Mitchell and Bernauer 1998; Victor, Raustiala, and Skolnikoff 1998; Young, Levy, and Osherenko 1999; Helm and Sprinz 2000; Miles et al. 2002; Underdal 2002b; A. Gupta and Falkner 2006.

12. For example, Stokke 2000; Velasquez 2000; Chambers 2001; Rosendal 2001a, 2001b; van Asselt, Gupta, and Biermann 2005; Oberthür and Gehring 2006c.

13. Note that the term *policy architecture* is more widely used.

14. To the extent that the environmental consequences of these nonenvironmental institutions are covered by environmental institutions at the same time, the problem of nonenvironmental institutions becomes a problem of institutional interaction and hence a problem of the architecture of environmental governance (see Gehring and Oberthür, chapter 6 in this volume, on the state of the art in this field).

15. See on the concept of scale Young 1994b; Gibson, Ostrom, and Ahn 2000; Alcock 2002; Lebel, Garden, and Imamura 2005; J. Gupta, chapter 7 in this volume.

16. This is largely linked to the theoretical strand of sociological institutionalism. See, among many others and with further references, March and Olsen 1989, 1996, 1998; Finnemore 1996b; Finnemore and Sikkink 1998; Barnett and Finnemore 1999.

17. See, for example, Princen and Finger 1994; Conca 1995; Princen, Finger, and Manno 1995; Wapner 1996; Charnovitz, 1997; Raustiala 1997; Arts 1998, 2002; Betsill and Corell 2001; Reinalda and Verbeek 2001; Van der Heijden 2002; Edwards and Zadek 2003; J. Gupta 2003; Van der Grijp and Brander 2004.

18. P. Haas 1992, 1994; Jasanoff 1996; Jäger 1998; Biermann 2001a, 2002a; Hisschemöller et al. 2001; Jasanoff and Long Martello 2004; R. Mitchell et al. 2006.

19. See, for example, Lee, Humphreys, and Pugh 1997; Clapp 1998; Cutler, Haufler, and Porter 1999; Sell 1999; Haufler 2000; D. Levy and Newell 2000, 2002; Rowlands 2001; D. Levy and Kolk 2002; Falkner 2003; Van der Woerd, Levy, and Begg 2005.

20. See the 2004 special issue of *Global Environmental Politics*, "Global Environmental Change and the Nation State," vol. 4 (1), in particular the introductory article by Biermann and Dingwerth.

21. On the democratic deficit of inter- and transnational politics and on attempts to conceptualize democratic governance on the transnational level, see, for instance, Archibugi and Held 1995; Commission on Global Governance 1995; Held 1995, 1997; South Centre 1996; Archibugi, Held, and Köhler 1998; Dryzek 1999; Scholte 2002; Dingwerth 2005, 2007.

22. See similarly Adger, Brown, and Hulme (2005), who write in their editorial to *Global Environmental Change* that a "more explicit concern with equity and justice will be important in furthering the study of global environmental change."

References

Abbott, K., and D. Snidal. 2000. "Hard and Soft Law in International Governance." *International Organization* 54:421–56.

Abu Amara, S. Forthcoming. "How to Protect Traditional Knowledge: 'Scaling Up' in TRIPS and the CBD." In *The Politics of Scale in Environmental Governance*, ed. J. Gupta and D. Huitema. IDGEC manuscript.

Adams, W. M., D. Brockington, J. Dyson, and B. Vira. 2003. "Managing Tragedies: Understanding Conflict over Common Pool Resources." *Science* 302:1915–16.

Adger, W. N., T. A. Benjaminsen, K. Brown, and H. Svarstad. 2001. "Advancing a Political Ecology of Global Environmental Discourses." *Development and Change* 32:681–715.

Adger, W. N., K. Brown, and M. Hulme. 2005. "Redefining Global Environmental Change" (editorial). *Global Environmental Change: Human and Policy Dimensions* 15:1–4.

Adger, W. N., T. Hughes, C. Folke, S. R. Carpenter, and J. Rockström. 2005. "Social-Ecological Resilience to Coastal Disasters." *Science* 309:1036–39.

Adger, W. N., J. Paavola, S. Huq, and M. J. Mace, eds. 2006. *Fairness in Adaptation to Climate Change*. Cambridge, MA: The MIT Press.

Agarwal, A., et al. 1992. *For Earth's Sake: A Report from the Commission on Developing Countries and Global Change*. Ottawa, Ontario: International Development Research Centre.

Aggarwal, V. K. 1983. "The Unraveling of the Multi-Fiber Arrangement, 1981: An Examination of International Regime Change." *International Organization* 37 (4): 617–46.

Agranoff, R. I., and M. McGuire. 2001. "Big Questions in Public Network Management Research." *Journal of Public Administration Research and Theory* 11:295–326.

Agrawal, A. 2002. "Common Resources and Institutional Sustainability." In *The Drama of the Commons*, ed. E. Ostrom, Dietz, N. Dolsak, P. C. Stern, S. Stonich, and U. W. Elke, 41–85. Washington, T. DC: National Academies Press.

————. 2005. *Governmentality: Technologies of Government and the Making of Subjects.* Durham, NC: Duke University Press.

Agrawal, A., and A. Chhatre. 2006. "Explaining Success on the Commons: Community Forest Governance in the Indian Himalaya." *World Development* 34:149–66.

Ahn, T., E. Ostrom, D. Schmidt, et al. 2001. "Cooperation in PD Games—Fear, Greed and History of Play." *Public Choice* 106:137–55.

Alcántara, C. H. de. 1998. "Uses and Abuses of the Concept of Governance." *International Social Science Journal* 155:105–13.

Alcock, F. 2002. "Scale Crisis and Sectoral Conflict: The Fisheries Development Dilemma." In *Global Environmental Change and the Nation State: Proceedings of the 2001 Berlin Conference on the Human Dimensions of Global Environmental Change,* ed. F. Biermann, R. Brohm, and K. Dingwerth. Potsdam, Germany: Potsdam Institute for Climate Impact Research.

Allison, H. E., and R. J. Hobbs. 2004. "Resilience, Adaptive Capacity, and the 'Lock-in Trap' of the Western Australian Agricultural Region." *Ecology and Society* 9 (1): 3.

Alston, P. 1997. "The Myopia of the Handmaidens: International Lawyers and Globalization." *European Journal of International Law* 8 (3): 435–48.

Anderies, J. M., P. Ryan, and B. H. Walker. 2006. "Loss of Resilience, Crisis, and Institutional Change: Lessons from an Intensive Agricultural System in Southeastern Australia." *Ecosystem* 9:865–78.

Anders, G. 2005. "Civil Servants in Malawi: Moonlighting, Kinship and Corruption in the Shadow of Good Governance." PhD manuscript, Law Faculty, Erasmus University, Rotterdam, The Netherlands.

Andersen, R. 2002. "The Time Dimension in International Regime Interplay." *Global Environmental Politics* 2 (3): 98–117.

————. 2008. *Governing Agrobiodiversity: Plant Genetics and Developing Countries.* Aldershot, UK: Ashgate (forthcoming).

Andonova, L. B., and M. A. Levy. 2003. "Franchising Global Governance: Making Sense of the Johannesburg Type II Partnerships." In *Yearbook of International Cooperation on Environment and Development,* ed. O. B. Thommessen, 19–32. London: Earthscan.

Andrée, P. 2005. "The Cartagena Protocol on Biosafety and Shifts in the Discourse of Precaution." *Global Environmental Politics* 5 (4): 25–46. (See http:// ideas.repec.org/a/tpr/glenvp/v5y2005i4p25-46.html)

Andresen, S. 2002. "The International Whaling Commission (IWC): More Failure than Success?" In *Environmental Regime Effectiveness: Confronting Theory with Evidence,* by E. L. Miles, A. Underdal, S. Andresen, J. Wettestad, J. B. Skjærseth, and E. M. Carlin, 375–403. Cambridge, MA: The MIT Press.

————, T. Skodvin, A. Underdal, and J. Wettestad, eds. 2000. *Science and Politics in International Environmental Regimes: Between Integrity and Involvement.* New York: Manchester University Press.

Ankersmit, W. 1998. *Water Supply and Sanitation in Developing Countries: Sectoral Policy Document Cooperation 12.* The Hague: Netherlands Development Assistance (NEDA).

Annan, K. 2004. "Secretary-General's Address to the General Assembly, New York, 21 September 2004." http://www.un.org/apps/sg/sgstats.asp?nid=1088.

Antweiler, W., B. R. Copeland, and M. S. Taylor. 2001. "Is Free Trade Good for the Environment?" *American Economic Review* 91 (4): 877–908.

Archibugi, D., and D. Held, eds. 1995. *Cosmopolitan Democracy: An Agenda for a New World Order.* Oxford: Polity Press.

Archibugi, D., D. Held, and M. Köhler, eds. 1998. *Re-imagining Political Community: Studies in Cosmopolitan Democracy.* Stanford, CA: Stanford University Press.

Arthur, B. W. 1999. "Complexity and the Economy." *Science* 284:107–9.

Arts, B. 1998. *The Political Influence of Global NGOs: Case Studies on the Climate Change and Biodiversity Conventions.* Utrecht, The Netherlands: International Books.

———. 2002. "Green Alliances of Business and NGOs: New Styles of Self-Regulation or Dead-End Roads?" *Corporate Social Responsibility and Environmental Management* 9:26–36.

Axelrod, M. 2006. "Saving Institutional Benefits: Path Dependence in International Law." Paper presented at the annual meeting of the International Studies Association, San Diego, California, March 22.

Axelrod, R. 1984. *The Evolution of Cooperation.* New York: Basic Books.

———. 1997. *The Complexity of Cooperation: Agent-Based Models of Cooperation and Collaboration.* Princeton, NJ: Princeton University Press.

Baden, J. A., and D. S. Noonan, eds. 1998. *Managing the Commons.* Bloomington: Indiana University Press.

Bardach, E. 1998. *Managerial Craftsmanship: Getting Agencies to Work Together.* Washington, DC: Brookings Institution Press.

Bardhan, P. 2002. "Decentralization of Governance and Development." *Journal of Economic Perspectives* 16 (4): 185–205.

Barlow, M., and T. Clarke. 2002. *Blue Gold: The Battle Against Corporate Theft of the World's Water.* London: Earthscan.

Barnett, M. N., and M. Finnemore. 1999. "The Politics, Power, and Pathologies of International Organizations." *International Organization* 53 (4): 699–732.

Barrett, S. 2003. *Environment and Statecraft: The Strategy of Environmental Treaty Making.* New York: Oxford University Press.

Baum, R. 1986. "Modernization and Legal Reform in Post-Mao China: The Rebirth of Socialist Legality." *Studies in Comparative Communism* 14 (2): 69–104.

Baumgartner, F., and B. Jones. 1991. "Agenda Dynamics and Policy Subsystems." *Journal of Politics* 53:1044–74.

Bebbington, A. 1997. "Social Capital and Rural Intensification: Local Organizations and Islands of Sustainability in the Rural Andes." *The Geographical Journal* 163:189–97.

Bellamy, I., ed. 2005. *International Society and Its Critics.* Oxford: Oxford University Press.

Benson, D., and A. Jordan. Forthcoming. "Subsidiarity as a 'Scaling Device' in Environmental Governance: The Case of the European Union." In *The Politics of Scale in Environmental Governance,* ed. J. Gupta and D. Huitema. IDGEC manuscript.

Berkes, F. 2002. "Cross-Scale Institutional Linkages: Perspectives from the Bottom Up." In *The Drama of the Commons,* ed. E. Ostrom, T. Dietz, N. Dolsak, P. C. Stern, S. Stonich, and U. W. Elke, 293–321. Washington, DC: National Academies Press.

———. 2006a. "Comments from the 'Knowledge Broker' on 'Global Change: Analysing Scale and Scaling in Environmental Governance,' by Joyeeta Gupta." Presentation at IDGEC Synthesis Conference, December 6–9, Bali, Indonesia.

———. 2006b. "From Community-Based Resource Management to Complex Systems: The Scale Issue and Marine Commons." *Ecology and Society* 11 (1). http://www.ecologyandsociety.org/vol11/iss1/art45/.

———. 2007. "Community-based Conservation in a Globalized World." *Proceedings of the National Academy of Sciences* 104 (39): 15188–93.

Berkes, F., J. Colding, and C. Folke. 2003. *Navigating Social-Ecological Systems: Building Resilience for Complexity and Change.* Cambridge: Cambridge University Press.

Berkes, F., and C. Folke, eds. 1998. *Linking Social and Ecological Systems: Management Practices and Social Mechanisms for Building Resilience.* Cambridge: Cambridge University Press.

Berkes, F., T. P. Hughes, R. S. Steneck, et al. 2006. "Globalization, Roving Bandits, and Marine Resources." *Science* 311:1557–58.

Bernauer, T. 1995. "The Effect of International Environmental Institutions: How We Might Learn More." *International Organization* 49 (2): 351–77.

Bernauer, T., and V. Koubi. 2006. "Political Determinants of Environmental Quality." SSRN (January). http://ssrn.com/abstract=882812.

Bernauer, T., and T. Siegfried. n.d. "Estimating International Policy Performance: Measurement Concept and Empirical Application to International Water Management in the Naryn/Syr Darya Basin." Unpublished paper.

Betsill, M. M., and E. Corell. 2001. "NGO Influence in International Environmental Negotiations: A Framework for Analysis." *Global Environmental Politics* 1 (4): 65–85.

———, eds. 2007. *NGO Diplomacy: The Influence of Nongovernmental Organizations in International Environmental Negotiations.* Cambridge, MA: The MIT Press.

Biermann, F. 1997. "Financing Environmental Policies in the South: Experiences from the Multilateral Ozone Fund." *International Environmental Affairs* 9 (3): 179–219.

———. 2001a. "Big Science, Small Impacts—in the South? The Influence of Global Environmental Assessments on Expert Communities in India." *Global Environmental Change* 11:297–309.

———. 2001b. "The Rising Tide of Green Unilateralism in World Trade Law: Options for Reconciling the Emerging North-South Conflict." *Journal of World Trade* 35 (3): 421–48.

———. 2002a. "Institutions for Scientific Advice: Global Environmental Assessments and Their Influence in Developing Countries." *Global Governance* 8 (2): 195–219.

———. 2002b. "Johannesburg plus 20: From International Environmental Policy to Earth System Governance." *Politics and the Life Sciences* 21 (2): 72–77.

———. 2002c. "Strengthening Green Global Governance in a Disparate World Society: Would a World Environmental Organization Benefit the South?" *International Environmental Agreements: Politics, Law and Economics* 2:297–315.

———. 2005. "Earth System Governance: The Challenge for Social Science." Inaugural lecture, Vrije Universiteit Amsterdam.

———. 2006. "Whose Experts? The Role of Geographic Representation in Global Environmental Assessments." In *Global Environmental Assessments: Information and Influence*, ed. R. B. Mitchell, W. C. Clark, D. W. Cash, and N. Dickson, 87–112. Cambridge, MA: The MIT Press.

———. 2007. " 'Earth System Governance' as a Crosscutting Theme of Global Change Research." *Global Environmental Change* 17:326–37.

Biermann, F., and S. Bauer. 2004. "Assessing the Effectiveness of Intergovernmental Organizations in International Environmental Politics." *Global Environmental Change: Human and Policy Dimensions* 14 (2): 189–93.

———, eds. 2005. *A World Environment Organization: Solution or Threat for Effective International Environmental Governance?* Aldershot, UK: Ashgate.

Biermann, F., and R. Brohm. 2005. "Implementing the Kyoto Protocol without the United States: The Strategic Role of Energy Tax Adjustments at the Border." *Climate Policy* 4 (3): 289–302.

Biermann, F., and K. Dingwerth. 2004. "Global Environmental Change and the Nation State." *Global Environmental Politics* 4 (1): 1–22.

Biermann, F., and B. Siebenhüner. 2007. "Managers of Global Change: The Core Findings of the MANUS Project on the Influence of International Bureaucracies." Global Governance Working Paper No. 25. Amsterdam: The Global Governance Project.

Blum, E. 1993. "Making Biodiversity Conservation Profitable: A Case Study of the Merck/INBIO Agreement." *Environment* 35 (4): 16–20, 38–45.

Board on Sustainable Development, Policy Division, National Research Council. 1999. *Our Common Journey: A Transition toward Sustainability.* Washington, DC: National Academies Press.

Bodansky, D. 1993. "The United Nations Framework Convention on Climate Change: A Commentary." *Yale Journal of International Law* 18:451–588.

———. 2004. *International Climate Efforts Beyond 2012: A Survey of Approaches.* Arlington, VA: Pew Center on Global Climate Change. http://www .test.earthscape.org/p1/ES16041/intclimate.pdf.

Bodin, Ö., B. Crona, and H. Ernstsson. 2006. "Social Networks in Natural Resource Management: What Is There to Learn from a Structural Perspective?" *Ecology and Society* 11 (2): r2.

Bodin, Ö., and J. Norberg. 2005. "Information Network Typologies for Enhanced Local Adaptive Management." *Environmental Management* 35 (2): 175–93.

Boehmer-Christiansen, S., and J. Skea. 1991. *Acid Politics.* London: Belhaven Press.

Boin, A., and P. t'Hart. 2003. "Public Leadership in Times of Crisis: Mission Impossible?" *Public Administration Review* 63 (5): 544–53.

Borrini-Feyerabend, G., M. Pimbert, M. T. Farvar, A. Kothari, and Y. Renard. 2004. *Sharing Power: Learning by Doing in Co-management of Natural Resources throughout the World.* Tehran, Iran: The Natural Resources Group and the Sustainable Agriculture and Rural Livelihoods Programme of the International Institute for Environment and Development (IIED) and the Collaborative Management Working Group (CMWG) of the IUCN Commission on Environmental, Economic and Social Policy (CEESP) of the World Conservation Union (IUCN).

Boyle, A. E. 1999. "Problems of Compulsory Jurisdiction and the Settlement of Disputes Relating to Straddling Fish Stocks." *International Journal of Marine and Coastal Law* 14 (1): 1–26.

Bozeman, A. 1960. *Politics and Culture in International History.* Princeton, NJ: Princeton University Press.

Brack, D. 2002. "Environmental Treaties and Trade: Multilateral Environmental Agreements and the Multilateral Trading System." In *Trade, Environment, and the Millennium.* 2nd ed. Ed. G. P. Sampson and W. Bradney Chambers, 321–52. Tokyo: United Nations University Press.

Brack, D., R. Falkner, and J. Goll. 2003. "The Next Trade War? GM Products, the Cartagena Protocol and the WTO." RIIA Briefing Paper No. 8 (September). London: Royal Institute of International Affairs.

Braybrooke, D., and C. E. Lindblom. 1963. *A Strategy for Decision.* New York: The Free Press.

Breitmeier, H., O. R. Young, and M. Zürn. 2006. *Analyzing International Environmental Regimes: From Case Study to Database.* Cambridge, MA: The MIT Press.

Brenner, N. 1998. "Between Fixity and Motion: Accumulation, Territorial Organization and the Historical Geography of Spatial Scales." *Environment and Planning D: Society and Space* 16:459–81.

Bromley, D. W. 1992. "The Commons, Common Property, and Environmental Policy." *Environmental and Resource Economics* 2 (1): 1–17.

Brown, K. 2003. "Integrating Conservation and Development: A Case of Institutional Misfit." *Frontiers in Ecology and the Environment* 1 (9): 479–87.

Brown Weiss, E. 1993. "International Environmental Issues and the Emergence of a New World Order." *Georgetown Law Journal* 81 (3): 675–710.

———, ed. 1997. *International Compliance with Nonbinding Accords.* Washington, DC: American Society of International Law.

Brown Weiss, E., and H. K. Jacobson, eds. 1998. *Engaging Countries: Strengthening Compliance with International Environmental Accords.* Cambridge, MA: The MIT Press.

Bryner, G. 1995. *Blue Skies, Green Politics: The Clean Air Act of 1990 and Its Implementation.* Washington, DC: Congressional Quarterly Books.

Bulkeley, H. Forthcoming. "Contesting Scale: Environmental Governance and the Geographies of Scale." In *The Politics of Scale in Environmental Governance*, ed. J. Gupta and D. Huitema. IDGEC manuscript.

Busch, P. O., and H. Jörgens. 2005. "The International Sources of Policy Convergence: Explaining the Spread of Environmental Policy Innovations." *Journal of European Public Policy* 12 (5): 860–84.

Büsscher, B., and W. Critchley. Forthcoming. "Indigenous Environmental Knowledge and Scale: Challenges in the Vertical Upscaling of Local Innovation in Sustainable Land Management within Africa." In *The Politics of Scale in Environmental Governance*, ed. J. Gupta and D. Huitema. IDGEC manuscript.

Capoor, K., and P. Ambrosi. 2006. *State and Trends of the Carbon Market 2006.* Washington, DC: World Bank.

Carmines, E. G., and R. A. Zeller. 1979. *Reliability and Validity Assessment.* Beverly Hills, CA: SAGE Publications.

Carothers, T. 1998. "The Rule of Law Revival." *Foreign Affairs* 77 (2): 95–106.

Cash, D. W. 2001. "'In Order to Aid in Diffusing Useful and Practical Information': Agricultural Extension and Boundary Organizations." *Science, Technology, and Human Values* 26:431–53.

Cash, D. W., W. N. Adger, F. Berkes, P. Garden, L. Lebel, P. Olsson, L. Pritchard, and O. Young. 2006. "Scale and Cross-Scale Dynamics: Governance and Information in a Multilevel World." *Ecology and Society* 11 (2): 8.

Cash, D. W., and S. C. Moser. 2000. "Linking Global and Local Scales: Designing Dynamic Assessment and Management Processes." *Global Environmental Change* 10:109–20.

Cashore, B. 2002. "Legitimacy and the Privatization of Environmental Governance: How Non-State Market-Driven (NSMD) Governance Systems Gain Rule-Making Authority." *Governance* 15:503–29.

Chalker, J. 2006. "Making the Investment Provisions of the Energy Charter Treaty Sustainable Development Friendly." *International Environmental Agreements: Politics, Law and Economics* 6 (4): 435–58.

Chambers, W. B., ed. 1998. *Global Climate Governance: Inter-linkages between the Kyoto Protocol and Other Multilateral Regimes.* Tokyo: United Nations University Press.

———, ed. 2001. *Inter-linkages: The Kyoto Protocol and the International Trade and Investment Regimes.* Tokyo: United Nations University Press.

Chambers, W. B., and J. F. Green, eds. 2005. *Reforming International Environmental Governance: From Institutional Limits to Innovative Reforms.* Tokyo: United Nations University Press.

Charnovitz, S. 1997. "Two Centuries of Participation: NGOs and International Governance." *Michigan Journal of International Law* 18:183–286.

———. 1998. "The World Trade Organization and the Environment." *Yearbook of International Environmental Law* 8:98–116.

———. 2003. *Trade and Climate: Potential Conflict and Synergies.* Washington, DC: Pew Center on Global Climate Change.

Chayes, A., and A. H. Chayes. 1991. "Compliance without Enforcement: State Behavior under Regulatory Treaties." *Negotiation Journal* 7 (3): 311–30.

———. 1993. "On Compliance." *International Organization* 47 (2): 175–205.

———. 1995. *The New Sovereignty: Compliance with International Regulatory Agreements.* Cambridge, MA: Harvard University Press.

Checkel, J. T. 1998. "The Constructivist Turn in International Relations Theory." *World Politics* 50 (3): 324–48.

———. 2001. "Why Comply? Social Learning and European Identity Change." *International Organization* 55:553–88.

Clapp, J. 1994. "Africa, NGOs, and the International Toxic Waste Trade." *Journal of Environment and Development* 2 (1): 17–46.

———. 1998. "The Privatization of Global Environmental Governance: ISO 14000 and the Developing World." *Global Governance* 4:295–316.

Clark, W. C. 1985. "Scales of Climate Impacts." *Climatic Change* 7 (1): 5–27.

Clark, W. C., P. J. Crutzen, and H. J. Schellnhuber. 2005. "Science for Global Sustainability: Toward a New Paradigm." Center for International Development Working Paper No. 120. Cambridge, MA: Harvard University.

Claude, I., Jr. 1988. *States and the Global System: Politics, Law, and Organization.* New York: St. Martin's.

Cleveland, C., R. Costanza, T. Eggertsson, L. Fortmann, B. Low, M. McKean, E. Ostrom, J. Wilson, and O. R. Young. 1996. "A Framework for Modeling the

Linkages between Ecosystems and Human Systems." Beijer Discussion Paper Series No. 76. Stockholm: Beijer International Institute of Ecological Economics.

Coffey, C. 2006. "The EU Habitats Directive: Enhancing Synergy with Pan-European Nature Conservation and with the EU Structural Funds." In *Institutional Interaction in Global Environmental Governance: Synergy and Conflict among International and EU Policies*, ed. S. Oberthür and T. Gehring, 233–58. Cambridge, MA: The MIT Press.

Cole, D. H. 2002. *Pollution and Property: Comparing Ownership Institutions for Environmental Protection*. Cambridge: Cambridge University Press.

Coleman, J. 1990. *Foundations of Social Theory*. Cambridge, MA: Belknap Press.

Commission on Global Governance. 1995. *Our Global Neighbourhood: The Report of the Commission on Global Governance*. Oxford: Oxford University Press.

Compagnon, D. Forthcoming. "Scaling and the Nation State in the Third World: Theoretical Gaps and Policy Consequences." In *The Politics of Scale in Environmental Governance*, ed. J. Gupta and D. Huitema. IDGEC manuscript.

Conca, K. 1995. "Greening the United Nations: Environmental Organizations and the UN System." *Third World Quarterly* 16 (3): 441–57.

———. 2006. *Governing Water: Contentious Transnational Politics and Global Institution Building*. Cambridge, MA: The MIT Press.

Connolly, B., and M. List. 1996. "Nuclear Safety in Eastern Europe and the Former Soviet Union." In *Institutions for Environmental Aid: Pitfalls and Promise*, ed. R. O. Keohane and M. A. Levy, 233–79. Cambridge, MA: The MIT Press.

Contreras, A. P. 2003. *The Kingdom and the Republic: Forest Governance and Political Transformation in Thailand and the Philippines*. Quezon City, Philippines: Ateneo de Manila University Press.

Contreras, A. P., L. Lebel, and S. Pasong. 2001. "The Political Economy of Tropical and Boreal Forests." IDGEC Scoping Report No. 3.

Conybeare, J. A. C. 1980. "International Organization and the Theory of Property Rights." *International Organization* 34:307–34.

Copeland, B. R., and M. S. Taylor. 2004. "Trade, Growth and the Environment." *Journal of Economic Literature* 42:7–71.

Cortell, A. P., and J. W. Davies. 1996. "How Do International Institutions Matter? The Domestic Impacts of International Rules and Norms." *International Studies Quarterly* 40 (4): 451–78.

Costanza, R. 1997. "Social Traps and Environmental Policy." *BioScience* 37 (6): 407–12.

Costanza, R., L. J. Graumlich, and W. Steffen, eds. 2005. *Sustainability or Collapse? An Integrated History and Future of People on Earth*. Cambridge, MA: The MIT Press.

Cowell, R. 2003. "Substitution and Scalar Politics: Negotiating Environmental Compensation in Cardiff Bay." *Geoforum* 34 (3): 343–58.

Cowles, M. G., ed. 2001. *Transforming Europe: Europeanization and Domestic Change.* Ithaca, NY: Cornell University Press.

Crepaz, M. L. 1995. "Explaining National Variations of Air Pollution Levels: Political Institutions and Their Impact on Environmental Policy-making." *Environmental Politics* 4 (3): 391–414.

Crona, B. 2006. "Supporting and Enhancing Development of Heterogeneous Ecological Knowledge among Resource Users in a Kenyan Seascape." *Ecology and Society* 11 (1): 32.

Cross, J. G., and M. J. Guyer. 1980. *Social Traps.* Ann Arbor: University of Michigan Press.

Crowder, L., et al. 2006. "Resolving Mismatches in U.S. Ocean Governance." *Science* 313:617–18.

Crutzen, P. J., and E. F. Stoermer. 2000. "The 'Anthropocene.'" *IGBP Newsletter* 41:17–18.

Cumming, G. S., D. H. M. Cumming, and C. L. Redman. 2006. "Scale Mismatches in Social-Ecological Systems: Causes, Consequences, and Solutions." *Ecology and Society* 11 (1): 14. http://www.ecologyandsociety.org/vol11/iss1/art14/.

Cutler, A. C., V. Haufler, and T. Porter, eds. 1999. *Private Authority and International Affairs.* Albany: State University of New York Press.

Dai, Xinyuan. 2005. "Why Comply? The Domestic Constituency Mechanism." *International Organization* 59:363–98.

Dalton, R. J. 2004. *Democratic Challenges, Democratic Choices: The Erosion of Political Support in Advanced Industrial Democracies.* Oxford: Oxford University Press.

Daly, H. 2003. "Ecological Economics: The Concept of Scale and Its Relation to Allocation, Distribution, and Uneconomic Growth." Presented at the CANSEE Conference, October 16–19, Alberta, Canada.

Datta, S., and V. Varalakshmi. 1999. "Decentralization: An Effective Method of Financial Management at the Grassroots." *Sustainable Development* 7:113–20.

Dauvergne, P. 1997. *Shadows in the Forest: Japan and the Politics of Timber in Southeast Asia.* Cambridge, MA: The MIT Press.

Deacon, R., and C. S. Norman. 2004. "Is the Environmental Kuznets Curve an Empirical Reality?" Departmental Working Paper, Department of Economics, University of California, Santa Barbara.

de la Torre Castro, M. 2006. "Beyond Regulations in Fisheries Management: The Dilemmas of the 'Beach Recorders' Bwana Dikos in Zanzíbar, Tanzania." In M. de la Torre Castro, "Humans and Seagrasses in East Africa." PhD thesis, Paper No. 6, Department of Systems Ecology, Stockholm University, Sweden.

Delmas, M., and O. R. Young, eds. Forthcoming. *Governing the Environment: Interdisciplinary Perspectives.* Cambridge: Cambridge University Press.

DeSombre, E. R. 2000. *Domestic Sources of International Environmental Policy.* Cambridge, MA: The MIT Press.

———. 2005. "Fishing under Flags of Convenience: Using Market Power to Increase Participation in International Regulation." *Global Environmental Politics* 5 (4): 73–94.

DeVellis, R. F. 2003. *Scale Development: Theory and Applications.* Thousand Oaks, CA: SAGE Publications.

Dhakal, S. 2006. "Response Comments to 'Global Change: Analysing Scale and Scaling in Environmental Governance,' by Joyeeta Gupta." Presentation at IDGEC Synthesis Conference, December 6–9, Bali, Indonesia.

Dietz, T., E. Ostrom, and P. C. Stern. 2003. "The Struggle to Govern the Commons." *Science* 302 (5652): 1907–12.

Dingwerth, K. 2005. "The Democratic Legitimacy of Public-Private Rule-Making: What Can We Learn from the World Commission on Dams?" *Global Governance* 11 (1): 65–83.

———. 2007. *The New Transnationalism: Transnational Governance and Democratic Legitimacy.* Basingstoke, UK: Palgrave Macmillan.

Dingwerth, K., and P. Pattberg. 2006. "Global Governance as a Perspective on World Politics." *Global Governance* 12 (2): 185–203.

Doty, H. D., and W. H. Glick. 1994. "Typologies as a Unique Form of Theory Building: Toward Increased Understanding and Modelling." *The Academy of Management Review* 19 (2): 230–51.

Downs, G. W. 2000. "Constructing Effective Environmental Regimes." *Annual Review of Political Science* 3:25–42.

Downs, G. W., D. M. Rocke, and P. N. Barsoom. 1996. "Is the Good News about Compliance Good News about Cooperation?" *International Organization* 50 (3): 379–406.

Dryzek, J. S. 1999. "Transnational Democracy." *The Journal of Political Philosophy* 7 (1): 30–51.

Duffield, J. 2007. "What Are International Institutions?" *International Studies Review* 9:1–22.

Duit, A., and V. Galaz. forthcoming. "Governance and Complexity: Emerging Challenges for Governance Theory." Accepted for publication in *Governance.*

Ebbin, S. A., ed. 2004. "Institutions and the Production of Knowledge for Environmental Governance: Empirical Evidence from Marine and Terrestrial Systems." Special issue, *International Environmental Agreements: Politics, Law and Economics* 4 (2).

———. 2005. "The Impact of the EEZ on Salmon Management in the US Pacific Northwest Region: An Examination of Institutional Change and Vertical Interplay." In *A Sea Change: The Exclusive Economic Zone and Governance*

Institutions for Living Marine Resources, ed. S. A. Ebbin, A. Håkon Hoel, and A. K. Sydnes. Dordrecht, The Netherlands: Springer Verlag.

Ebbin, S. A., A. Håkon Hoel, and A. K. Sydnes, eds. 2005. *A Sea Change: The Exclusive Economic Zone and Governance Institutions for Living Marine Resources*. Dordrecht, The Netherlands: Springer Verlag.

Eckersley, R. 2004. "The Big Chill: The WTO and Multilateral Environmental Agreements." *Global Environmental Politics* 4 (2): 24–50.

Edwards, M., and S. Zadek. 2003. "Governing the Provision of Global Public Goods: The Role and Legitimacy of Nonstate Actors." In *Providing Global Public Goods: Managing Globalization*, ed. I. Kaul, P. Conceicao, K. Le Goulven, and R. U. Mendoza, 200–224. New York, Oxford: Oxford University Press.

Ehrlich, Paul R. and Ann H. Ehrlich (1990). *The Population Explosion*. New York: Simon and Schuster.

Eisenack, K. 2003. "Qualitative Viability Analysis of a Bio-Socio-Economic System." In *Working Papers of 17th Workshop on Qualitative Reasoning*, ed. P. Salles and B. Bredeweg, 63–70. Brasilia. http://www.pik-potsdam.de/~eisenack/downloads/QR-2003.pdf.

Eisenack, K., J. Kropp, and H. Welsch. 2006. "A Qualitative Dynamical Modelling Approach to Capital Accumulation in Unregulated Fisheries." *Journal of Economic Dynamics and Control* 30:2613–36.

Eising, R., and B. Kohler-Koch. 2000. "Introduction: Network Governance in the European Union." In *The Transformation of Governance in the European Union*, ed. B. Kohler-Koch and R. Eising. London and New York: Routledge.

Elkins, Z., and B. Simmons. 2005. "On Waves, Clusters, and Diffusion: A Conceptual Framework." *The Annals of the American Academy of Political and Social Science* 598:33–51.

Ellickson, R. C. 1991. *Order without Law: How Neighbors Settle Disputes*. Cambridge, MA: Harvard University Press.

Ellis, F., and S. Biggs. 2001. "Evolving Themes in Rural Development 1950s–2000s." *Development Policy Review* 19 (4): 437–48.

Elmqvist, T., C. Folke, M. Nystroem, G. Peterson, J. Bengtsson, B. Walker, and J. Norberg. 2003. "Response Diversity, Ecosystem Change, and Resilience." *Frontiers in Ecology and the Environment* 1 (9): 488–94.

Esty, D. C. 1994. *Greening the GATT: Trade, Environment, and the Future*. Washington, DC: International Institute for Economics.

———. 2006. "Good Governance at the Supranational Scale: Globalizing Administrative Law." *Yale Law Journal* 115 (7): 1490–1562.

Evans, P. 1996. "Government Action, Social Capital, and Development: Reviewing the Evidence on Synergy." *World Development* 24 (6): 178–209.

Fairbrass, J., and A. Jordan. 2001. "Protecting Biodiversity in the European Union: National Barriers and European Opportunities?" *Journal of European Public Policy* 8 (4): 499–518.

Falkner, R. 2003. "Private Environmental Governance and International Relations: Exploring the Links." *Global Environmental Politics* 3:72–87.

Fearon, J. D. 1991. "Counterfactuals and Hypothesis Testing in Political Science." *World Politics* 43 (2): 169–95.

Fearon, J., and A. Wendt. 2002. "Rationalism v. Constructivism: A Skeptical View." In *Handbook of International Relations*, ed. W. Carlsnaes, T. Risse, and B. A. Simmon, 52–72. London: SAGE Publications.

Finkelstein, L. S. 1995. "What Is Global Governance?" *Global Governance* 1 (3): 367–72.

Finnemore, M. 1993. "International Organizations as Teachers of Norms: The United Nations Educational, Scientific, and Cultural Organization and Science Policy." *International Organization* 47 (4): 565–97.

Finnemore, M., and K. Sikkink. 1998. "International Norm Dynamics and Political Change." *International Organization* 52 (4): 887–917.

Fiori, S. 2006. "The Emergence of Institutions in Hayek's Theory: Two Views or One?" *Constitutional Political Economy* 17:49–66.

Fisher, R. 1981. *Improving Compliance with International Law*. Charlottesville: University Press of Virginia.

Fitzpatrick, P. 2003. "'Gods Would be Needed ...': American Empire and the Rule of (International) Law." *Leiden Journal of International Law* 16 (3): 429–66.

Folke, C. 2006. "Resilience: The Emergence of a Perspective for Social-Ecological System Analyses." *Global Environmental Change* 16 (3): 253–67.

Folke, C., S. Carpenter, B. Walker, et al. 2004. "Regime Shifts, Resilience, and Biodiversity in Ecosystem Management." *Annual Review of Ecology, Evolution, and Systematics* 35:557–81.

Folke, C., J. Colding, and F. Berkes. 2003. "Synthesis: Building Resilience and Adaptive Capacity in Social-Ecological Systems." In *Navigating Social-Ecological Systems: Building Resilience for Complexity and Change*, ed. F. Berkes, J. Colding, and C. Folke. Cambridge: Cambridge University Press.

Folke, C., T. Hahn, P. Olsson, et al. 2005. "Adaptive Governance of Social-Ecological Systems." *Annual Review of Environment and Resources* 30:441–73.

Folke, C., L. Pritchard, F. Berkes, J. Colding, and U. Svedin. 1998. "The Problem of Fit between Ecosystems and Institutions." IHDP Working Paper No. 2. Bonn, Germany: International Human Dimensions Programme.

Franck, T. M. 1995. *Fairness in International Law and Institutions*. Oxford: Oxford University Press.

French, D. 2002. "The Role of the State and International Organizations in Reconciling Sustainable Development and Globalization." *International Environmental Agreements: Politics, Law and Economics* 2 (2): 135–50.

Friedman, T. L. 2006. *The World Is Flat: The Globalized World in the Twenty-first Century*. London: Penguin Books.

Friibergh Workshop on Sustainability Science. 2000. Statement of the Friibergh Workshop on Sustainability Science, Friibergh, Sweden, October 11–14, 2000. http://sustsci.aaas.org/content.html?contentid=774.

Fukuyama, F. 1992. *The End of History and the Last Man.* New York: Avon Books.

Gadgil, M., P. Olsson, F. Berkes, and C. Folke. 2003. "Exploring the Role of Local Ecological Knowledge for Ecosystem Management: Three Case Studies." In *Navigating Social-Ecological Systems: Building Resilience for Complexity and Change*, ed. F. Berkes, J. Colding, and C. Folke. Cambridge: Cambridge University Press.

Gadgil, M., P. R. Seshagiri Rao, G. Utkarsh, P. Pramod, and A. Chhatre. 2000. "New Meanings for Old Knowledge: The People's Biodiversity Registers Programme." *Ecological Applications* 10:1307–17.

Galaz, V. 2006. "The Rönne and Em Rivers, Sweden: Resilience, Networks and Bargaining Power in Water Management." In *Networks and Institutions in Natural Resource Management*, ed. E. Falleth and Y. Rydin. Cheltenham, UK: Edward Elgar Publishing.

Garden, P., L. Lebel, F. Viseskul, C. Chirangworapat, and M. Prompanyo. 2006. "The Consequences of Institutional Interplay and Density on Local Governance in Northern Thailand." USER Working Paper WP-2006-03, Chiang Mai University, Thailand.

Gareau, B. Forthcoming. "The Global Scale: Global Political Economy Theory and Application to the Montreal Protocol." In *The Politics of Scale in Environmental Governance*, ed. J. Gupta and D. Huitema. IDGEC manuscript.

Gehring, T. 1994. *Dynamic International Regimes: Institutions for International Environmental Governance.* Frankfurt am Main: P. Lang.

———. 2007. "Einflussbeziehungen zwischen internationalen Institutionen im Spannungsfeld von Handel und Umwelt: Von gegenseitiger Störung zur institutionalisierten Arbeitsteilung." In *Politik und Umwelt*, ed. K. Jakob, F. Biermann, P. O. Busch, and K. H. Feindt. Special issue of *Politische Vierteljahresschrift* 39:94–114.

Gehring, T., and S. Oberthür. 2004. "Exploring Regime Interaction: A Framework of Analysis." In *Regime Consequences: Methodological Challenges and Research Strategies*, ed. A. Underdal and O. R. Young, 247–69. Dordrecht, The Netherlands: Kluwer Academic Publishers.

———. 2006. "Comparative Empirical Analysis and Ideal Types of Institutional Interaction." In *Institutional Interaction in Global Environmental Governance: Synergy and Conflict among International and EU Policies*, ed. S. Oberthür and T. Gehring, 307–71. Cambridge, MA: The MIT Press.

George, A., and A. Bennett. 2005. *Case Studies and Theory Development in the Social Sciences.* Cambridge, MA: The MIT Press.

Gibson, C. C., E. Ostrom, and T. K. Ahn. 2000. "The Concept of Scale and the Human Dimensions of Global Change." *Ecological Economics* 32:217–39.

Gibson, C. C., J. T. Williams, and E. Ostrom. 2005. "Local Enforcement and Better Forests." *World Development* 33:273–84.

Gill, P. 2004. "Securing the Globe: Intelligence and the Post 9/11 Shift from 'Liddism' to 'Drainism.'" *Intelligence and National Security* 19 (3): 467–89.

Gillespie, A. 2002. "Forum Shopping in International Environmental Law: The IWC, CITES, and the Management of Cetaceans." *Ocean Development and International Law* 33 (1): 17–56.

Glantz, M. H., ed. 1976. *The Politics of Natural Disaster: The Case of the Sahel Drought.* New York: Praeger.

———. 1990. "Does History Have a Future? Forecasting Climate Change Effects by Analogy." *Fisheries* 15 (6): 39–44.

Glasbergen, P., F. Biermann, and A. Mol. 2007. *Partnerships, Governance and Sustainable Development: Reflections on Theory and Practice.* Cheltenham, UK: Edward Elgar Publishing.

Gleick, P., et al. 2003. "The Privatization of Water and Water Systems." In *The World's Water 2002–2003: The Biennial Report on Freshwater Resources*, 33–56. Washington, DC: Island Press.

Global Water Partnership. 2000. "Integrated Water Resources Management." TAC Background Papers, No. 4. Stockholm: Global Water Partnership.

González, A. A., and R. Nigh. 2005. "Smallholder Participation and Certification of Organic Farm Products in Mexico." *Journal of Rural Studies* 21:449–60.

Gordenker, L., and T. G. Weiss. 1996. "Pluralizing Global Governance: Analytical Approaches and Dimensions." In *NGOs, the UN, and Global Governance*, ed. T. G. Weiss and L. Gordenker, 17–47. Boulder, CO: Lynne Rienner.

Gordon, L., M. Dunlop, and B. Foran. 2003. "Land Cover Change and Water Vapour Flows: Learning from Australia." *Philosophical Transactions: Biological Sciences* 358 (1440): 1973–84.

Gough, C., and S. Shackley. 2001. "The Respectable Politics of Climate Change: The Epistemic Communities and NGOs." *International Affairs* 77 (2): 329–45.

Grant, R. W., and R. O. Keohane. 2005. "Accountability and Abuses of Power in World Politics." *American Political Science Review* 99 (1): 29–44.

Grossman, G. M., and A. B. Krueger. 1995. "Economic Growth and the Environment." *Quarterly Journal of Economics* 110 (2): 353–75.

Gunderson, L. H. 1999. "Resilience, Flexibility and Adaptive Management: Antidotes for Spurious Certitude?" *Conservation Ecology* 3:7.

———. 2000. "Ecological Resilience—in Theory and Application." *Annual Review of Ecology and Systematics* 31:425–39.

———. 2003. "Adaptive Dancing: Interactions between Social Resilience and Ecological Crises." In *Navigating Social-Ecological Systems: Building Resilience for Complexity and Change*, ed. F. Berkes, J. Colding, and C. Folke. Cambridge: Cambridge University Press.

Gunderson, L. H., and C. S. Holling, eds. 2002. *Panarchy: Understanding Transformations in Human and Natural Systems.* Washington, DC: SCOPE/Island Press.

Gunderson, L. H., and L. Pritchard Jr. 2002. *Resilience and the Behavior of Large-Scale Systems.* Washington, DC: Island Press.

Gupta, A. 2004. "When Global Is Local: Negotiating Safe Use of Biotechnology." In *Earthly Politics: Local and Global in Environmental Governance,* ed. S. Jasanoff and M. Long Martello, 127–48. Cambridge, MA: The MIT Press.

———. 2006. "Problem Framing in Assessment Processes: The Case of Biosafety." In *Global Environmental Assessments: Information and Influence,* ed. R. B. Mitchell, W. C. Clark, D. W. Cash, and N. Dickson, 57–86. Cambridge, MA: The MIT Press.

Gupta, A., and R. Falkner. 2006. "The Influence of the Cartagena Protocol on Biosafety: Comparing Mexico, China and South Africa." *Global Environmental Politics* 6 (4): 23–44.

Gupta, J. 1995. "The Global Environment Facility in Its North-South Context." *Environmental Politics* 4 (1): 19–43.

———. 1997. *The Climate Change Convention and Developing Countries: From Conflict to Consensus?* Dordrecht, The Netherlands: Kluwer Academic Publishers.

———. 2003. "The Role of Non-state Actors in International Environmental Affairs." *Heidelberg Journal of International Law* 63 (2): 459–86.

———. 2005. "Who's Afraid of Global Warming?" Inaugural address as Professor of Climate Change: Policy and Law, Vrije Universiteit Amsterdam, October 21. ISBN 90-90201-43-2.

———. 2007. "Climate Change: Legal Challenges with Respect to Developing Countries." *Yearbook of International Environmental Law* 16: 114–153.

———. Forthcoming. "Constructing Scale in the Climate Change Negotiations." In *The Politics of Scale in Environmental Governance,* ed. J. Gupta and D. Huitema. IDGEC manuscript.

Gupta, J., and M. Hisschemöller. 1997. "Issue-Linkages: A Global Strategy Towards Sustainable Development." *International Environmental Affairs* 9 (4): 289–308.

Gupta, J., and D. Huitema, eds. Forthcoming. *The Politics of Scale in Environmental Governance,* Amsterdam: Institute of Environmental Studies, Free University of Amsterdam. IDGEC Manuscript.

Gupta, J., and H. van Asselt. 2006. "Helping Operationalise Article 2: A Transdisciplinary Methodological Tool for Evaluating When Climate Change Is Dangerous." *Global Environmental Change* 16 (1): 83–94. http://authors.elsevier.com/sd/article/S0959378005000762.

Gupta, J., D. Weber, H. van Asselt, B. van der Veen, and F. Sindico. 2003. "Relationship between Domestic, Supranational, and International Policymaking and Law: The Problem of Scale." Report of the Workshop on Multi-level Environmental Governance, Institute for Environmental Studies, Vrije Universiteit Amsterdam, December.

Guston, D. H. 1999. "Stabilizing the Boundary between Politics and Science: The Role of the Office of Technology Transfer as a Boundary Organization." *Social Studies of Science* 29:87–112.

———. 2001. "Boundary Organizations in Environmental Policy and Science: An Introduction." *Science, Technology & Human Values* 26 (4): 399–408.

Haas, Ernst B. 1990. *When Knowledge Is Power: Three Models of Change in International Organizations.* Berkeley: University of California Press.

Haas, P. M. 1989. "Do Regimes Matter? Epistemic Communities and Mediterranean Pollution Control." *International Organization* 43 (3): 377–403.

———. 1992a. "Banning Chlorofluorocarbons: Epistemic Community Efforts to Protect Stratospheric Ozone." *International Organization* 46:187–224.

———. 1992b. "Introduction: Epistemic Communities and International Policy Coordination." *International Organization* 46 (1): 1–36.

———. 2004. "Addressing the Global Governance Deficit." *Global Environmental Politics* 4 (4): 1–15.

Haas, P. M., R. O. Keohane, and M. A. Levy, eds. 1993. *Institutions for the Earth: Sources of Effective International Environmental Protection.* Cambridge, MA: The MIT Press.

Haftel, Y. Z., and A. Thompson. 2006. "The Independence of International Organizations." *Journal of Conflict Resolution* 50:253–75.

Hahn, T. Forthcoming. "Lessons from the Sub-global Assessments: Scaling Down the Ecosystem Approach of the CBD to Local Policies and Processes." In *The Politics of Scale in Environmental Governance*, ed. J. Gupta and D. Huitema. IDGEC manuscript.

Hahn, T., P. Olsson, C. Folke, et al. 2006. "Trust-building, Knowledge Generation and Organizational Innovations: The Role of Bridging Organization for Adaptive Comanagement of a Wetland Landscape around Kristianstad, Sweden." *Human Ecology* 34:573–92.

Hall, R. B., and T. J. Biersteker, eds. 2002. *The Emergence of Private Authority in Global Governance.* Cambridge: Cambridge University Press.

Halls, A. S., R. I. Arthur, D. Bartley, et al. 2005. "Guidelines for Designing Data Collection and Sharing Systems for Co-managed Fisheries. Part 1: Practical Guide." FAO Fisheries Technical Paper No. 494/1. Rome: FAO.

Harbaugh, W., A. Levinson, and D. Wilson. 2000. *Re-examining the Empirical Evidence for an Environmental Kuznets Curve.* Cambridge, MA: National Bureau of Economic Research.

Hardin, G. 1968. "The Tragedy of the Commons." *Science* 162:1243–48.

Hardin, R. 1982. *Collective Action.* Baltimore, MD: Johns Hopkins University Press.

Harris, M. 1998. *Lament for an Ocean: The Collapse of the Atlantic Cod Fishery; a True Crime Story.* Toronto: McClelland and Stewart.

Hart, H. L. A. 1961. *The Concept of Law.* Oxford: Oxford University Press.

Hasenclever, A., P. Mayer, and V. Rittberger. 1997. *Theories of International Regimes.* Cambridge: Cambridge University Press.

Haufler, V. 2000. "Private Sector International Regimes." In *Non-state Actors and Authority in the Global System*, ed. R. A. Higgott, G. R. D. Underhill, and A. Bieler, 121–37. London and New York: Routledge.

Hayek, F. A. 1973. *Law, Legislation, and Liberty.* Vol. 1, *Rules and Order.* Chicago: University of Chicago Press.

Hedström, P., and R. Swedberg. 1998. *Social Mechanisms: An Analytical Approach to Social Theory.* Cambridge: Cambridge University Press.

Held, D. 1999. "The Transformation of Political Community." In *Democracy's Edges*, ed. I. Shapiro and C. Hacker-Cordón. Cambridge: Cambridge University Press.

———. 2000. "Regulating Globalization? The Reinvention of Politics." *International Sociology* 15 (2): 394–408.

———. 2004. "Democratic Accountability and Political Effectiveness from a Cosmopolitan Perspective." *Government and Opposition* 39 (2): 364–91.

Helm, C., and D. Sprinz. 2000. "Measuring the Effectiveness of International Environmental Regimes." *Journal of Conflict Resolution* 44 (5): 630–52.

Henriksen, T., G. Honneland, and A. Sydnes, eds. 2006. *Law and Politics in Ocean Governance.* Leiden, The Netherlands: Martinus Nijhoff.

Héritier, A. 1999. *Policy-making and Diversity in Europe: Escaping Deadlock.* Cambridge: Cambridge University Press.

Héritier, A., C. Knill, and S. Mingers. 1996. *Ringing the Changes in Europe.* Berlin and New York: Walter de Gruyter.

Herr, R. A., and E. Chia. 1995. "The Concept of Regime Overlap: Toward Identification and Assessment." In *Overlapping Maritime Regimes: An Initial Reconnaissance*, ed. B. Davis, 11–26. Hobart, Australia: Antarctic CRC and Institute of Antarctic and Southern Ocean Studies.

Higgott, R. A., G. D. Underhill, and A. Bieler, eds. 1999. *Non-state Actors and Authority in the Global System.* London: Routledge.

Hirst, P. 2000. "Democracy and Governance." In *Debating Governance: Authority, Steering and Democracy*, ed. J. Pierre, 13–35. Oxford: Oxford University Press.

Hirst, P., and G. Thompson. 1995. "Globalization and the Future of the Nation State." *Economy and Society* 24 (3): 408–42.

Hisschemöller, M., R. Hoppe, W. N. Dunn, and J. R. Ravetz, eds. 2001. *Knowledge, Power, and Participation in Environmental Policy Analysis.* New Brunswick and London: Transaction Publishers.

Hoel, M. 1997. "Coordination of Environmental Policy for Transboundary Environmental Problems?" *Journal of Public Economics* 66 (2): 199–224.

Hoel, A. H. 2000. "Performance of Exclusive Economic Zones." IDGEC Scoping Report No. 2.

Hoel, A. H., and I. Kvalvik. 2006. "The Allocation of Scarce Resources: The Case of Fisheries." *Marine Policy* 30:347–56.

Hoffmann, M. J. 2005. *Ozone Depletion and Climate Change.* Albany: State University of New York Press.

Holling, C. S., ed. 1978. *Adaptive Environmental Assessment and Management.* New York: John Wiley and Sons.

———. 1986. "The Resilience of Terrestrial Ecosystems: Local Surprise and Global Change." In *Sustainable Development of the Biosphere*, ed. W. C. Clark and R. E. Munn, 292–317. Cambridge: Cambridge University Press.

Holling, C. S., and G. K. Meffe. 1996. "Command and Control and the Pathology of Natural Resource Management." *Conservation Biology* 10 (2): 328–37.

Hood, C., and D. Heald, eds. 2006. *Transparency: The Key to Better Governance?* Oxford: Oxford University Press.

Hooghe, L., and G. Marks. 2003. "Unraveling the Central State, but How? Types of Multi-level Governance." *American Political Science Review* 97 (2): 233–43.

Hooper, B. 2005. *Integrated River Basin Governance: Learning from International Experience.* London: IWA Publishing.

Hovi, J. 2004. "Causal Mechanisms and the Study of Regime Effectiveness." In *Regime Consequences: Methodological Challenges and Research Strategies*, ed. A. Underdal and O. R. Young, 71–86. Dordrecht, The Netherlands: Kluwer Academic Publishers.

Hovi, J., D. F. Sprinz, and A. Underdal. 2003a. "The Oslo-Potsdam Solution to Measuring Regime Effectiveness: Critique, Response, and the Road Ahead." *Global Environmental Politics* 3 (3): 74–96.

Hovi, J., D. F. Sprinz, and A. Underdal. 2003b. "Regime Effectiveness and the Oslo-Potsdam Solution: A Rejoinder to Oran Young." *Global Environmental Politics* 3 (3): 105–7.

Hovi, J., and D. Sprinz. 2006. "The Limits of the Law of the Least Ambitious Program." *Global Environmental Politics* 6:28–42.

Huitema, D., and H. Bressers. Forthcoming. "Scaling Water Governance: The Case of the Implementation of the European Water Framework Directive in the Netherlands." In *The Politics of Scale in Environmental Governance*, ed. J. Gupta and D. Huitema. IDGEC manuscript.

Huitric, M. 2005. "Lobster and Conch Fisheries of Belize: A History of Sequential Exploitation." *Ecology and Society* 107:21.

Humphreys, D. 2005. "The Elusive Quest for a Global Forests Convention." *RECIEL* 14 (1): 1–10.

Imperial, M. T. 1999. "Institutional Analysis and Ecosystem-Based Management: The Institutional Analysis and Development Framework." *Environmental Management* 24:449–65.

———. 2005. "Using Collaboration as a Governance Strategy: Lessons from Six Watershed Management Programs." *Administration and Society* 37:281–320.

International Maritime Organization. 2006. "New International Rules to Allow Storage of CO_2 in Seabed Adopted." IMO Press Briefing 43/2006, November 8. London: International Maritime Organization.

Isaac, G. E., and W. A. Kerr. 2003. "Genetically Modified Organisms and Trade Rules: Identifying Important Challenges for the WTO." *World Economy* 26 (1): 29–42.

Jackson, J. B. C. 1997. "Reefs since Columbus." *Coral Reefs* 16 (Supplement 1): S23–S32.

———. 2001. "What Was Natural in the Coastal Oceans?" *Proceedings of the National Academy of Sciences* 98:5411–18.

Jackson, J. B. C., M. X. Kirby, W. H. Berger, K. A. Bjorndal, L. W. Botsford, B. J. Bourque, R. H. Bradbury, R. Cooke, J. Erlandson, J. A. Estes, et al. 2001. "Historical Overfishing and the Recent Collapse of Coastal Ecosystems." *Science* 293:629–37.

Jacobson, H. K., and E. Brown Weiss. 1998. "Assessing the Records and Designing Strategies to Engage Countries." In *Engaging Countries: Strengthening Compliance with International Environmental Accords*, ed. E. Brown Weiss and H. K. Jacobson, 511–54. Cambridge, MA: The MIT Press.

Jacquemont, F., and A. Caparrós. 2002. "The Convention on Biological Diversity and the Climate Change Convention 10 Years after Rio: Towards a Synergy of the Two Regimes?" *Review of European Community and International Environmental Law* 11 (2): 139–80.

Jäger, J. 1998. "Current Thinking on Using Scientific Findings in Environmental Policy Making." *Environmental Modeling and Assessment* 3:143–53.

Jahiel, A. R. 2006. "China, the WTO, and Implications for the Environment." *Environmental Politics* 15:310–29.

Jänicke, M., and H. Weidner, eds. 1997. *National Environmental Policies: A Comparative Study of Capacity-Building.* Berlin: Springer.

Janssen, M., Ö. Bodin, J. M. Anderies, et al. 2006. "Toward a Network Perspective of the Study of Resilience in Social-Ecological Systems." *Ecology and Society* 11 (1): 15.

Janssen, M. A., and E. Ostrom. 2006. "Resilience, Vulnerability, and Adaptation: A Cross-cutting Theme of the International Human Dimensions Programme on Global Environmental Change" (editorial). *Global Environmental Change* 16 (3): 237–39.

Jasanoff, S. 1996. "Science and Norms in Global Environmental Regimes." In *Earthly Goods: Environmental Change and Social Justice*, ed. J. Reppy, 173–97. Ithaca, NY: Cornell University Press.

Jasanoff, S., and M. Long Martello, eds. 2004. *Earthly Politics: Local and Global in Environmental Governance*. Cambridge, MA: The MIT Press.

Jervis, R. 1997. *System Effects: Complexity in Political and Social Life*. Princeton, NJ: Princeton University Press.

Jones, C., W. S. Hesterly, and S. P. Borgatti. 1997. "A General Theory of Network Governance: Exchange Conditions and Social Mechanisms." *The Academy of Management Review* 22 (4): 911–945.

Jordan, A., R. K. W. Wurzel, and A. Zito. 2005. "The Rise of 'New' Policy Instruments in Comparative Perspective: Has Governance Eclipsed Government?" *Political Studies* 53:477–96.

Kahneman, D. 2003. "Maps of Bounded Rationality: Psychology for Behavioral Economics." *American Economic Review* 93:449–75.

Kahneman, D., and A. Tversky. 2000. *Choices, Values, and Frames*. Cambridge: Cambridge University Press.

Kanie, N., and P. M. Haas, eds. 2004. *Emerging Forces in Environmental Governance*. Tokyo: United Nations University Press.

Karlsson, S. 2000. "Multilayered Governance: Pesticides in the South—Environmental Concerns in a Globalized World." PhD dissertation, Department of Water and Environmental Studies, Linköping University, Sweden.

Kates, R. W., W. C. Clark, R. Corell, J. M. Hall, C. C. Jaeger, I. Lowe, J. McCarthy, H.-J. Schellnhuber, B. Bolin, N. M. Dickson, S. Faucheux, et al. 2001. "Sustainability Science." *Science* 292 (5517): 641–42.

Kates, R. W., and T. M. Parris. 2003. "Long-term Trends and a Sustainability Transition." *Proceedings of the National Academy of Sciences* 100 (14): 8062–67.

Kates, R. W., T. M. Parris, and A. A. Leiserowitz. 2005. "What Is Sustainable Development? Goals, Indicators, Values, and Practice." *Environment: Science and Policy for Sustainable Development* 47 (3): 8–21.

Katzenstein, P., ed. 1996. *The Culture of National Security: Norms and Identity in World Politics*. New York: Columbia University Press.

Katzenstein, P. J., R. O. Keohane, and S. Krasner. 1998. "International Organization and the Study of World Politics." *International Organization* 52:645–85.

Katzner, T. E. 2005. "Corruption—a Double-edged Sword for Conservation? A Response to Smith and Walpole." *Oryx* 39 (3): 260–62.

Kaul, I., and K. Le Goulven. 2003. "Institutional Options for Providing Global Public Goods." In *Providing Global Public Goods: Managing Globalization*, ed. I. Kaul, P. Conceicao, K. Le Goulven, and R. U. Mendoza, 371–409. New York and Oxford: Oxford University Press.

Keck, M., and K. Sikkink. 1988. *Activists beyond Borders: Advocacy Networks in International Politics*. Ithaca, NY: Cornell University Press.

Keohane, R. O. 1984. *After Hegemony: Cooperation and Discord in the World Political Economy*. Princeton, NJ: Princeton University Press.

———. 1993. "The Analysis of International Regimes: Towards a European-American Research Programme." In *Regime Theory and International Relations*, ed. V. Rittberger, 23–45. Oxford: Clarendon Press.

Keohane, R. O., P. M. Haas, and M. A. Levy. 1993. "The Effectiveness of International Environmental Institutions." In *Institutions for the Earth*, ed. P. M. Haas, R. O. Keohane, and M. A. Levy, 3–24. Cambridge, MA: The MIT Press.

Keohane, R. O., and M. A. Levy, eds. 1996. *Institutions for Environmental Aid: Pitfalls and Promise*. Cambridge, MA: The MIT Press.

Keohane, R. O., and L. L. Martin. 2003. "Institutional Theory, Endogeneity, and Delegation." In *Progress in International Relations Theory*, ed. C. Elman and M. Elman, 71–107. Cambridge, MA: The MIT Press.

Keohane, R. O., and J. S. Nye. 2003. "Redefining Accountability for Global Governance." In *Governance in a Global Economy*, ed. M. Kahler and D. A. Lake, 386–412. Princeton, NJ, and Oxford: Princeton University Press.

Kickert, W. J. M., E.-H. Klijn, and J. F. M. Koppenjan, eds. 1997. *Managing Complex Networks: Strategies for the Public Sector*. London: SAGE Publications.

Kickert, W. J. M., and J. F. M. Koppenjan. 1997. "Public Management and Network Management: An Overview." In *Managing Complex Networks: Strategies for the Public Sector*, ed. W. J. M. Kickert, E.-H. Klijn, and J. F. M. Koppenjan. London: SAGE Publications.

Kim, L. 1998. "Crisis Construction and Organizational Learning: Capability Building in Catching-up at Hyundai Motor." *Organizational Science* 9 (4): 506–21.

King, L. A. 1997. "Institutional Interplay: Research Questions." Background paper prepared for the IDGEC project, University of Vermont.

Kingdon, J. W. 1995. *Agendas, Alternatives, and Public Policies*. 2nd ed. New York: Harper Collins.

Kingsbury, B. 1995. "The Tuna-Dolphin Controversy, the World Trade Organization, and the Liberal Project to Reconceptualize International Law." In *Yearbook of International Environmental Law*, ed. G. Handl, 1–40. Oxford: Clarendon Press.

Kinzig, A. P., P. Ryan, M. Etienne, H. Allison, T. Elmqvist, and B. H. Walker. 2006. "Resilience and Regime Shifts: Assessing Cascading Effects." *Ecology and Society* 11 (1): 20. http://www.ecologyandsociety.org/vol11/iss1/art20/.

Klijn, E.-H., and J. F. M. Koppenjan. 2004. *Managing Uncertainties in Networks: A Network Approach to Problem Solving and Decision Making*. London: Routledge.

Klijn, E.-H., and G. R. Teisman. 1997. "Strategies and Games in Networks." In *Managing Complex Networks: Strategies for the Public Sector*, ed. W. J. M. Kickert, E.-H. Klijn, and J. F. M. Koppenjan, 98–118. London: SAGE Publications.

Knight, J., and D. North. 1997. "Explaining the Complexity of Institutional Change." In *The Political Economy of Property Rights*, ed. D. L. Weimer, 349–54. Cambridge: Cambridge University Press.

Komesar, N. K. 2001. *Law's Limits: The Rule of Law and the Supply and Demand of Rights*. Cambridge: Cambridge University Press.

Kooiman, J. ed. 1993. *Modern Governance: New Government-Society Interactions*. London: SAGE Publications.

———. 2003. *Governing as Governance*. London: SAGE Publications.

Koremenos, B., C. Lipson, and D. Snidal. 2001. "The Rational Design of International Institutions." *International Organization* 55 (4): 761–99.

Koskenniemi, M. 1990. "The Politics of International Law." *European Journal of International Law* 11 (1): 4–32.

Krasner, S. D. 1982a. "Regimes and the Limits of Realism: Regimes as Autonomous Variables." *International Organization* 36:497–510.

———. 1982b. "Structural Causes and Regime Consequences: Regimes as Intervening Variables." *International Organization* 36:185–205.

———. 1983. "Structural Causes and Regime Consequences: Regimes as Intervening Variables." In *International Regimes*, ed. S. D. Krasner, 1–21. Ithaca, NY: Cornell University Press.

Kratochwil, F., and J. G. Ruggie. 1986. "International Organization: A State of the Art or an Art of the State." *International Organization* 40:753–75.

Krott, M., and N. D. Hasanagas. 2006. "Measuring Bridges between Sectors: Causative Evaluation of Cross-sectorality." *Forest Policy and Economics* 8:555–63.

Kruger, J. 2005. "From SO_2 to Greenhouse Gases: Trends and Events Shaping Future Emissions Trading Programs in the United States." RFF DP 05-20. Washington, DC: Resources for the Future.

Krupnik, I., and D. Jolly, eds. 2002. *The Earth Is Faster Now: Indigenous Observations of Arctic Environmental Change*. Fairbanks, AK: Arctic Research Consortium of the United States.

Kuhn, T. 1962. *The Structure of Scientific Revolutions*. Chicago: University of Chicago Press.

Lafferty, W. M., and J. Meadowcroft, eds. 1996. *Democracy and the Environment: Problems and Prospects*. Cheltenham, UK: Edward Elgar Publishing.

Lambin, E., H. J. Geist, and E. Lepers. 2003. "Dynamics of Land-Use and Land-Cover Change in Tropical Forests." *Annual Review of Environmental Resources* 28:205–41.

Lambin, E. F., and H. J. Geist, eds. 2006. *Land-Use and Land-Cover Change: Local Processes and Global Impacts*. Berlin: Springer.

Lanchbery, J. 2006. "The Convention on International Trade in Endangered Species of Wild Fauna and Flora (CITES): Responding to Calls for Action from

Other Nature Conservation Regimes." In *Institutional Interaction in Global Environmental Governance: Synergy and Conflict among International and EU Policies*, ed. S. Oberthür and T. Gehring, 157–79. Cambridge, MA: The MIT Press.

Larson, A. 1992. "Network Dyads in Entrepreneurial Settings: A Study of the Governance of Exchange Relationships." *Administrative Science Quarterly* 37 (1): 76–104.

Laurance, W. F. 2004. "The Perils of Pay-off: Corruption as a Threat to Global Biodiversity." *Trends in Ecology and Evolution* 19 (8): 399–401.

Lavranos, N. 2006. "The MOX Plant and Ijzeren Rijn Disputes: Which Court Is the Supreme Arbiter?" *Leiden Journal of International Law* 19 (1): 223–46.

Lebel, L. 2004. "The Politics of Scale in Environmental Assessment." USER Working Paper, WP-2004-07, Chiang Mai University, Thailand.

———. 2005. "Institutional Dynamics and Interplay: Critical Processes for Forest Governance and Sustainability in the Mountain Regions of Northern Thailand." In *Global Change and Mountain Regions: An Overview of Current Knowledge*, ed. U. M. Huber, H. K. M. Bugmann, and M. A. Reasoner, 531–40. Berlin: Springer.

———. 2006. "Comments from the 'Knowledge Broker' on 'Global Change: Analysing Scale and Scaling in Environmental Governance,' by Joyeeta Gupta." Presentation at IDGEC Synthesis Conference, December 6–9, Bali, Indonesia.

Lebel, L., J. M. Anderies, B. Campbell, C. Folke, S. Hatfield-Dodds, T. P. Hughes, and J. Wilson. 2006. "Governance and the Capacity to Manage Resilience in Regional Social-Ecological Systems." *Ecology and Society* 11 (1): 19.

Lebel, L., P. Garden, and M. Imamura. 2005. "Politics of Scale, Position, and Place in the Governance of Water Resources in the Mekong Region." *Ecology and Society* 10 (2): 18. http://www.ecologyandsociety.org/vol10/iss2/art18/.

Lebel, L., and M. Imamura. Forthcoming. "Water Governance at Multiple Levels and Scales in the Mekong Region." In *The Politics of Scale in Environmental Governance*, ed. J. Gupta and D. Huitema. IDGEC manuscript.

Lee, Kai. 1993a. *Compass and Gyroscope*. Washington, DC: Island Press.

———. 1993b. "Greed, Scale Mismatch, and Learning." *Ecological Applications* 3:560–64.

———. 2006. "Urban Sustainability and the Limits of Classical Environmentalism." *Environment & Urbanization* 18 (1): 9–22.

Lee, K., D. Humphreys, and M. Pugh. 1997. "Privatisation in the United Nations System: Patterns of Influence in Three Intergovernmental Organizations." *Global Society* 11(3) (January): 339–59.

Levin, S. 1998. "Ecosystems and the Biosphere as Complex Adaptive Systems." *Ecosystems* 1:431–36.

Levy, D. L., and A. Kolk. 2002. "Strategic Responses to Global Climate Change: Conflicting Pressures on Multinationals in the Oil Industry." *Business and Politics* 4:275–300.

Levy, D. L., and P. J. Newell. 2000. "Oceans Apart? Business Responses to Global Environmental Issues in Europe and the United States." *Environment* 42:8–20.

Levy, D. L., and P. J. Newell. 2002. "Business Strategy and International Environmental Governance." *Global Environmental Politics* 2:84–101.

Levy, M. A. 1993. "European Acid Rain: The Power of Tote-Board Diplomacy." In *Institutions for the Earth*, ed. P. M. Haas, R. O. Keohane, and M. A. Levy, 75–132. Cambridge, MA: The MIT Press.

Levy, M. A., O. R. Young, and M. Zürn. 1995. "The Study of International Regimes." *European Journal of International Relations* 1:267–330.

Linden, E. 2006. *The Winds of Change: Climate, Weather, and the Destruction of Civilizations*. New York: Simon and Schuster.

Lipson, C. 1991. "Why Are Some International Agreements Informal?" *International Organization* 45:495–538.

Litfin, K. T. 1994. *Ozone Discourses: Science and Politics in Global Environmental Cooperation*. New York: Columbia University Press.

Liu, J., T. Dietz, S. R. Carpenter, M. Alberti, C. Folke, E. Moran, A. N. Pell, P. Deadman, T. Kratz, et al. 2007. "Complexity of Coupled Human and Natural Systems." *Science* 317:1513–16.

Lofdahl, C. L. 2002. *Environmental Impact of Globalization and Trade*. Cambridge, MA: The MIT Press.

Low, B., E. Ostrom, C. Simon, and J. Wilson. 2003. "Redundancy and Diversity: Do They Influence Optimal Management?" In *Navigating Social-Ecological Systems*, ed. F. Berkes, J. Colding, and C. Folke, 83–114. Cambridge: Cambridge University Press.

Lowndes, V., and C. Skelcher. 1998. "The Dynamics of Multi-organizational Partnerships: An Analysis of Changing Modes of Governance." *Public Administration* 76:313–33.

Lundqvist, L. J. 1980. *The Hare and the Tortoise*. Ann Arbor: University of Michigan Press.

———. 2001. "Implementation from Above: The Ecology of Power in Sweden's Environmental Governance." *Governance* 14:319–37.

———. 2004. *Sweden and Ecological Governance: Straddling the Fence*. Manchester and New York: Manchester University Press.

Malayang, B. S., T. Hahn, and P. Kumar. 2006. "Responses to Ecosystem Changes and to Their Impacts on Human Well-being: Lessons from Sub-global Assessments." In Millennium Ecosystem Assessment, *Ecosystems and Human Well-being: Multiscale Assessment*, vol. 4, 205–28. Washington, DC: Island Press.

March, J. G., and J. P. Olsen. 1989. *Rediscovering Institutions: The Organizational Basis of Politics*. New York: The Free Press.

————. 1995. *Democratic Governance*. New York: The Free Press.

————. 1996. "Institutional Perspectives on Political Institutions." *Governance* 9 (3): 247–64.

————. 1998. "The Institutional Dynamics of International Political Orders." *International Organization* 52 (4): 943–69.

————. 2006. "Elaborating the 'New Institutionalism.'" In *The Oxford Handbook of Political Institutions*, ed. R. A. W. Rhodes, S. A. Binder, and B. A. Rockman, 3–20. Oxford: Oxford University Press.

Marshall, G. R. 2005. *Economics for Collaborative Environmental Management: Renegotiating the Commons*. London: Earthscan.

Martin, L. L. 1993. "The Rational State Choice of Multilateralism." In *Multilateralism Matters: The Theory and Praxis of an Institutional Form*, ed. J. G. Ruggie, 91–121. New York: Columbia University Press.

Mayntz, R. 1993. "Governing Failures and the Problem of Governability: Some Comments on a Theoretical Paradigm." In *Modern Governance: New Government-Society Interactions*, ed. J. Kooiman. London: SAGE Publications.

McGinnis, M. D., ed. 2000. *Polycentric Games and Institutions*. Ann Arbor: University of Michigan Press.

McGinnis, M. V., J. Woolley, and J. Gamman. 1999. "Bioregional Conflict Resolution: Rebuilding Community in Watershed Planning and Organizing." *Environmental Management* 24 (1): 1–12.

McIntosh, R. J. 2000. "Social Memory in Mande." In *The Way the Wind Blows: Climate, History, and Human Action*, ed. R. J. McIntosh, J. A. Tainter, and S. K. McIntosh, 141–80. New York: Columbia University Press.

McLain, R., and R. Lee. 1996. "Adaptive Management: Promises and Pitfalls." *Journal of Environmental Management* 20:437–48.

Meadows, D. H., D. L. Meadows, and J. Randers. 1992. *Beyond the Limits: Confronting Global Collapse, Envisioning a Sustainable Future*. Post Mills, VT: Chelsea Green.

Mearsheimer, J. J. 1994/1995. "The False Promise of International Institutions." *International Security* 19 (3): 5–49.

Meier, G. M. 2001. "Introduction: Ideas for Development." In *Frontiers of Development Economics: The Future in Perspective*, ed. G. M. Meier and J. Stiglitz, 1–12. A World Bank Publication. New York: Oxford University Press.

Meinke, B. 2002. *Multi-Regime-Regulierung. Wechselwirkungen zwischen globalen und regionalen Umweltregimen*. Wiesbaden: Deutscher Universitäts-Verlag.

Meyerson, L. A., and J. K. Reaser. 2003. "Bioinvasions, Bioterrorism, and Biosecurity." *Frontiers in Ecology and the Environment* 1 (6): 307–14.

Midlarsky, M. I. 1998. "Democracy and the Environment: An Empirical Assessment." *Journal of Peace Research* 35 (3): 341–62.

Miles, E. L. 2006. "Principles for Designing International Environmental Institutions." Paper presented at the IDGEC Synthesis Conference, December 6–9, Bali, Indonesia.

Miles, E. L., A. Underdal, S. Andresen, J. Wettestad, J. B. Skjærseth, and E. M. Carlin, eds. 2002. *Environmental Regime Effectiveness: Confronting Theory with Evidence*. Cambridge, MA: The MIT Press.

Millennium Ecosystem Assessment. 2005. *Ecosystems and Human Well-being: Synthesis*. Washington, DC: Island Press.

Miller, M. L., and L. Gunderson. 2004. "Biological and Cultural Camouflage: The Challenges of Seeing the Harmful Invasive Species Problem and Doing Something about It." In *Harmful Invasive Species: Legal Responses*, ed. M. Miller and R. Fabian. Environmental Law Institute. SSRN (September 2003). http://ssrn.com/abstract=452982.

Mitchell, J. K. 2006. "The Primacy of Partnership: Scoping a New National Disaster Recovery Policy." *The Annals of the American Academy of Political and Social Science* 604: 228–55.

Mitchell, R. B. 1993. "Compliance Theory: A Synthesis." *Review of European Community and International Environmental Law* 2 (4): 327–34.

———. 1994a. *Intentional Oil Pollution at Sea: Environmental Policy and Treaty Compliance*. Cambridge, MA: The MIT Press.

———. 1994b. "Regime Design Matters: Intentional Oil Pollution and Treaty Compliance." *International Organization* 48 (3): 425–58.

———. 2002 "A Quantitative Approach to Evaluating International Environmental Regimes." *Global Environmental Politics* 2 (4): 58–83.

———. 2003a. "International Environmental Agreements: A Survey of Their Features, Formation, and Effects." *Annual Review of Environment and Resources* 28:429–61.

———. 2003b. "The Relative Effects of Environmental Regimes: A Qualitative Comparison of Four Acid Rain Protocols." Paper presented at the 2003 International Studies Association Conference, Portland, Oregon.

———. 2004. "A Quantitative Approach to Evaluating International Environmental Regimes." In *Regime Consequences: Methodological Challenges and Research Strategies*, ed. A. Underdal and O. R. Young, 121–49. Dordrecht, The Netherlands: Kluwer Academic Publishers.

———. 2005. "Flexibility, Compliance and Norm Development in the Climate Regime." In *Implementing the Climate Regime: International Compliance*, ed. J. H. Olav, S. Stokke, and G. Ulfstein, 65–83. London: Earthscan.

———. 2006. "Problem Structure, Institutional Design, and the Relative Effectiveness of International Environmental Agreements." *Global Environmental Politics* 6 (3): 72–89.

———. 2007. "Compliance Theory: Compliance, Effectiveness, and Behavior Change in International Environmental Law." In *The Oxford Handbook of*

International Environmental Law, ed. J. Brunée, D. Bodansky, and E. Hey, 893–921. Cambridge: Oxford University Press.

Mitchell, R. B., and T. Bernauer. 1998. "Empirical Research on International Environmental Policy: Designing Qualitative Case Studies." *Journal of Environment and Development* 7 (1): 4–31.

Mitchell, R. B., W. C. Clark, D. Cash, and N. M. Dickson, eds. 2006. *Global Environmental Assessments: Information and Influence*. Cambridge, MA: The MIT Press.

Mitchell, R. B., and P. Keilbach. 2001. "Reciprocity, Coercion, or Exchange: Symmetry, Asymmetry and Power in Institutional Design." *International Organization* 55 (4): 891–917.

Morgenstern, R. D., and W. A. Pizer, eds. 2007. *Reality Check: The Nature and Performance of Voluntary Environmental Programs in the United States, Europe, and Japan*. Washington, DC: RFF Press.

Munich Re. 2006. *Weather Catastrophes and Climate Change: Is There Still Hope for Us?* Münchener Rückversicherungs Gesellschaft. Munich: PG Verlag.

Munton, D., M. Soroos, E. Nikitina, and M. A. Levy. 1999. "Acid Rain in Europe and North America." In *The Effectiveness of International Environmental Regimes*, ed. O. R. Young, 155–247. Cambridge, MA: The MIT Press.

Murdoch, J. C., T. Sandler, and K. Sargent. 1997. "A Tale of Two Collectives: Sulphur versus Nitrogen Oxides Emission Reductions in Europe." *Economica* 64:281–301.

Najam, A., M. Papa, and N. Taiyab. 2006. *Global Environmental Governance: A Reform Agenda*. Winnipeg, MB, Canada: International Institute for Sustainable Development.

Neuendorf, K. A. 2002. *The Content Analysis Guidebook*. Thousand Oaks, CA: SAGE Publications.

Newman, L., and A. Dale. 2005. "Network Structure, Diversity, and Proactive Resilience Building: A Response to Tompkins and Adger." *Ecology and Society* 10 (1): r2.

New York Times. 2006. "Outsize Profits, and Questions, in Effort to Cut Warming Gases." December 21: A1.

North, Douglass C. 1990. *Institutions, Institutional Change and Economic Performance*. Cambridge: Cambridge University Press.

Oberthür, S. 2001. "Linkages between the Montreal and Kyoto Protocols: Enhancing Synergies between Protecting the Ozone Layer and the Global Climate." *International Environmental Agreements: Politics, Law and Economics* 1 (3): 357–77.

———. 2002. "Clustering of Multilateral Environmental Agreements: Potentials and Limitations." *International Environmental Agreements: Politics, Law and Economics* 2 (4): 317–40.

————. 2003. "Institutional Interaction to Address Greenhouse Gas Emissions from International Transport: ICAO, IMO and the Kyoto Protocol." *Climate Policy* 3 (3): 191–205.

————. 2006. "The Climate Change Regime: Interactions with ICAO, IMO, and the EU Burden-Sharing Agreement." In *Institutional Interaction in Global Environmental Governance: Synergy and Conflict among International and EU Policies*, ed. S. Oberthür and T. Gehring, 53–77. Cambridge, MA: The MIT Press.

Oberthür, S., and T. Gehring. 2006a. "Conceptual Foundations of Institutional Interaction." In *Institutional Interaction in Global Environmental Governance: Synergy and Conflict among International and EU Policies*, ed. S. Oberthür and T. Gehring, 19–51. Cambridge, MA: The MIT Press.

————. 2006b. "Institutional Interaction in Global Environmental Governance: The Case of the Cartagena Protocol and the World Trade Organization." *Global Environmental Politics* 6 (2): 1–31.

————, eds. 2006c. *Institutional Interaction in Global Environmental Governance: Synergy and Conflict among International and EU Policies*. Cambridge, MA: The MIT Press.

Oberthür, S., and H. E. Ott. 1999. *The Kyoto Protocol: International Climate Policy for the 21st Century*. Berlin: Springer.

Olson, M., Jr. 1965. *The Logic of Collective Action*. Cambridge, MA: Harvard University Press.

————. 1982. *The Rise and Decline of Nations: Economic Growth, Stagflation, and Social Rigidities*. New Haven, CT: Yale University Press.

Olsson, P., C. Folke, and F. Berkes. 2004. "Adaptive Comanagement for Building Resilience in Social-Ecological Systems." *Environmental Management* 34:75–90.

Olsson, P., L. H. Gunderson, S. R. Carpenter, P. Ryan, L. Lebel, C. Folke, and C. S. Holling. 2006. "Shooting the Rapids: Navigating Transitions to Adaptive Governance of Social-Ecological Systems." *Ecology and Society* 11 (1): 18.

Onuf, N. G. 1989. *World of Our Making: Rules and Rule in Social Theory and International Relations*. Columbia: University of South Carolina Press.

Osborne, S. P., ed. 2000. *Public-Private Partnerships: Theory and Practice in International Perspective*. London: Routledge.

Ostrom, E. 1990. *Governing the Commons: The Evolution of Institutions for Collective Acton*. Cambridge: Cambridge University Press.

————. 1996. "Crossing the Great Divide: Coproduction, Synergy, and Development." *World Development* 24 (6): 1073–87.

————. 1998. "Scales, Polycentricity, and Incentives: Designing Complexity to Govern Complexity." In *Protection of Global Biodiversity: Converging Strategies*, ed. L. D. Guruswamy and J. A. McNeely, 149–67. Durham, NC: Duke University Press.

———. 1999. "Coping with Tragedies of the Commons." *Annual Review of Political Science* 2:493–535.

———. 2005. *Understanding Institutional Diversity.* Princeton, NJ: Princeton University Press.

Ostrom, E., J. Burger, C. B. Field, R. B. Norgaard, and D. Policansky. 1999. "Revisiting the Commons: Local Lessons, Global Challenges." *Science* 284:278–82.

Ostrom, E., T. Dietz, N. Dolsak, P. C. Stern, S. Stonich, and U. W. Elke, eds. 2002. *The Drama of the Commons.* Washington, DC: National Academies Press.

Ostrom, E., J. Walker, and R. Gardner. 1992. "Covenants With and Without a Sword: Self-governance Is Possible." *American Political Science Review* 86:404–17.

Oye, K. A. 1985. "Explaining Cooperation under Anarchy: Hypotheses and Strategies." *World Politics* 38 (1): 1–24.

———, ed. 1986. *Cooperation under Anarchy.* Princeton, NJ: Princeton University Press.

Paarlberg, R. L. 1993. "Managing Pesticide Use in Developing Countries." In *Institutions for the Earth: Sources of Effective International Environmental Protection,* ed. P. Haas, R. O. Keohane, and M. Levy, 309–50. Cambridge, MA: The MIT Press.

Paavola, J., and W. N. Adger. 2006. "Fair Adaptation to Climate Change." *Ecological Economics* 56 (4): 594–609.

Pahl-Wostl, C., J. Gupta, and D. Petry. 2006. "Governance and the Global Water System: The Need to Adopt a Global Perspective on Water Issues." Presentation at the GWSP Water Governance Workshop, June 20–23, Bonn, Germany.

Palmer, A., B. Chaytor, and J. Werksman. 2006. "Interaction between the World Trade Organization and International Environmental Regimes." In *Institutional Interaction in Global Environmental Governance: Synergy and Conflict among International and EU Policies,* ed. S. Oberthür and T. Gehring, 181–204. Cambridge, MA: The MIT Press.

Parks, B. C., and J. Timmons Roberts. 2006. "Globalization, Vulnerability to Climate Change, and Perceived Injustice." *Society and Natural Resources* 19 (4): 337–55.

Parson, E. A. 2003. *Protecting the Ozone Layer: Science and Strategy.* New York: Oxford University Press.

Parson, E. A., and W. C. Clark. 1995. "Sustainable Development as Social Learning: Theoretical Perspectives and Practical Challenges for the Design of a Research Program." In *Barriers and Bridges to the Renewal of Ecosystems and Institutions,* ed. L. H. Gunderson, C. S. Holling, and S. S. Light. New York: Columbia University Press.

Parto, S. 2004. "Sustainability and the Local Scale: Squaring the Peg?" *International Journal of Sustainable Development* 7 (1): 76–97.

Pascual, M., X. Rodo, S. P. Ellner, R. Colwell, and M. J. Bouma. 2000. "Cholera Dynamics and El Niño-Southern Oscillation." *Science* 289:1766–69.

Pasong, S., and L. Lebel. 2000. "Political Transformation and the Environment in Southeast Asia." *Environment* 42 (8): 8–19.

Pattberg, P. 2005. "The Institutionalization of Private Governance: How Business and Non-profits Agree on Transnational Rules." *Governance: An International Journal of Policy, Administration, and Institutions* 18 (4): 589–610.

———. 2006. "Private Governance and the South: Lessons from Global Forest Politics." *Third World Quarterly* 27 (4): 579–93.

———. 2007. *Private Institutions and Global Governance: The New Politics of Environmental Sustainability.* Cheltenham, UK, and Northampton, MA: Edward Elgar Publishing.

Pauly, D., V. Christensen, J. Dalsgaard, R. Froese, and F. Torres Jr. 1998. "Fishing Down Marine Food Webs." *Science* 279 (5352): 860–63.

Pauwelyn, J. 2003. *Conflict of Norms in Public International Law: How WTO Law Relates to Other Rules of International Law.* Cambridge: Cambridge University Press.

Peterson, G. D. 2002. "Forest Dynamics in the Southeastern United States: Managing Multiple Stable States." In *Resilience and the Behavior of Large-Scale Systems,* ed. L. H. Gunderson and L. Pritchard Jr., 227–48. Washington, DC: SCOPE/Island Press.

Peterson, G. D., T. D. Beard Jr., B. E. Beisner, E. M. Bennett, S. R. Carpenter, G. S. Cumming, C. L. Dent, and T. D. Havlicek. 2003. "Assessing Future Ecosystem Services: A Case Study of the Northern Highlands Lake District, Wisconsin." *Conservation Ecology* 7 (3): 1. http://www.consecol.org/vol7/iss3/art1/.

Petrella, R. 2001. *The Water Manifesto: Arguments for a World Water Contract.* London: Zed Books.

Pierre, J. 1999. "Models of Urban Governance: The Institutional Dimension of Urban Politics." *Urban Affairs Review* 34 (3): 372–96.

———, ed. 2000. *Debating Governance: Authority, Steering, and Democracy.* Oxford: Oxford University Press.

Pierre, J., and G. B. Peters. 2005. *Governing Complex Societies: Trajectories and Scenarios.* Basingstoke, UK: Palgrave Macmillan.

Pierson, P. 2000. "The Limits of Design: Explaining Institutional Origins and Change." *Governance* 13 (4): 475–99.

———. 2003. "Big, Slow-Moving, and ... Invisible." In *Comparative Historical Analysis in the Social Sciences,* ed. J. Mahoney and D. Rueschemeyer, 177–207. Cambridge: Cambridge University Press.

Pomeroy, R. S., B. D. Ratnera, S. J. Halla, J. Pimoljindab, and V. Vivekanandan. 2006. "Coping with Disaster: Rehabilitating Coastal Livelihoods and Communities." *Marine Policy* 30:786–93.

Pontecorvo, C. M. 1999. "Interdependence between Global Environmental Regimes: The Kyoto Protocol on Climate Change and Forest Protection." *Zeitschrift für ausländisches öffentliches Recht und Völkerrecht* 59 (3): 709–49.

Prakash, A., and M. Potoski. 2006. "Racing to the Bottom? Trade Environmental Governance, and ISO 14001." *American Journal of Political Science* 50:350–64.

Princen, T., and M. Finger. 1994. *Environmental NGOs in World Politics: Linking the Local and the Global.* London: Routledge.

Princen, T., M. Finger, and J. Manno. 1995. "Nongovernmental Organizations in World Environmental Politics." *International Environmental Affairs* 7 (1): 42–58.

Princen, T., M. Maniates, and K. Conca. 2002. *Confronting Consumption.* Cambridge, MA: The MIT Press.

Przeworski, A. 2004. "Institutions Matter?" *Government and Opposition* 39:527–40.

Putnam, R. 1988. "Diplomacy and Domestic Politics: The Logic of Two-Level Games." *International Organization* 42:427–60.

Ragin, C. C. 1987. *The Comparative Method.* Berkeley, Los Angeles, and London: University of California Press.

Rajesh, D. N., and L. Lebel. 2006. "Land Policy, Tenure, and Use: Institutional Interplay at the Rural-Forest Interface in Thailand." USER Working Paper WP-2006-01, Chiang Mai University, Thailand.

Rapoport, A. 1960. *Fights, Games, and Debates.* Ann Arbor: University of Michigan Press.

Rausser, G., L. Simon, and H. Ameden. 2000. "Public–Private Alliances in Biotechnology: Can They Narrow the Knowledge Gaps between Rich and Poor?" *Food Policy* 25: 499–513.

Raustiala, K. 1997. "States, NGOs, and International Environmental Institutions." *International Studies Quarterly* 42 (4): 719–40.

Raustiala, K., and D. G. Victor. 1998. "Conclusions." In *The Implementation and Effectiveness of International Environmental Commitments,* ed. D. G. Victor, K. Raustiala, and E. B. Skolnikoff, 659–707. Cambridge, MA: The MIT Press.

———. 2004. "The Regime Complex for Plant Genetic Resources." *International Organization* 55:277–309.

Raymond, L. 2003. *Private Rights in Public Resources: Equity and Property Allocation in Market-Based Environmental Policy.* Washington, DC: RFF Press.

Reeve, R. 2002. *Policing International Trade in Endangered Species: The CITES Treaty and Compliance*. London: Royal Institute of International Affairs and Earthscan.

Reinalda, B., and B. Verbeek. 2001. "Theorising Power Relations between NGOs, Inter-governmental Organisations and States." In *Non-state Actors in International Relations*, ed. B. Reinalda, 145–59. Aldershot, UK: Ashgate.

Reinicke, W. H. 1998. *Global Public Policy: Governing without Government?* Washington, DC: Brookings Institution Press.

Reinicke, W. H., and F. Deng, eds. 2000. *Critical Choices: The United Nations, Networks, and the Future of Global Governance*. Ottawa, ON, Canada: International Development Research Centre.

Ribot, J. C. 2004. *Waiting for Democracy: The Politics of Choice in Natural Resource Decentralization*. Washington, DC: World Resources Institute.

Riker, W. 1962. *The Theory of Political Coalitions*. New Haven, CT: Yale University Press.

Ringquist, E. J., and T. Kostadinova. 2005. "Assessing the Effectiveness of International Environmental Agreements: The Case of the 1985 Helsinki Protocol." *American Journal of Political Science* 49 (1): 86–102.

Rio Declaration on Environment and Development. 1992. "Report of the UN Conference on Environment and Development, Rio de Janeiro, June 1992." Also reported in 31 ILM 876.

Risse, T. 2000. "'Let's Argue!': Communicative Action in World Politics." *International Organization* 54 (1): 1–39.

Risse, T., S. Ropp, and K. Sikkink, eds. 1999. *The Power of Human Rights*. Cambridge: Cambridge University Press.

Rockström, J., L. Gordon, C. Folke, M. Falkenmark, and M. Engwall. 1999. "Linkages among Water Vapor Flows, Food Production, and Terrestrial Ecosystem Services." *Conservation Ecology* 3 (2): 5. http://www.consecol.org/vol3/iss2/art5.

Rodrik, D. 2006. "Goodbye Washington Consensus, Hello Washington Confusion? A Review of the World Bank's *Economic Growth in the 1990s: Learning from a Decade of Reform*." *Journal of Economic Literature* 44: 973–87.

Rogers, K., and H. Biggs. 1999. "Integrating Indicators, Endpoints and Value Systems in Strategic Management of the Rivers of the Kruger National Park." *Freshwater Biology* 41:439–51.

Rolén, M., H. Sjöberg, and U. Svedin. 1997. *International Governance on Environmental Issues*. Dordrecht, The Netherlands: Kluwer Academic Publishers.

Rosamond, B. 2000. *Theories of European Integration*. Basingstoke, UK: Palgrave Macmillan.

Rosen, R. 1986. "On Information and Complexity." In *Biomathematics*, vol. 16, *Complexity, Language and Life: Mathematical Approaches*, ed. J. L. Casti and A. Karlqvist, 174–96. Berlin: Springer.

Rosenau, J. N. 1995. "Governance in the Twenty-first Century." *Global Governance* 1 (1): 13–43.

Rosenau, J. N., and E. Czempiel, eds. 1992. *Governance without Government: Order and Change in World Politics.* Cambridge: Cambridge University Press.

Rosendal, G. K. 2001a. "Impacts of Overlapping International Regimes: The Case of Biodiversity." *Global Governance* 7 (1): 95–117.

———. 2001b. "Overlapping International Regimes: The Case of the Intergovernmental Forum on Forests (IFF) Between Climate Change and Biodiversity." *International Environmental Agreements: Politics, Law and Economics* 1 (4): 447–68.

———. 2006. "The Convention on Biological Diversity: Tensions with the WTO TRIPS Agreement over Access to Genetic Resources and the Sharing of Benefits." In *Institutional Interaction in Global Environmental Governance: Synergy and Conflict among International and EU Policies*, ed. S. Oberthür and T. Gehring, 79–102. Cambridge, MA: The MIT Press.

Rosenthal, U., and A. Kouzmin. 1997. "Crises and Crisis Management: Toward Comprehensive Government Decision Making." *Journal of Administration Research and Theory* 2:277–304.

Roughgarden, J., and F. Smith. 1996. "Why Fisheries Collapse and What to Do about It." *Proceedings of the National Academy of Sciences* 93 (10): 5078–83.

Rowlands, I. H. 2001. "Transnational Corporations and Global Environmental Politics." In *Non-state Actors in World Politics*, ed. W. Wallace. Basingstoke, UK: Palgrave Macmillan.

Ruggie, J. G. 1982. "International Regimes, Transactions, and Change: Embedded Liberalism in the Postwar Economic Order." *International Organization* 36:379–415.

Ruitenbeek, J., and C. Cartier. 2001. "The Invisible Wand: Adaptive Co-management as an Emergent Strategy in Complex Bio-economic Systems." Occasional Paper 34, Center for International Forestry Research, Bogor, Indonesia.

Rutherford, M. 1994. *Institutions in Economics: The Old and the New Institutionalism.* Cambridge: Cambridge University Press.

Sabatier, P. 1998. "An Advocacy Coalition Framework of Policy Change and the Role of Policy-Oriented Learning Therein." *Policy Sciences* 21:129–68.

Safrin, S. 2002. "Treaties in Collision? The Biosafety Protocol and the World Trade Organization Agreements." *The American Journal of International Law* 96 (3): 606–28.

Sandler, T. 2004. *Global Collective Action.* Cambridge: Cambridge University Press.

Santiso, C. 2001. "Good Governance and Aid Effectiveness: The World Bank and Conditionality." *The Georgetown Public Policy Review* 7 (1): 1–22.

———. 2005. "Governance Conditionality and the Reform of Multilateral Development Finance: The Role of the Group of Eights." *G8 Governance* 7: 2–36.

Scharpf, F. 1997. *Games Real Actors Play: Actor-Centered Institutionalism in Policy Research*. Boulder, CO: Westview Press.

Schattschneider, E. E. 1960/1975. *The Semisovereign People: A Realist's View of Democracy in America*. Hinsdale, IL: Dryden Press.

Scheffer, M., S. Carpenter, J. Foley, C. Folke, and B. Walker. 2001. "Catastrophic Shifts in Ecosystems." *Nature* 413:591–696.

Scheffer, M., F. Westley, and W. A. Brock. 2003. "Slow Response of Societies to New Problems: Causes and Costs." *Ecosystems* 6:493–502.

Schelling, T. C. 1960. *The Strategy of Conflict*. New York and Oxford: Oxford University Press.

———. 1978. *Micromotives and Macrobehavior*. New York: W. W. Norton.

———. 1998. "Social Mechanisms and Social Dynamics." In *Social Mechanisms: An Analytical Approach to Social Theory*, ed. P. Hedström and R. Swedberg, 32–44. Cambridge: Cambridge University Press.

Schellnhuber, H.-J. 1998. "Earth System Analysis: The Scope of the Challenge." In *Earth System Analysis: Integrating Science for Sustainability*, ed. H.-J. Schellnhuber and V. Wenzel, 3–195. Berlin: Springer.

———. 1999. "Earth System Analysis and the Second Copernican Revolution." *Nature* 402 (Millennium Supplement, December 2), C19–C23.

———, ed. 2006. *Avoiding Dangerous Climate Change*. Cambridge: Cambridge University Press.

Schellnhuber, H.-J., P. Crutzen, W. C. Clark, M. Claussen, and H. Held. 2004. *Earth System Analysis for Sustainability*. Cambridge, MA: The MIT Press.

Schellnhuber, H.-J., and D. Sahagian. 2002. "The Twenty-three GAIM Questions." *Global Change Newsletter* (International Geosphere-Biosphere Programme) 49: 20–21.

Schellnhuber, H.-J., and V. Wenzel. 1998. "Preface." In *Earth System Analysis: Integrating Science for Sustainability*, ed. H.-J. Schellnhuber and V. Wenzel, vii–xvi. Berlin: Springer.

Schiffman, H. 1999. "The Southern Bluefin Tuna Case: ITLOS Hears Its First Fishery Dispute." *Journal of International Wildlife Law and Policy* 2 (3): 1–15.

Schneider, H. 1999. "Participatory Governance for Poverty Reduction." *Journal of International Development* 11:521–34.

Schneider, L., J. Graichen, and N. Matz. 2005. Implications of the Clean Development Mechanism under the Kyoto Protocol on Other Conventions: The Case of HFC-23 Destruction." *elni Review* 1:41–52.

Schneider, S. H. 2004. "Abrupt Non-linear Climate Change, Irreversibility and Surprise." *Global Environmental Change* 14:245–58.

Schneider, S. H., and T. L. Root. 1995. "Ecological Implications of Climate Change Will Include Surprises." *Biodiversity and Conservation* 5:1109–19.

Scholte, J. A. 2002. "Civil Society and Democracy in Global Governance." *Global Governance* 8 (3): 281–304.

Schouten, M., and K. Schwartz. 2006. "Water as a Political Good: Implications for Investments." *International Environmental Agreements: Politics, Law and Economics* 6 (4): 407–21.

Schrijver, N. 2001. "On the Eve of Rio plus Ten: Development: The Neglected Dimension in the International Law on Sustainable Development." Dies Natalis 2001, Institute of Social Studies, The Hague, The Netherlands.

Schroeder, H. 2008. "Biosafety and Trade through the Lens of Institutional Interplay." In *Institutional Interplay: The Case of Biosafety*, ed. O. R. Young, W. B. Chambers, J. A. Kim, and C. ten Have. Tokyo: United Nations University Press.

Schusler, T. M., D. J. Decker, and M. J. Pfeffer. 2003. "Social Learning for Collaborative Natural Resource Management." *Society and Natural Resources* 15:309–26.

Scott, W. R. 1995. *Institutions and Organizations*. Thousand Oaks, CA: SAGE Publications.

Scruggs, L. 1999. "Institutions and Environmental Performance in Seventeen Western Democracies." *British Journal of Political Science* 29 (1): 1–31.

Sebenius, J. K., M. Grubb, A. Magalhaes, and S. Subak. 1992. "Sharing the Burden: Fair Allocation and Equity." In *Confronting Climate Change: Risks, Implications and Responses*, ed. I. Mintzer, 305–22. Cambridge: Cambridge University Press.

Selden, T., and D. Song. 1995. "Neoclassical Growth, the J Curve for Abatement, and the Inverted U Curve for Pollution." *Journal of Environmental Economics and Management* 29:162–68.

Selin, H., and S. D. VanDeveer. 2003. "Mapping Institutional Linkages in European Air Pollution Politics." *Global Environmental Politics* 3 (3): 14–46.

Sell, S. K. 1999. "Multinational Corporations as Agents of Change: The Globalization of Intellectual Property Rights." In *Private Authority and International Affairs*, ed. C. Cutler, V. Haufler, and T. Porter. Albany: State University of New York Press.

Sewell, G., M. Wasson, and Y. Yamagata. 2000. "The Institutional Dimensions of Carbon Management." IDGEC Scoping Report No. 1.

Shafik, N. 1994. "Economic Development and Environmental Quality: An Econometric Analysis." *Oxford Economic Papers* 46 (0): 757–73.

Shany, Y. 2003. The Competing Jurisdictions of International Courts and Tribunals. Oxford: Oxford University Press.

Shaw, N., and A. Cosbey. 1994. "GATT, the World Trade Organization, and Sustainable Development." *International Environmental Affairs* 6:245–72.

Shelton, D. 1991. "Human Rights, Environmental Rights, and the Right to Environment." *Stanford Journal of International Law* 28:103–38.

Shiva, V. 1992. "Why GEF Is an Inadequate Institute for UNCED." Earth Summit Briefings, No. 19, Third World Network, Penang, Malaysia.

———. 2002. *Water Wars: Privatization, Pollution and Profit*. Cambridge, MA: South End Press.

Siebenhüner, B. 2004. "Social Learning and Sustainability Science: Which Role Can Stakeholder Participation Play?" *International Journal of Sustainable Development* 7 (2): 146–63.

———. 2005. "The Role of Social Learning on the Road to Sustainability." In *Governance and Sustainability*, ed. J. N. Rosenau, E. U. von Weizsäcker, and U. Petschow, 86–99. Sheffield, UK: Greenleaf Publishing.

Siegfried, T., and T. Bernauer. 2006. "Estimating the Performance of International Regulatory Regimes." CIS Working Paper 16/2006, ETH/Universität Zürich, Zürich.

Simon, H. A. 1957. *Administrative Behavior*. New York: The Free Press.

———. 1972. "Theories of Bounded Rationality." In *Decision and Organization*, ed. C. B. McGuire and R. Radner, 161–76. Amsterdam: North-Holland.

Skjærseth, J. B. 2000. *North Sea Cooperation: Linking International and Domestic Pollution Control*. Manchester, UK: Manchester University Press.

———. 2006. "Protecting the Northeast Atlantic: One Problem, Three Institutions." In *Institutional Interaction in Global Environmental Governance: Synergy and Conflict among International and EU Policies*, ed. S. Oberthür and T. Gehring, 102–25. Cambridge, MA: The MIT Press.

Skjærseth, J. B., O. S. Stokke, and J. Wettestad. 2006. "Soft Law, Hard Law, and Effective Implementation of International Environmental Norms." *Global Environmental Politics* 6 (3): 104–20.

Smouts, M.-C. 1998. "The Proper Use of Governance in International Relations." *International Social Science Journal* 155:81–89.

Snidal, D. 1985. "The Game Theory of International Politics." *World Politics* 38 (1): 25–57.

———. 1994. "The Politics of Scope: Endogenous Actors, Heterogeneity and Institutions." *Journal of Theoretical Politics* 6 (4): 449–72.

Social Learning Group. 2001. *Learning to Manage Global Environmental Risks*, vol. 2. Cambridge, MA: The MIT Press.

Sorensen, E. 2002. "Democratic Theory and Network Governance." *Administrative Theory and Praxis* 24 (4): 693–720.

Sornarajah, M. 2006. "A Law for Need or a Law for Greed: Finding the Lost Law in the International Law of Foreign Investment." *International Environmental Agreements: Politics, Law and Economics* 6 (4): 329–57.

South Centre. 1996. *For a Strong and Democratic United Nations: A South Perspective on UN Reform*. Geneva: South Centre.

Spierenburg, M. Forthcoming. "The Nature of Cultural Discourses, Cultural Discourses on Nature." In *The Politics of Scale in Environmental Governance*, ed. J. Gupta and D. Huitema. IDGEC manuscript.

Spierenburg, M., and H. Wels. Forthcoming. "'Scaling Up' and 'Scaling Down': Transfrontier Conservation Areas and Community Participation; the Case of the Great Limpopo Park, Southern Africa." In *The Politics of Scale in Environmental Governance*, ed. J. Gupta and D. Huitema. IDGEC manuscript.

Sprinz, D. F., and C. Helm. 1999. "The Effect of Global Environmental Regimes: A Measurement Concept." *International Political Science Review* 20 (4): 359–69.

———. 2000. "Measuring the Effectiveness of International Environmental Regimes." *Journal of Conflict Resolution* 45:630–52.

Sprinz, D. F., J. Hovi, R. B. Mitchell, and A. Underdal. 2004. "Separating and Aggregating Regime Effects." Paper presented at the International Studies Association Conference, Montreal, Canada (March).

Steffen, W., A. Sanderson, P. D. Tyson, J. Jäger, P. M. Matson, B. Moore III, F. Oldfield, K. Richardson, H. J. Schellnhuber, B. L. Turner II, and R. J. Wasson. 2004. *Global Change and the Earth System: A Planet under Pressure*. New York: Springer.

Stegner, W. 1954. *Beyond the Hundredth Meridian: John Wesley Powell and the Second Opening of the West*. Boston: Houghton Mifflin.

Stein, A. 1982. "Coordination and Collaboration Regimes in an Anarchic World." *International Organization* 36:299–324.

Steinbruner, J. D. 1976. *The Cybernetic Theory of Decisions*. Princeton, NJ: Princeton University Press.

Stern, D. I. 2004. "The Rise and Fall of the Environmental Kuznets Curve." *World Development* 32:1419–39.

Stern, E. 1997. "Crisis and Learning: A Conceptual Balance Sheet." *Journal of Contingencies and Crisis Management* 5 (2): 69–86.

Stewart, D. O. 2007. *The Summer of 1787*. New York: Simon and Schuster.

Stewart, M., and P. Collett. 1998. "Accountability in Community Contributions to Sustainable Development." In *Community and Sustainable Development*, ed. D. Warburton, 52–67. London: Earthscan.

Stoker, G. 1998. "Governance as Theory." *International Social Science Journal* 155:17–28.

Stokke, O. S. 2000. "Managing Straddling Stocks: The Interplay of Global and Regional Regimes." *Ocean and Coastal Management* 43 (2/3): 205–34.

———, ed. 2001a. *Governing High Seas Fisheries: The Interplay of Global and Regional Regimes*. Oxford: Oxford University Press.

———. 2001b. "The Interplay of International Regimes: Putting Effectiveness Theory to Work." Report No. 14, The Fridtjof Nansen Institute, Norway.

———. 2007. "A Legal Regime for the Arctic? Interplay with the Law of the Sea Convention." *Marine Policy* 31 (4): 402–8.

———. Forthcoming. "International Governance and Fisheries Management: Determining and Explaining the Effectiveness of International Regimes." Lysaker, Norway: The Fridtjof Nansen Institute.

Stokke, O. S., and C. Coffey. 2006. "Institutional Interplay and Responsible Fisheries: Combating Subsidies, Developing Precaution." In *Institutional Interaction in Global Environmental Governance: Synergy and Conflict among International and EU Policies*, ed. S. Oberthür and T. Gehring, 127–50. Cambridge, MA:

Stokke, O. S., and G. Hønneland, eds. 2007. *International Cooperation and Arctic Governance: Regime Effectiveness and Northern Region Building*. London: Routledge.

Stokke, O. S., and D. Vidas. 1996. *Governing the Antarctic: The Effectiveness and Legitimacy of the Antarctic Treaty System*. Cambridge: Cambridge University Press.

Stoll-Kleemann, S., and T. O'Riordan. 2002. "From Participation to Partnership in Biodiversity Protection: Experience from Germany and South Africa." *Society and Natural Resources* 15:161–77.

Stone, D. 2002. *Policy Paradox: The Art of Political Decision Making*. Rev. ed. New York: W. W. Norton.

Strange, S. 1983. "*Cave! hic dragones*: A Critique of Regime Analysis." In *International Regimes*, ed. Stephen D. Krasner, 337–54. Ithaca, NY: Cornell University Press.

Streeter, C. 1992. "Redundancy in Organizational Systems." *Social Service Review* March: 6(1):97–111.

Stubbs, M., and M. Lemon. 2001. "Learning to Network and Networking to Learn: Facilitating the Process of Adaptive Management in a Local Response to the UK's National Air Quality Strategy." *Environmental Management* 27:321–34.

Subak, S. 1993. "Assessing Emissions: Five Approaches Compared." In *The Global Greenhouse Regime: Who Pays?* ed. P. Hayes and K. Smith, 51–69. London: Earthscan.

Sugiyama, T., ed. 2005. "Scenarios for the Global Climate Regime after 2012." Special issue, *International Environmental Agreements* 5.

Svedin, U. 1991. "The Contextual Features of the Economy-Ecology Dialogue." In *Linking the Natural Environment and the Economy: Essays from the Eco-Eco Group*, ed. C. Folke and T. Kåberg. Dordrecht, The Netherlands: Kluwer Academic Publishers.

Svedin, U., T. O'Riordan, and A. Jordan. 2001. "Multilevel Governance for the Sustainability Transition." In *Globalism, Localism and Identity*, ed. T. O'Riordan, 43–60. London: Earthscan.

Swedish Government Inquiry. 2005. *Sverige och tsunamin: kastastrofkommis-sionens rapport* [Sweden and the Tsunami]. SOU 2005: 104. Stockholm: Lind & Co.

Swyngedouw, E. 2004. "Scaled Geographies: Nature, Place and the Politics of Scale." In *Scale and Geographic Inquiry: Nature, Society and Method*, ed. E. Sheppard and R. McMaster, 129–53. Oxford: Blackwell.

Tainter, J. 2004. "Problem Solving: Complexity, History, Sustainability." *Population and Environment* 22 (1): 3–41.

Tarasofsky, R. G. 1997. "Ensuring Compatibility between Multilateral Environmental Agreements and GATT/WTO." *Yearbook of International Environmental Law* 7:52–74.

Tay, S. 1999. "The Southeast Asian Fires and Haze: A Challenge to International Environmental Law and Sustainable Development." *Georgetown International Environmental Law Review*, Winter 241–305.

Thompson, L., B. Jago, L. Fernandes, and J. Day. 2004. "Barriers to Communication: How These Critical Aspects Were Addressed during the Public Participation for the Rezoning of the Great Barrier Reef Marine Park." Staff Paper, GBRMPA. http://www.gbrmpa.gov.au/_data/assets/pdf_file/0016/8251/Breaking _through_the_barriers_15April0420FINAL.pdf.

Tienhaara, K. 2006. "What You Don't Know Can Hurt You: Investor-Disputes and the Protection of the Environment in Developing Countries." *Global Environmental Politics* 6 (4): 73–100.

———. Forthcoming. "Not Seeing the Forest or the Trees: Global Forest Production and the Production of Scale." In *The Politics of Scale in Environmental Governance*, ed. J. Gupta and D. Huitema. IDGEC manuscript.

Tietenberg, T. 2002. "The Tradable Permits Approach to Protecting the Commons: What Have We Learned?" In *The Drama of the Commons*, ed. E. Ostrom, T. Dietz, N. Dolsak, P. C. Stern, S. Stonich, and U. W. Elke, 197–232. Washington, DC: National Academies Press.

Torras, M., and J. K. Boyce. 1998. "Income, Inequality, and Pollution: A Reassessment of the Environmental Kuznets Curve." *Ecological Economics* 25:147–60.

Tóth, F., ed. 1999. *Fair Weather? Equity Concerns in Climate Change*. London: Earthscan.

Trebuil, G., B. Shinawatra-Ekasingh, F. Bousquet, and C. Thong-Ngam. 2002. "Multi-agent Systems Companion Modeling for Integrated Watershed Management: A Northern Thailand Experience." In the proceedings of the 3rd International Conference on Montage Mainland Southeast Asia (MMSEA 3), Lijiang, Yunnan, China, 14.

True, James L., Bryan D. Jones, and Frank R. Baumgartner. 1999. "Punctuated Equilibrium Theory." In *Theories of the Policy Process*, ed. P. Sabbatier. Boulder, CO: Westview Press.

Turner, B. L., W. C. Clark, R. W. Kates, J. F. Richards, J. T. Mathews, and W. B. Meyer, eds. 1990. *The Earth as Transformed by Human Action: Global and Regional Changes in the Biosphere over the Past 300 Years.* Cambridge: Cambridge University Press.

Turton, A. R., B. Schreiner, and J. Leestemaker. 2000. "Feminization as a Critical Component of the Changing Hydrosocial Contract." *Water Science and Technology* 43 (4): 155–64.

Tversky, A., and D. Kahnemann. 1981. "The Framing of Decision and Rational Choice." *Science* 211:453–58.

———. 1984. "Choices, Values, and Frames." *American Psychologist* 39 (4): 341–50.

Underdal, A. 1992. "The Concept of Regime 'Effectiveness.'" *Cooperation and Conflict* 27 (3): 227–40.

———. 1994. "Progress in the Absence of Substantive Joint Decisions? Notes on the Dynamics of Regime Formation Processes." In *Climate Change and the Agenda for Research*, ed. Ted Hanisch, 113–30. Boulder, CO: Westview Press.

———. 1998. "Explaining Compliance and Defection: Three Models." *European Journal of International Relations* 4 (1): 5–30.

———. 2002a. "Conclusions: Patterns of Regime Effectiveness." In *Environmental Regime Effectiveness: Confronting Theory with Evidence*, by E. L. Miles, A. Underdal, S. Andresen, J. Wettestad, J. B. Skjærseth, and E. M. Carlin, 433–65. Cambridge, MA: The MIT Press.

———. 2002b. "One Question, Two Answers." In *Environmental Regime Effectiveness: Confronting Theory with Evidence*, by E. L. Miles, A. Underdal, S. Andresen, J. Wettestad, J. B. Skjærseth, and E. M. Carlin, 3–45. Cambridge, MA: The MIT Press.

———. 2004. "Methodological Challenges in the Study of Regime Effectiveness." In *Regime Consequences: Methodological Challenges and Research Strategies*, ed. A. Underdal and O. R. Young, 27–48. Dordrecht, The Netherlands: Kluwer Academic Publishers.

Underdal, A., and K. Hanf, eds. 2000. *International Environmental Agreements and Domestic Politics: The Case of Acid Rain.* Aldershot, UK: Ashgate.

Underdal, A., and O. R. Young, eds. 2004. *Regime Consequences: Methodological Challenges and Research Strategies.* Dordrecht, The Netherlands: Kluwer Academic Publishers.

United Nations. 2002. "Plan of Implementation of the World Summit on Sustainable Development." http://www.un.org/esa/sustdev/documents/WSSD_POI_PD/English/WSSD_PlanImpl.pdf.

United Nations Environment Programme. 2000. *Global Environmental Outlook.* Nairobi: United Nations Environment Programme.

Upadhyay, B. 2003. "Water, Poverty and Gender: Review of Evidences from Nepal, India and South Africa." *Water Policy* 5 (5–6): 503–11.

van Asselt, H. 2006. "Climate and . . . : Legal and Political Aspects of Resolving Conflicts and Enhancing Synergies between International Agreements Related to Climate Change." Paper presented at the IDGEC Synthesis Conference, Bali, Indonesia, December 6–9.

van Asselt, H., F. Biermann, and J. Gupta. 2004. "Interlinkages of Global Climate Governance." In *Beyond Climate: Options for Broadening Climate Policy*, ed. M. Kok and H. de Coninck, 221–46. Netherlands Research Programme on Climate Change, Report 500036001. Bilthoven: RIVM.

van Asselt, H., J. Gupta, and F. Biermann. 2005. "Advancing the Climate Agenda: Exploiting Material and Institutional Linkages to Develop a Menu of Policy Options." *Review of European Community and International Environmental Law* 14 (3): 255–64.

Van de Kerkhof, M. "Making a Difference: On the Constraints of Consensus Building and the Relevance of Deliberation in Stakeholder Dialogues." *Policy Sciences* 39:279–99.

Van den Bergh, J. Forthcoming. "Scale in Economic Policy Analysis: On Micro-Foundations, Space and Groups." In *The Politics of Scale in Environmental Governance*, ed. J. Gupta and D. Huitema. IDGEC manuscript.

Van der Grijp, N. M., and L. Brander. 2004. "The Role of the Business Community and Civil Society." In *Sustainability Labelling and Certification*, ed. M. Campins Eritja, 223–43. Madrid: Marcial Pons.

Van der Heijden, H.-A. 2002. "Political Parties and NGOs in Global Environmental Politics." *International Political Science Review* 23:187–201.

Van der Woerd, K. F., D. Levy, and K. Begg, eds. 2005. *Corporate Responses to Climate Change*. Sheffield, UK: Greenleaf Publishing.

Van Kersbergen, K., and F. van Waarden. 2004. "'Governance' as a Bridge between Disciplines: Cross-disciplinary Inspiration Regarding Shifts in Governance and Problems of Governability, Accountability and Legitimacy." *European Journal of Political Research* 43:143–71.

Velasquez, J. 2000. "Prospects for Rio + 10: The Need for an Inter-linkages Approach to Global Environmental Governance." *Global Environmental Change* 10 (4): 307–12.

Vermaat, J., and A. Gilbert. Forthcoming. "Ecological Perspectives on Scale: From Sermon and Band Aid." In *The Politics of Scale in Environmental Governance*, ed. J. Gupta and D. Huitema. IDGEC manuscript.

Victor, D. G., J. C. House, and S. Joy. 2004. "A Madisonian Approach to Climate Policy." *Science* 309:1820–21.

Victor, D. G., K. Raustiala, and E. B. Skolnikoff, eds. 1998. *The Implementation and Effectiveness of International Environmental Commitments: Theory and Practice*. Cambridge, MA: The MIT Press.

Vidas, D. 2000a. "Emerging Law of the Sea Issues in the Antarctic Maritime Area: A Heritage for the New Century?" *Ocean Development and International Law* 31 (1/2): 197–222.

————, ed. 2000b. *Protecting the Polar Marine Environment: Law and Policy for Pollution Prevention*. Cambridge: Cambridge University Press.

Vitousek, P. M., H. Mooney, J. Lubchenko, and J. Melillo. 1997. "Human Domination of Earth's Ecosystems." *Science* 277:494–99.

Vogel, D. 1986. *National Styles of Regulation*. Ithaca, NY, and London: Cornell University Press.

————. 1995. *Trading Up*. Cambridge, MA: Harvard University Press.

von Moltke, K. 1997. "Institutional Interactions: The Structure of Regimes for Trade and the Environment." In *Global Governance: Drawing Insights from the Environmental Experience*, ed. Oran R. Young, 247–72. Cambridge, MA: The MIT Press.

————. 2005. "Clustering International Environmental Agreements as an Alternative to a World Environment Organization." In *A World Environment Organization: Solution or Threat for Effective International Environmental Governance?* ed. F. Biermann and S. Bauer, 175–204. Aldershot, UK: Ashgate.

von Stein, J. 2005. "Do Treaties Constrain or Screen? Selection Bias and Treaty Compliance." *American Political Science Review* 99:611–22.

von Weizsäcker, E. U., O. Young, and M. Finger, eds. 2005. *Limits to Privatization: How to Avoid Too Much of a Good Thing*. London: Earthscan.

Wälde, T. (2006). Investment Arbitration and Sustainable Development: Good Intentions or Effective Results?; Commentary on James Chalker's paper on Making the Investment Provisions of the Energy Charter Treaty Sustainable Development Friendly, International Environmental Agreements: Politics, Law and Economics, 6(4): 459–466.

Walsh, V. M. 2004. *Global Institutions and Social Knowledge*. Cambridge, MA: The MIT Press.

Walters, C. 1986. *Adaptive Management of Renewable Resources*. New York: Macmillan.

Waltz, K. N. 1979. *Theory of International Politics*. Reading, MA: Addison-Wesley.

Wapner, P. 1996. *Environmental Activism and World Civic Politics*. Albany: State University of New York Press.

Ward, H. 2006. "International Linkages and Environmental Sustainability: The Effectiveness of the Regime Network." *Journal of Peace Research* 43:149–66.

Watts, A. 1993. "The International Rule of Law." *German Yearbook of International Law* 36:15–45.

Waugh, W. L. 2006. "The Political Costs of Failure in the Katrina and Rita Disasters." *Annals of the American Academy of Political and Social Science* 604:10–25.

Webler, T., H. Kastenholz, and O. Renn. 1995. "Public Participation in Impact Assessment: A Social Learning Perspective." *Environmental Impact Assessment Review* 15:443–63.

Weisslitz, M. 2002. "Rethinking the Equitable Principle of Common but Differentiated Responsibility: Differential versus Absolute Norms of Compliance and Contribution in the Global Climate Change Context." *Colorado Journal of International Environmental Law and Policy* 13:473.

Wendt, A. 1987. "The Agent-Structure Problem in International Relations Theory." *International Organization* 41 (3): 335–70.

———. 1999. *Social Theory of International Politics*. Cambridge: Cambridge University Press.

———. 2001. "Driving with the Rearview Mirror: On the Rational Science of Institutional Design." *International Organization* 55:1019–49.

Werksman, J., and C. Santoro. 1998. "Investing in Sustainable Development: The Potential Interaction between the Kyoto Protocol and the Multilateral Agreement on Investment." In *Global Climate Governance: Inter-linkages between the Kyoto Protocol and Other Multilateral Regimes*, ed. W. B. Chambers, 59–76. Tokyo: Global Environmental Information Centre.

Westley, F. 1995. "Governing Design: The Management of Social Systems and Ecosystems Management." In *Barriers and Bridges to Renewal of Ecosystems and Institutions*, ed. L. H. Gunderson, C. S. Holling, and S. S. Light, 391–427. New York: Columbia University Press.

———. 2002. "The Devil in the Dynamics: Adaptive Management on the Front Lines." In *Panarchy: Understanding Transformations in Human and Natural Systems*, ed. L. H. Gunderson and C. S. Holling, 333–60. Washington, DC: Island Press.

Westley, F., S. R. Carpenter, W. A. Brock, C. S. Holling, and L. H. Gunderson. 2002. "Why Systems of People and Nature Are Not Just Social and Ecological Systems." In *Panarchy: Understanding Transformations in Systems of Humans and Nature*, ed. L. H. Gunderson and C. S. Holling, 3–119. Washington, DC: Island Press.

Wettestad, J. 2002. "The Convention on Long-Range Transboundary Air Pollution (CLRTAP)." In *Environmental Regime Effectiveness*, by E. L. Miles, A. Underdal, S. Andresen, J. Wettestad, J. B. Skjærseth, and E. M. Carlin, 197–221. Cambridge, MA: The MIT Press.

Whalley, J., and B. Zissimos. 2001. "What Could a World Environmental Organization Do?" *Global Environmental Politics* 1 (1): 29–34.

Whittell, G. 2002. "Albright Attacks US Foreign Policy as Schizophrenic." *The Times*, May 21: 8.

Wilbanks, T. J., and R. W. Kates. 1999. "Global Change in Local Places: How Scale Matters." *Climatic Change* 43:601–28.

Wilhite, D. A. 1996. "A Methodology for Drought Preparedness." *Natural Hazards* 13:229–52.

Wilson, J. A. 2006. "Matching Social and Ecological Systems in Complex Ocean Fisheries." *Ecology and Society* 11 (1): 9.

Winter, G., ed. 2006. *Multilevel Governance of Global Environmental Change: Perspectives from Science, Sociology and the Law*. Cambridge: Cambridge University Press.

———. Forthcoming. "Dangerous Chemicals: In Search of the Proper Level in Multi-level Global Governance." In *The Politics of Scale in Environmental Governance*, ed. J. Gupta and D. Huitema. IDGEC manuscript.

Wirth, D. A. 1991. "Legitimacy, Accountability, and Partnership: A Model for Advocacy on Third World Environmental Issues." *Yale Law Journal* 100 (8): 2645–66.

Wolfrum, R., and N. Matz. 2003. *Conflicts in International Environmental Law*. Berlin: Springer.

Wollenberg, E., D. Edmunds, and L. Buck. 2000. "Using Scenarios to Make Decisions about the Future: Anticipatory Learning for the Adaptive Co-management of Community Forests." *Landscape and Urban Planning* 47:65–77.

Wondolleck, J. M., and S. L. Yaffee. 2000. *Making Collaboration Work: Lessons from Innovation in Natural Resource Management*. Washington, DC: Island Press.

World Bank, UNDP and UNEP (1991). GEF Brochure. Washington, DC: GEF.

World Commission on Environment and Development. 1987. *Our Common Future*. Oxford: Oxford University Press.

Worm, B., E. B. Barbier, N. Beaumont, et al. 2006. "Impacts of Biodiversity Loss on Ocean Ecosystem Services." *Science* 314:787–90.

Yamin, F., and J. Depledge. 2004. *The International Climate Change Regime: A Guide to Rules, Institutions and Procedures*. Cambridge: Cambridge University Press.

Young, O. R., ed. 1975. *Bargaining: Formal Theories of Cooperation*. Urbana: University of Illinois Press.

———. 1979. *Compliance and Public Authority: A Theory with International Applications*. Baltimore, MD: Johns Hopkins University Press.

———. 1982a. "Regime Dynamics: The Rise and Fall of International Regimes." *International Organization* 36 (2): 277–97.

———. 1982b. *Resource Regimes: Natural Resources and Social Institutions*. Berkeley: University of California Press.

———. 1989a. *International Cooperation: Building Regimes for Natural Resources and the Environment*. Ithaca, NY: Cornell University Press.

———. 1989b. "The Politics of International Regime Formation: Managing Natural Resources and the Environment." *International Organization* 43 (3): 349–75.

———. 1989c. "The Power of Institutions: Why International Regimes Matter." In *International Cooperation: Building Regimes for Natural Resources and the Environment*, 58–80. Ithaca, NY: Cornell University Press.

———. 1992. "The Effectiveness of International Institutions: Hard Cases and Critical Variables." In *Governance without Government: Change and Order in*

World Politics, ed. J. N. Rosenau and E.-O. Czempiel, 160–94. New York: Cambridge University Press.

——. 1994a. *International Governance: Protecting the Environment in a Stateless Society*. Ithaca, NY: Cornell University Press.

——. 1994b. "The Problem of Scale in Human/Environment Relations." *Journal of Theoretical Politics* 6:429–47.

——. 1996. "Institutional Linkages in International Society: Polar Perspectives." *Global Governance* 2 (1): 1–24.

——, ed. 1997. *Global Governance: Drawing Insights from the Environmental Experience*. Cambridge, MA: The MIT Press.

——. 1998. *Creating Regimes: Arctic Accords and International Governance*. Ithaca, NY: Cornell University Press.

——. 1999a. *The Effectiveness of International Environmental Regimes: Causal Connections and Behavioral Mechanisms*. Cambridge, MA: The MIT Press.

——. 1999b. *Governance in World Affairs*. Ithaca, NY: Cornell University Press.

——. 2001a. "Environmental Ethics in International Society." In *Ethics in International Affairs*, ed. J. Coicaud and D. Warner, 161–93. Tokyo: United Nations University Press.

——. 2001b. "Inferences and Indices: Evaluating the Effectiveness of International Environmental Regimes." *Global Environmental Politics* 1:99–121.

——. 2001c. "Political Leadership and Regime Formation: On the Development of Institutions in International Society." *International Organization* 45 (3): 281–308.

——. 2002a. "Are Institutions Intervening Variables or Basic Causal Forces? Causal Clusters vs. Causal Chains in International Society." In *Millennium Reflections on International Studies*, ed. M. Brecher and F. Harvey, 176–91. Ann Arbor: University of Michigan Press.

——. 2002b. *The Institutional Dimensions of Environmental Change: Fit, Interplay, and Scale*. Cambridge, MA: The MIT Press.

——. 2002c. "Institutional Interplay: The Environmental Consequences of Cross-Scale Interactions." In *The Drama of the Commons*, ed. E. Ostrom, T. Dietz, N. Dolsak, P. C. Stern, S. Stonich, and U. W. Elke, 263–91. Washington, DC: National Academies Press.

——. 2002d. "Matching Institutions and Ecosystems: The Problem of Fit." Paper for the seminar series "Economy of the Environment," Institut du développement durable et des relations internationales (Iddri), Paris.

——. 2003a. "Determining Regime Effectiveness: A Commentary on the Oslo-Potsdam Solution." *Global Environmental Politics* 3 (3): 97–104.

——. 2003b. "International Governance: The Role of Institutions in Causing and Confronting Environmental Problems." *International Environmental Agreements* 3:377–93.

————. 2005a. "Regime Theory and the Quest for Global Governance." In *Contending Perspectives on Global Governance*, ed. A. D. Ba and M. J. Hoffmann. London: Routledge.

————. 2005b. "Why Is There No Unified Theory of Environmental Governance?" In *Handbook of Global Environmental Politics*, ed. Peter Dauvergne, 170–84. Cheltenham, UK: Edward Elgar Publishing.

————. 2006. "Vertical Interplay among Scale-Dependent Environmental and Resource Regimes." *Ecology and Society* 11 (1): article 27.

————. 2008. "Institutional Coordination through Inter-linkages." In *Institutional Interplay: Biosafety and Trade*, ed. O. R. Young, W. B. Chambers, J. A. Kim, and C. ten Have. Tokyo: United Nations University Press.

Young, O. R., with contributions from A. Agrawal, L. A. King, P. H. Sand, A. Underdal, and M. Wasson. 1999/2005. Institutional Dimensions of Global Environmental Change (IDGEC) Science Plan. Bonn: IHDP Report Nos. 9, 16.

Young, O., F. Berkhout, G. C. Gallopin, et al. 2006. "The Globalization of Socio-ecological Systems: An Agenda for Scientific Research." *Global Environmental Change* 16 (3): 304–16.

Young, O. R., W. B. Chambers, J. A. Kim, and C. ten Have, eds. 2008. *Institutional Interplay: Biosafety and Trade*. Tokyo: United Nations University Press.

Young, O. R., E. Lambin, et al. 2006. "A Portfolio Approach to Analyzing Complex Human-Environment Interactions: Institutions and Land Use." *Ecology and Society* 11 (2): article 31.

Young, O. R., and M. A. Levy. 1999. "The Effectiveness of International Environmental Regimes." In *The Effectiveness of International Environmental Regimes*, ed. Oran R. Young, 1–32. Cambridge, MA: The MIT Press.

Young, O. R., M. A. Levy, and G. Osherenko, eds. 1999. *Effectiveness of International Environmental Regimes: Causal Connections and Behavioral Mechanisms*. Cambridge, MA: The MIT Press.

Young, O. R., and G. Osherenko, eds. 1993. *Polar Politics: Creating International Environmental Regimes*. Ithaca, NY: Cornell University Press.

Zaelke, D., D. Kaniaru, and E. Kruzikova, eds. 2005. *Making Law Work: Environmental Compliance and Sustainable Development*. 2 vols. London: Cameron May.

Zebich-Knos, M. 1997. "Preserving Biodiversity in Costa Rica: The Case of the Merck-INBio Agreement." *Journal of Environment and Development* 6 (2): 180–86.

Zito, A. R. 2001. "Epistemic Communities, Collective Entrepreneurship and European Integration." *Journal of European Public Policy* 8 (4): 585–603.

Zürn, M., and C. Joerges, eds. 2005. *Law and Governance in Postnational Europe: Compliance beyond the Nation-State*. Cambridge: Cambridge University Press.

Contributors

Frank Biermann Professor and Head, Department of Environmental Policy Analysis, Institute for Environmental Studies, Vrije Universiteit Amsterdam; Chair, IHDP Earth System Governance Scientific Planning Committee

Carl Folke Stockholm Resilience Centre, Stockholm University; Beijer International Institute of Ecological Economics

Victor Galaz Stockholm Resilience Centre, Stockholm University

Thomas Gehring Professor of International Politics, Faculty of Social Science, Economics and Management, Otto-Friedrich-University Bamberg

Joyeeta Gupta Professor in Policy and Law on the Environment, Vrije Universiteit Amsterdam

Thomas Hahn Stockholm Resilience Centre, Stockholm University

Leslie A. King Vice President, Academic, Vancouver Island University

Ronald B. Mitchell Department of Political Science, University of Oregon, Eugene

Sebastian Oberthür Academic Director, Institute for European Studies of the Vrije Universiteit Brussels

Per Olsson Stockholm Resilience Centre, Stockholm University

Heike Schroeder Tyndall Research Fellow, Environmental Change Institute, Oxford University Centre for the Environment

Uno Svedin Swedish Research Council for Environment, Agricultural Sciences & Spatial Planning (Formas)

Simon Tay Chairman, Singapore Institute of International Affairs; Chairman, Singapore Environment Agency; Professor, International Law, National University of Singapore

Arild Underdal Department of Political Science, University of Oslo

Oran R. Young Donald Bren School of Environmental Science and Management, University of California, Santa Barbara; Chair, IHDP Scientific Committee

Index